Lecture Notes in Civil Engineering

Volume 230

Lecture Notes in Civil Engineering (LNCE) publishes the latest developments in Civil Engineering - quickly, informally and in top quality. Though original research reported in proceedings and post-proceedings represents the core of LNCE, edited volumes of exceptionally high quality and interest may also be considered for publication. Volumes published in LNCE embrace all aspects and subfields of, as well as new challenges in, Civil Engineering. Topics in the series include:

- Construction and Structural Mechanics
- Building Materials
- Concrete, Steel and Timber Structures
- Geotechnical Engineering
- Earthquake Engineering
- Coastal Engineering
- Ocean and Offshore Engineering; Ships and Floating Structures
- Hydraulics, Hydrology and Water Resources Engineering
- Environmental Engineering and Sustainability
- Structural Health and Monitoring
- Surveying and Geographical Information Systems
- Indoor Environments
- Transportation and Traffic
- Risk Analysis
- Safety and Security

To submit a proposal or request further information, please contact the appropriate Springer Editor:

- Pierpaolo Riva at pierpaolo.riva@springer.com (Europe and Americas);
- Swati Meherishi at swati.meherishi@springer.com (Asia - except China, and Australia, New Zealand);
- Wayne Hu at wayne.hu@springer.com (China).

All books in the series now indexed by Scopus and EI Compendex database!

More information about this series at https://link.springer.com/bookseries/15087

Hong-Hu Zhu · Ankit Garg ·
Askar Zhussupbekov · Li-Jun Su
Editors

Advances in Geoengineering along the Belt and Road

Proceedings of 1st Belt and Road Webinar Series on Geotechnics, Energy and Environment 2021

Springer

Editors
Hong-Hu Zhu
School of Earth Sciences and Engineering
Nanjing University
Nanjing, China

Askar Zhussupbekov
Department of Design of Buildings
and Structures
L. N. Gumilyov Eurasian National
University
Nur-Sultan, Kazakhstan

Ankit Garg
Department of Civil
and Environmental Engineering
Shantou University
Shantou, Guangdong, China

Li-Jun Su
Institute of Mountain Hazards
and Environment
Chinese Academy of Sciences
Chengdu, China

ISSN 2366-2557 ISSN 2366-2565 (electronic)
Lecture Notes in Civil Engineering
ISBN 978-981-16-9962-7 ISBN 978-981-16-9963-4 (eBook)
https://doi.org/10.1007/978-981-16-9963-4

This Springer imprint is published by the registered company Springer Nature Singapore Pte Ltd.
The registered company address is: 152 Beach Road, #21-01/04 Gateway East, Singapore 189721, Singapore

Foreword

On behalf of the International Society of Environmental Geotechnology (ISEG), I would like to thank Profs. Hong-Hu Zhu, Ankit Garg, Askar Zhussupbekov, Li-Jun Su, and the organizing committee for holding the 1st Belt and Road Webinar Series on Geotechnics, Energy, and Environment.

ISEG was founded in 2000. During the past two decades, the parent group of individuals and organizations has organized 15 symposia and conferences across the world, seeking to promote scientific cooperation among countries, organizations, and individuals on geo-environmental issues on a global scale.

The Belt and Road (B & R) Initiative is China's proposal to build a Silk Road Economic Belt and a 21st Century Maritime Silk Road in cooperation with related countries, which was unveiled by Chinese President Xi Jinping during his visits to Central and Southeast Asia in September and October 2013. One major aim of the B & R Initiative is to develop intellectual capacity and establish strong research and education cooperation among various countries and regions. This B & R Webinar Series focuses on "Geotechnics, Energy, and Environment" and is being held with the aim of connecting young scholars among B & R countries. I would like to express my sincere congratulations to the organizing committee for holding this webinar series. As one of the sponsors, ISEG will continue to support such international academic activities. We welcome and encourage researchers around the world to actively participate in such international forums and to bring in their competence and ideas whether coming from academic level, the industrial environment, or from public authorities and political organizations.

Bin Shi

Bin Shi

Foreword

My congratulations to organizers for successfully organizing this webinar series. As understood, the Belt and Road Initiative (BRI) is a global infrastructure initiative established under the leadership of Honorable President of China, President Xi Jinping. This project was originally announced as the "Silk Road Economic Belt" during an official visit to Kazakhstan in September 2013. One of the major aims of BRI is to develop intellectual capacity and establish strong research and education cooperation among various countries and regions. I am delighted to see diversity among topics as well as speakers who come from a number of countries/continents listed in BRI (Canada, Italy, Russia, Singapore, China, Japan, Ukraine, Nigeria, Ethiopia, Iran, Kuwait, Uzbekistan, Kyrgyzstan, Kazakhstan, Tajikistan, Thailand, Malaysia, and New Zealand). Such events and exchange of ideas may not be easily feasible in a more physical environment due to the significant requirement of funding and logistics support. However, the pandemic has given us a good opportunity to explore the organization of gatherings in a digital environment. Of course, we can never fully capture the advantages of conferences that are held in a physical environment. Nevertheless, the theme of the webinar series is highly relevant for research toward sustainable development of all countries. The aim of the webinar series was to connect and understand the research of scholars from different countries. What is next?

Now, the next step is to convert this exchange and understanding into long-term cooperation. I urge participants and speakers to explore opportunities for joint projects and student/faculty exchange. There are many opportunities listed by the National Natural Science Foundation (NSFC) at both the central and provincial levels every year for joint international project applications, which includes academic visits/exchange. I would be glad to see the building of teams from this webinar series for conducting long-term research cooperation. From an administrative point of view at Shantou University, we strongly support international research and education cooperation among BRI countries.

Once again, thank you all for making this webinar series and its proceedings a great success.

Shui-Long Shen

Preface

This volume is the collection of manuscripts submitted to the 1st Belt and Road Webinar Series on Geotechnics, Energy, and Environment, which was held online from March 27 to May 29, 2021. The manuscripts have been submitted by authors from various countries and regions along the Belt and Road, including China, Singapore, Malaysia, Kazakhstan, Tajikistan, Kyrgyzstan, India, Kuwait, Iran, Italy, Russia, Ukraine, Canada, and Hong Kong SAR (China). This webinar series aimed to connect young scholars working in the research areas of geotechnics, energy, and environment across countries listed in the Belt and Road Initiative (BRI). The BRI is a global infrastructure project connectivity established by the People's Republic of China. One of the aims of the BRI is to develop intellectual capacity as well as education and research cooperation among countries for mutual sustainable development.

Due to the global COVID-19 pandemic, the organizing committee adopted a new mode of international academic exchange and held a series of webinars without any registration fees. The webinar series consisting of 25 sessions lasted for two months. Considering the time differences between various countries and regions, two–three sessions were held every week.

This webinar series was organized by the International Society for Environmental Geotechnical Engineering (ISEG), TC305 Technical Committee of the International Society for Soil Mechanics and Geotechnical Engineering (ISSMGE), and the Chinese Subsociety for Intelligent Monitoring in Geoengineering of Chinese Society for Rock Mechanics and Engineering (CSRME). The organizing committee includes faculty members from Nanjing University (Prof. Hong-Hu Zhu and Dr. Cheng-Cheng Zhang), Shantou University (Dr. Ankit Garg), Eurasian National University (Prof. Askar Zhussupbekov), and the Institute of Mountain Hazards and Environment, Chinese Academy of Sciences (Prof. Li-Jun Su).

There were five keynote presentations and 20 invited presentations in the webinar series from more than 22 countries. These presentations were categorized into three themes, i.e., geotechnics, energy, and environment, with rich and diverse content and a wide range of coverage. Around 1200 delegates around the world

Fig. 1. Guest lecturers and organizing committee members

participated in this webinar series, and the invited presentations were highly appreciated. The theme covers geological disaster prevention and control, underground space development, engineering monitoring and early warning, infrastructure operation and maintenance, geo-environmental engineering, coastal and offshore engineering, energy security and sustainable development, etc. Lively exchanges and discussions were carried out on research topics of common concern, which surpassed the barrier of the pandemic and helped in deepening mutual understanding and friendship.

The opening ceremony was held on March 27, 2021. The chairman of ISEG, Prof. Bin Shi (Nanjing University), congratulated the initiative and delivered a warm speech. Subsequently, Prof. Feng Da Hsuan, Honorary President of Belt and Road Research Institute at Hainan University, China, and former Vice President of Research and Development at the University of Texas at Dallas, USA, gave a keynote lecture titled "China's Millennium Transformation: The Belt and Road Initiative." He discussed the origin and goals of the BRI, predicting that it will undoubtedly promote the prosperity of all countries.

The closing ceremony of the webinar series was held on May 29, 2021. Prof. Askar Zhussupbekov (Eurasian National University), Past Vice President of ISSMGE for Asia, Chair of TC305 on Geotechnical Infrastructures for Megacities and New Capitals, and President of Kazakhstan Geotechnical Society, suggested enhancing further academic, cultural exchanges, and mutual understanding among scholars of the BRI. He sincerely thanked all the experts for their presentations and enthusiastic participants.

Fig. 2. Closing ceremony

Prof. Shui-Long Shen, Dean of College of Engineering, Shantou University, congratulated the organizers for the successful organization of the webinar series, which promotes international communication related to geotechnical, energy, and the environment. He pointed out that the distinctive feature of this webinar series is its wide coverage and strong participation. It is unprecedented in terms of the country of the participants and the broadness and depth of the topics. He looked forward to closer cooperation between the invited speakers and participants in the future.

Finally, Prof. Hong-Hu Zhu, Chair of the organizing committee, gave a brief concluding speech. He expressed his heartfelt thanks to the organizing committee members for their hard work in the past two months. The organizers will summarize the experience of hosting the webinar series and continue to promote China's cooperation along the BRI in the research areas of geotechnics, energy, and environment.

The editors would like to thank all the distinguished authors contributing to this volume for the high scientific quality and diversity of their manuscripts, successfully addressing the main topics of the 1st Belt and Road Webinar Series on Geotechnics, Energy, and Environment. Special thanks are given to Dr. Cheng-Cheng Zhang (Nanjing University) for his hard work as Secretary-General of the organizing committee and to Ms. Yu-Xin Gao (Nanjing University) for her assistance in editing the proceedings. The editors would also like to acknowledge the timely help provided by Dr. Meghna Sharma (IIT Indore), Dr. Nitin Tiwari (IIT Bombay), and Dr. Chandra Bogireddy (Vardhaman Engineering College). Finally, the editors are particularly grateful to Dr. Mengchu Huang, Karthikeyan Durairaj, and the whole Springer team for their full support and cooperation at various stages of the preparation and production of this volume.

Organization

Organizers of Webinar Series

Hong-Hu Zhu	Nanjing University, China
Ankit Garg	Shantou University, China
Askar Zhussupbekov	Eurasian National University, Kazakhstan
Li-Jun Su	Institute of Mountain Hazards and Environment, Chinese Academy of Sciences, China

Contents

About the Editors

Prof. Hong-Hu Zhu is Professor of Engineering Geology and Geotechnics and Dean of the Institute of Earth Exploration and Sensing at Nanjing University, China. He holds a Ph.D. degree in Geotechnical Engineering from the Hong Kong Polytechnic University, Hong Kong SAR, China. During 2014 and 2015, he was Visiting Scholar in the Department of Engineering, the University of Cambridge, UK. His research interests lie primarily in fiber optic sensing-based geoengineering monitoring and stability evaluation of geohazards, with a particular focus on interfacial behaviors. In the past decade, he has co-authored two books, ten patents, and over 100 international journal and conference papers. He won the first-class prize of the National Scientific and Technological Progress Award of China (sixth rank) in January 2019.

Dr. Ankit Garg is currently Associate Professor at the Department of Civil and Environmental Engineering, Shantou University, China. He obtained his Ph.D. from Hong Kong University of Science and Technology under the Hong Kong Ph.D. Fellowship Scheme and B. Tech from IIT Guwahati (2006–2010). He was former Assistant Professor at IIT, Guwahati (2015–2017) and part-time World Bank Consultant (2016–2017) for monitoring transport infrastructure projects in Assam, India. He is also Recipient of the Young Alumni Achiever Award. His research focuses on the development and utilization of sustainable materials (biochar, fiber), including vegetation for geo-environmental applications. He has been recently awarded the prestigious "Telford Premium Prize" from the British Civil Engineers Association for publication in Geotechnique Letters, ICE, UK. He is Recipient of the Young Alumni Achiever Award from IIT Guwahati, India. His research group has also been cited (Refer to Green Infrastructure and Water: An Analysis of Global Research by Caparrós-Martínez et al. 2020) for most number of publications in Web of Science. He is also currently Guest Editor for special issues related to IoT and AI in geotechnical engineering in the Journal of Rock Mechanics and Geotechnical Engineering. Besides, he was awarded the Talented Youth Scientist Program, the Young Doctor Award from the Ministry of Education Guangdong, and the International Scientist Exchange Program. Furthermore, he currently presides over the National Natural Science Foundation of China (NSFC) Youth Project on "Vegetation for soil remediation." He organized the 1st Indo-China webinar series in 2020, whose proceedings have been published in Springer.

Prof. Askar Zhussupbekov is Professor at Saint-Petersburg State Architecture and Civil Engineering University and Director of Geotechnical Institute, Department of Design of Buildings and Structures, Eurasian National University. He is Scientific Consultant on piling works of such projects as the second generation plant and in Tengiz (Caspian Sea coast) and Karabotan, Kashagan (Atyrau), where the clients are PFD company (USA), AGIP (Italy); the International Airport Project in Astana (new capital), the clients are Asian Pacific (Japan) and Alsim Alarko (Turkey); Buildings for the USA Embassy, the client is Fluor Caspian Services, Ltd (USA) and other mega projects on problematical soil ground of Kazakhstan, like as EXPO 2017, Abu-Dhabi Plaza (Astana). Starting in 2013, Askar Zhussupbekov has been appointed as Chair of T305 (ISSMGE) "Geotechnical Infrastructure of Mega cities and new capitals." He organized several workshops at Eurasian National University related to this TC305 activity. He is also Active Member of ATC-3, ATC-10, ATC-19, RSSMGE, and IALT.

Prof. Askar is Past Vice President of ISSMGE for Asia and President of Kazakhstan Geotechnical Society.

Prof. Li-Jun Su is Professor and Deputy Director-General of the Institute of Mountain Hazards and Environment, Chinese Academy of Sciences. He is also Deputy Director-General of the China–Pakistan Joint Research Center on Earth Sciences. He obtained his Ph.D. at the Hong Kong Polytechnic University in 2006. After that, he joined the Xi'an University of Architecture and Technology as Associate Professor. From 2008 to 2011, he worked at the University of Wollongong, Australia, as Research Fellow. He has 15 years experience in research and teaching in geotechnical engineering. He conducted numerous pull-out tests of soil nails to investigate the mechanism of a soil nail in stabilizing slopes. He is currently working on the mechanism and hazard risk analysis of landslides under the support of several national research funds. He has recently developed a seepage model that can consider the preferential flow of rainfall water in a slope. This model overcame the difficulty of the current seepage model in simulating deep-seated landslides. He developed comprehensive geophysical methods to generate the geological structure and underground water conditions of deep-seated landslides. His research interests also include the mechanism and risk of seismic landslides by conducting shaking table tests and developing new evaluation models. He has published more than 100 papers in top journals and international conferences. His wide research experience in soil and rock mechanics, engineering geology, and geophysical methods is very important for conducting multi-discipline research.

Distributed Fiber Optic Sensing in Pile Load Tests: Technological Development and Applications

Hong-Hu Zhu[1,2(✉)], Jing Wang[1], Bin Shi[1], and Guang-Qing Wei[3]

[1] School of Earth Sciences and Engineering, Nanjing University, Nanjing, China
zhh@nju.edu.cn
[2] Institute of Earth Exploration and Sensing, Nanjing University, Nanjing, China
[3] Suzhou Nanzee Sensing Technology Co., Ltd., Suzhou, China

Abstract. Piles enable a safe and convenient solution for providing adequate bearing capacity for civil infrastructures, which can be individually customized to changing soil conditions. Fully instrumented piles with advanced monitoring techniques can offer new insights into the bearing capacity and structural integrity of piles in field tests. Recently distributed fiber optic sensing (DFOS) technologies provide a powerful tool for geotechnical monitoring by enabling distributed and automatic strain measurement along fiber optic (FO) cables. This paper presents the DFOS-based pile monitoring system, which helps to quantify and refine each step during pile testing. According to the construction characteristics of precast piles, bored concrete piles and steel piles, various installation methods of FO cables on piles are introduced, respectively. Two case studies are illustrated to show the capability of DFOS in monitoring the performance of axially and horizontally loaded piles.

Keywords: Distributed fiber optic sensing (DFOS) · Field instrumentation · Pile load tests · Geotechnical monitoring · Interfacial behavior

1 Introduction

With the continuous acceleration of urbanization in Belt and Road countries and regions, super high-rise buildings, deep underground structures, large bridges, and other civil infrastructure are being built in major cities. Piles are common deep foundations used to transmit loads through shallow soil of low bearing capacity to deep soil or rock strata [1, 2]. However, as piles are embedded in ground soils, there is great difficulty in detecting their behavior, integrity, and load transfer properties. The existing detection methods cannot meet the increasing demand of engineering practice.

Static pile load tests are the basic and reliable method to understand the actual performance of piles in the field and obtain the relevant geotechnical parameters. For instance, the compressive, pullout and horizontal bearing capacity of piles can be determined based on the load-displacement curves, which are very important for optimizing pile design schemes. Appropriate instrumentation is required to get reliable measurements

H.-H. Zhu et al. (Eds.): BRWSG 2021, LNCE 230, pp. 1–11, 2022.
https://doi.org/10.1007/978-981-16-9963-4_1

so that the deformation and loading conditions of the instrumented piles can be evaluated [3–5]. Conventional sensors include extensometers, load cells, and vibration-wire strain gauges. However, these point-type sensors can only provide discrete information on the pile performance. The installation is tedious and may affect the pile integrity due to a large number of communication cables. In addition, conventional sensors often require substantial protection measures.

In recent years, a critical advance in geotechnical instrumentation is the wide applications of distributed fiber optic sensing (DFOS) technologies, which bring a new concept and methodology for measuring the behaviors of shallow and deep foundations due to their continuous nature [6–9]. In the past three decades, a number of DFOS technologies have been successfully used for pile monitoring, such as the fiber Bragg grating (FBG) [10–12], Brillouin optical time-domain reflectometry (BOTDR) [13–20], Brillouin optical time-domain analysis (BOTDA) [21–23], Brillouin optical frequency-domain analysis (BOFDA) [24, 25], and optical frequency domain reflectometry (OFDR) [26]. Among them, the Brillouin scattering-based technologies can perform fully distributed strain sensing of pile behavior. In the UK, the BOTDR technique was used to monitor the performance of a secant pile wall subjected to multiple props during construction of an adjacent basement, which can also obtain the axial force and a bending moment of the pile and then obtain the axial and lateral movement of the pile [15]. Distributed fiber optic sensors can also be accomplished to obtain strain profiles along piles during pile driving, which is helpful to reveal the failure mechanism of displacement piles [19, 24]. In China, the quasi-distributed FBG sensors were packaged into various types to monitor pile deformation, internal force and earth pressure near the pile [12, 21]. Using FBG sensing systems, elaborate strain data along the entire length of a pile can be performed with high reliability, which can be further processed to provide detailed information concerning pile behavior and load transfer properties [10]. These two types of DFOS technologies for measuring pile internal forces and deformation have the advantages of abundant monitoring information, high viability, easy implementation, accurate and reliable data, and good long-term stability.

This paper presents the basics of the DFOS-based pile monitoring system and introduces detailed sensor installation methods. Two case studies of using DFOS to measure pile strains during vertical and lateral loading are illustrated.

2 Fiber Optic Monitoring System of Piles

In order to effectively carry out pile testing, a distributed fiber optic monitoring system is developed. The framework of the monitoring system is shown in Fig. 1. The system is divided into four parts, including the measurement subsystem, the data acquisition subsystem, the data transmission and storage subsystem, and the data processing subsystem. As shown in Fig. 1, the stress and deformation characteristics of a pile are measured using fiber optic (FO) cables laid on the pile body. The data acquisition subsystem obtains the distributed information of temperature and strain of the pile shaft. Then, the data transmission subsystem connects the information to the network through the data transmission line and store the data in computers. Finally, the results of data processing and analysis are obtained in the data processing subsystem.

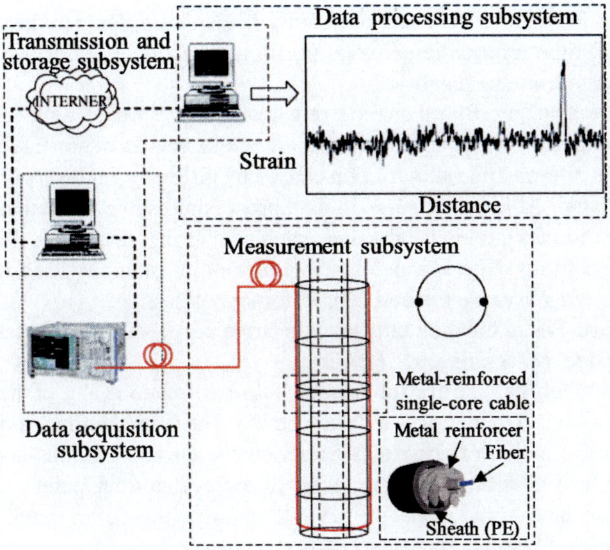

Fig. 1. Fiber optic monitoring system of piles

2.1 Measurement Subsystem

This subsystem comprises various sensors, including the distributed strain and temperature sensing cables, FBG strain sensors, FBG load cells, and FBG displacement sensors. The deformation compatibility between the FO cables and the pile body must be ensured so that the detected parameters can reflect the real state of the pile. If connections of FO cables are needed in the field, the optical fiber fusion splicer will be used. Using the distributed and long-distance monitoring characteristics of DFOS technologies, the monitoring data of multiple piles can be collected at one time, which greatly saves testing time.

2.2 Data Acquisition Subsystem and Data Transmission and Storage Subsystem

The subsystem comprises various optical fiber demodulation instruments and equipment, such as BOTDR, BOTDA, OFDR and FBG demodulators. During the test, the monitoring data of the sensing cables and FBG sensors under various loading conditions are obtained by the demodulation instrument. The collected data can be transmitted to the monitoring station automatically through the data transmission and storage subsystem. In general, the methods for data transmission are either wireless or wired.

2.3 Data Processing Subsystem

Unlike traditional point measurement technologies, the DFOS technology can obtain the distribution of strain and temperature along piles. The amount of monitoring data obtained in one test is tremendous. The DFOS-based pile performance evaluation system provides a powerful tool for the analysis and design of piles. The data processing subsystem mainly consider the following parts:

(a) Data read in section. This section mainly reads strain, temperature, wavelength, amplitude, time information, spatial positioning matching information and other relevant data from the database.
(b) Data processing part. Based on the strain, temperature, wavelength and other information obtained by the demodulator, pile strain and axial force, soil settlement around the pile, and pile skin friction caused by different soil layers on the pile side are calculated. At the same time, in data processing, some mathematical methods and means are used to revise the data, such as filtering, denoising, numerical integration and fitting. The wavelet analysis method is generally used for denoising, and the moving average method is used for smoothing.
(c) Output part. The output contents include strain curve, pile shaft axial force curve, friction value, $Q \sim s$ curve, $s \sim lgt$ curve.
(d) Diagnostic analysis. The maximum allowable deformation value of the pile shaft set in advance is used to identify abnormal points. Through identification, the relevant position information of abnormal points on the curve of strains and axial force is output in a specific way. According to these abnormal points, qualitative or quantitative analysis is made on the pile quality, pile deformation and damage mechanism.

3 Field Installation of Fiber Optic Sensors

Different types of piles have different working performance and load transfer mechanisms. Thus, the sensor installation and layout methods are quite different. In the following sections, the installation methods for precast piles, bored concrete piles and steel pipe piles are introduced.

3.1 Precast Pile

FO cables can be symmetrically embedded on two opposite sides along the axial direction of precast piles. In the process of slotting, firstly, it needs to determine the layout path of cables and mark it with an ink line. The cutting machine is used for slotting along the line. The recommended width of the groove is 2.0 mm–4.0 mm, and the depth of the groove may be 5.0 mm–8.0 mm. It passes through a U-shaped circuit at 50 cm from the pile bottom to ensure that the bending radius is greater than 100 mm. Use various tools such as a blower, brush and rag to remove dust from the groove and grind and straighten the uneven position in the groove.

Secondly, put the FO cable or FBG sensor into the design position in the groove after dispensing and fixing one end, properly pre-pull the fiber (e.g. 1000 $\mu\varepsilon$ in tensile strain), and then dispensing and fixing the other end. At the same time, it needs to glue the cable loop and set it in the slot. After the cable is implanted, the epoxy resin shall be coated, and the epoxy resin shall be repaired and leveled with a hot-air gun to ensure that all bubbles are eliminated. After 12 h, the epoxy resin glue is solidified, the other side of the pile shaft is constructed according to this method, and the bottom U-loop optical fiber is welded. Finally, the data transmission cable should be properly protected. The groove to embed the cables does not reach the pile top in most cases, so a suitable protection system should be provided to prevent the breakage of cables during pile pressing (Fig. 2).

Fig. 2. Layout of FO cables on a precast pile

3.2 Bored Pile

In order to monitor the pile behavior during pile load tests, it is necessary to ensure that the strain sensing cable is closely connected with the pile to make the deformation of the two consistent as much as possible. The quality of strain sensing FO cable determines the survival rate of the cable and affects the accuracy of monitoring results. Referring to the measuring principle of traditional reinforcement meters, the armored FO cables are laid on the two main reinforcements symmetrical to the reinforcement cage. The FO cables are fixed on the main reinforcement using the fixed-point bonding method, with a fixed point every 30–50 cm. In order to ensure that the sensing cables is laid vertically in a pile, a certain prestraining should be applied on the FO cables during fixation. The sensing cables laid in a pile are normally U-shaped. The laying position shall be close to the side of the main reinforcement of the reinforcement cage as far as possible to reduce the damage to the cable caused by concrete grouting. Then, the reinforcement cage with FO cables is inserted in the borehole, and then the concrete is poured in it slowly. In this way, the deformation of the cable is consistent with that of the pile. In order to prevent damage to the sensing optical cable during construction, the cables at the pile head should be specially protected (Fig. 3).

3.3 Steel Pipe Pile

For steel pipe piles, strain sensing FO cables with steel strands are generally installed on the steel pipe surface using welding and cementation. Then the pile is slowly driven into the soil layer. The installation of FO cables is divided into six steps: grinding in a pile, laying of FO cables, epoxy bonding, aluminum foil covering, channel steel covering, and pile head treatment, as shown in Fig. 4.

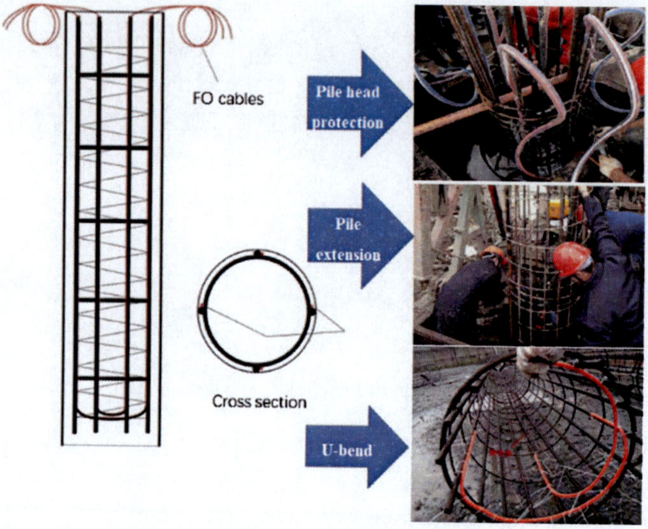

Fig. 3. Layout of FO cables on a bored pile

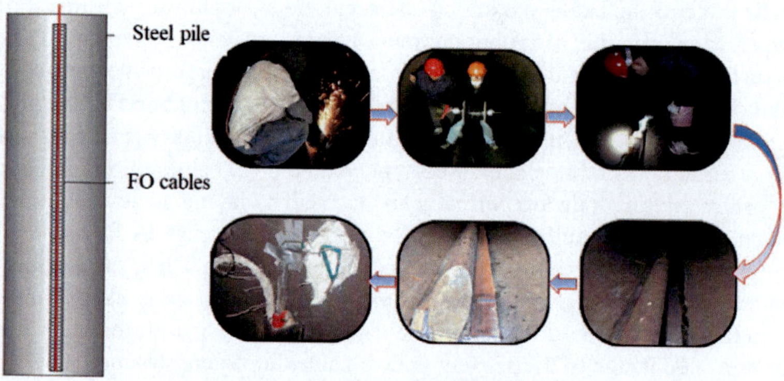

Fig. 4. Layout of FO cables on a steel pipe pile

4 Case Studies

4.1 Case Study 1: Vertical Static Load Test

The static load tests of a permeable pipe pile were performed in a construction site in Zhenjiang City, Jiangsu Province, China. The groundwater table was located about 2.0 m below the ground surface. The site is underlain consecutively with a plain fill layer, silty clay layer, silt layer and muddy, silty clay layer. The plain fill layer is mainly composed of silty clay, which is partially mixed with silt. The silt layer is locally composed of fine sand. The outer diameters and wall thickness of the test pile were 500 and 125 mm, respectively, and the pile length was 12 m. The bare FBG strain sensing arrays were installed for monitoring the axial strains of the test pile. Installation of the sensors is shown in Fig. 5. The FBG sensors were symmetrically embedded on two opposite sides of the test pile along the depth. Most FBG sensors were placed at 1.0 m intervals along the pile. The distance from the toe or the pile head to the nearest FBG sensors was 1.5 m to avoid the boundary effect. The static load tests were carried out in accord with the Chinese Code on Pile Testing.

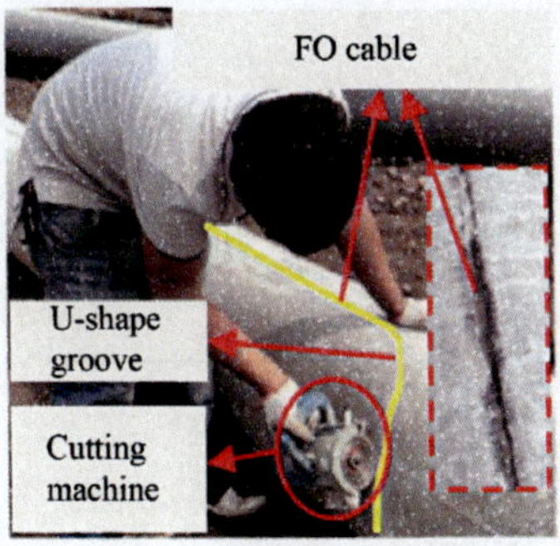

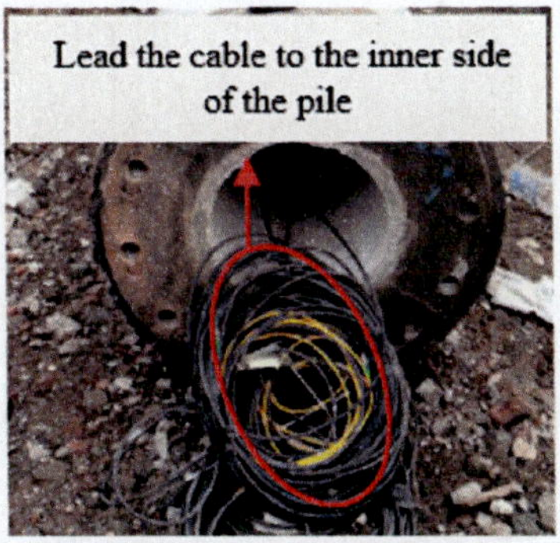

Fig. 5. Sensors installation in the field (adapted from [12]): (a) u-shape for installation of FBG sensors; (b) protection of fiber optical cables at the pile head.

The axial force profiles of the pile were calculated from strains multiplied by the pile axial rigidity, *EA*, and using $E = 3.8 \times 10^4$ N/mm^2. The value of *E* was obtained following the concrete strength of piles (C80). The axial force distribution curves of the permeable pipe pile in three static load tests are plotted in Fig. 6. It can be seen that the axial force increased steadily with the increase of load and gradually attenuated along the depth direction, showing an inverted triangle, which was the same as the distribution characteristics of the axial force of the pile shaft of a normal pipe pile. Because the pile side friction of each soil layer was different under the pile top load, the pile shaft axial force decreased nonlinearly along the depth. Under loading, the axial force changed significantly within 0–7 m of the pile shaft (miscellaneous fill, silty clay and silty soil layer), while the axial force transfer slowed down within 7–12 m (muddy, silty clay layer). The attenuation rate of axial force in the silt and silty clay layer was faster than that in the muddy, silty clay layer.

4.2 Case Study 2: Horizontal Static Load Test

At a construction site, a PHC pile with a diameter of 500 mm and a length of 15 m was built. In order to determine the deformation of pile top under free state horizontal load and determine the horizontal bearing capacity of the single pile, the horizontal loading test of the PHC pile was carried out, as shown in Fig. 7. In the test, FO cables were implanted in the PHC pile to couple it with the pile shaft for deformation, and the internal force and deformation of the pile shaft were measured. The pile top was sunk 40 cm below the ground, and the horizontal loading point was 1 m below the pile

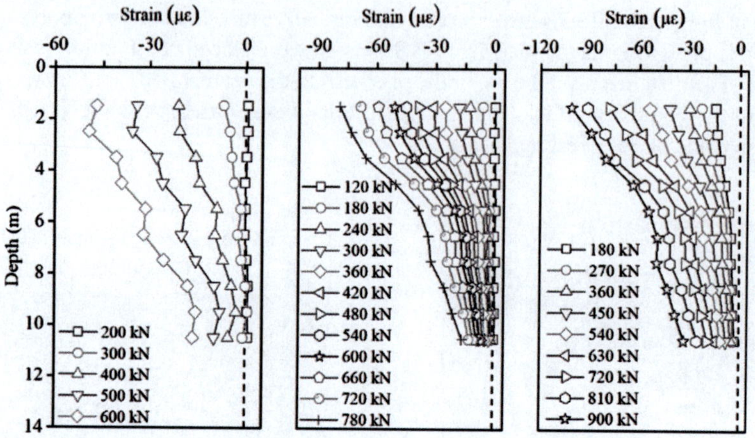

Fig. 6. Development of the pile axial strains with depth

top and 1.4 m below the ground. The horizontal hydraulic jack was used for applying horizontal loads in stages, 20 kN for each stage. A total of 11 steps were carried out. In the horizontal load test, the BOTDA optical fiber demodulator was used to collect strain data. According to the strain data, the deflection of the pile shaft was deduced and calculated.

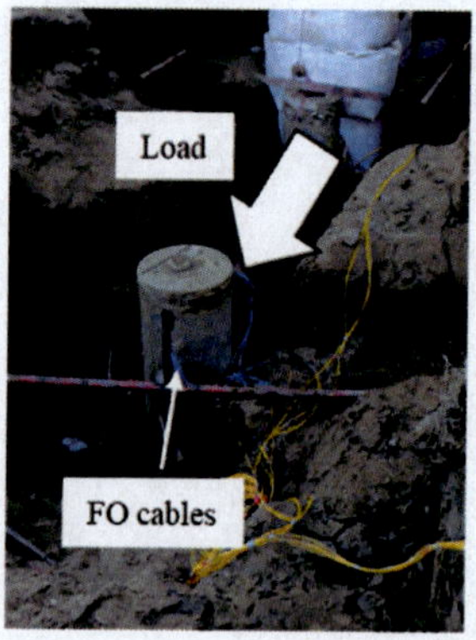

Fig. 7. Horizontal load test of a pile in the field

It can be seen from Fig. 8 that under the action of horizontal load, the displacement of the pile head increased with the increase of load. The zero point of pile displacement moves downward with the increase of load. It shows that the depth of the pile shaft affected by the horizontal load is increasing. The pile deflection changed obviously between 0–4 m. When the pile shaft was less than 4 m, the pile shaft deflection was small. The horizontal load mainly affected the soil in the upper 0–4 m. In this test, when the horizontal load reached the maximum horizontal load of 220 kN, the maximum influence

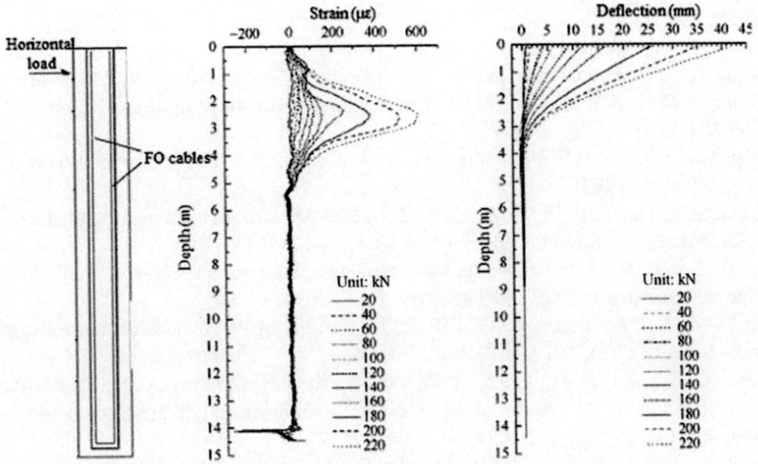

Fig. 8. Pile strains and deflection based on BOTDA

depth was about 6 m. The peak point (maximum strain point) of the pile strain curve increased with the increase of load, and the position of the peak point moved to the lower part of the pile. The results also show the superiority of BOTDA distributed monitoring, which can directly and accurately reflect the pile deformation and soil influence depth under horizontal load.

5 Conclusion

This paper presents the DFOS-based pile monitoring system and its field applications. The monitoring system is divided into four parts, including the measurement subsystem, data acquisition subsystem, data transmission and storage subsystem, and data processing subsystem. According to the construction characteristics of precast piles, bored concrete piles and steel pipe piles, various installation methods of FO cables or sensors are used. Two case studies are introduced to illustrate the performance of this pile monitoring system during vertical and horizontal static load tests. It proves that the distributed FO data can provide reliable information on the load transfer mechanism and bearing capacity. An available and reliable set of monitoring data along the whole pile length allows an estimation of the shaft friction development during testing, which is very important for optimizing pile design schemes.

Acknowledgments. This research was funded by the National Key Research and Development Program of China (Grant No. 2018YFC1505104) and the National Natural Science Foundation of China (Grant No. 42077235).

References

1. Poulos, H.G.: Pile behaviour: theory and application. Geotechnique **39**(3), 365–415 (1989)
2. Randolph, M.F.: Science and empiricism in pile foundation design. Géotechnique **53**(10), 847–875 (2003)
3. Coop, M.R., Wroth, C.P.: Field studies of an instrumented model pile in clay. Géotechnique **39**(4), 679–696 (1989)
4. Jardine, R.J., Zhu, B.T., Foray, P., Yang, Z.X.: Measurement of stresses around closed-ended displacement piles in sand. Géotechnique **63**(1), 1–17 (2013)
5. Seo, H., Prezzi, M., Salgado, R.: Instrumented static load test on rock-socketed micropile. J. Geotech. Geoenviron. Eng. **139**(12), 2037–2047 (2013)
6. Zhu, H.-H., Shi, B., Zhang, C.-C.: FBG-based monitoring of geohazards: current status and trends. Sensors **17**(3), 452 (2017)
7. Zhang, C.-C., Zhu, H.-H., Chen, D.-D., Xu, X.-Y., Shi, B., Chen, X.-P.: Feasibility study of anchored fiber-optic strain-sensing arrays for monitoring soil deformation beneath model foundation. Geotech. Test. J. **42**(4), 966–984 (2019)
8. Cao, D.F., Zhu, H.-H., Guo, C.C., Wu, J.H., Fatahi, B.: Investigating the hydro-mechanical properties of calcareous sand foundations using distributed fiber optic sensing. Eng. Geol. **295**, 106440 (2021)
9. Cao, D.F., Zhu, H.-H., Wu, B., Wang, J.C., Shukla, S.K.: Investigating temperature and moisture profiles of seasonally frozen soil under different land covers using actively heated fiber Bragg grating sensors. Eng. Geol. **290**, 106197 (2021)
10. Lee, W., Lee, W.J., Lee, S.B., Salgado, R.: Measurement of pile load transfer using the Fiber Bragg Grating sensor system. Can. Geotech. J. **41**(6), 1222–1232 (2004)
11. Doherty, P., et al.: Field validation of fibre Bragg grating sensors for measuring strain on driven steel piles. Geotech. Lett. **5**(2), 74–79 (2015)
12. Wang, J., Zhu, H.H., Mei, G.X., Xiao, T., Liu, Z.Y.: Field monitoring of bearing capacity efficiency of permeable pipe pile in clayey soil: a comparative study. Measurement **186**, 110151 (2021)
13. Klar, A., et al.: Distributed strain measurement for pile foundations. ICE Proc. Geotech. Eng. **159**(3), 135–144 (2006)
14. Wei, G.Q., Shi, B., Jia, J.X.: Application of distributed optical fiber sensing to testing inner force of prefabricated piles. Chin. J. Geotech. Eng. **31**(6), 911–916 (2009). (In Chinese)
15. Mohamad, H., Soga, K., Pellew, A., Bennett, P.J.: Performance monitoring of a secant-piled wall using distributed fiber optic strain sensing. J. Geotech. Geoenviron. Eng. **137**(12), 236–1243 (2011)
16. Lu, Y., Shi, B., Wei, G.Q., Chen, S.E., Zhang, D.: Application of a distributed optical fiber sensing technique in monitoring the stress of precast piles. Smart Mater. Struct. **21**(11), 115011 (2012)
17. Rui, Y., Kechavarzi, C., O'Leary, F., Barker, C., Nicholson, D., Soga, K.: Integrity testing of pile cover using distributed fibre optic sensing. Sensors **17**(12), 2949 (2017)
18. Pelecanos, L., et al.: Distributed fiber-optic monitoring of an Osterberg-cell pile test in London. Geotech. Lett. **7**(2), 152–160 (2017)
19. Pelecanos, L., et al.: Distributed fiber optic sensing of axially loaded bored piles. J. Geotech. Geoenviron. Eng. ASCE **144**(3), 04017122 (2018)
20. Gao, L., Cao, Y., Liu, H.L., Zhao, Z.X., Tu, W.J.: Experiment and numerical study on the monitoring of super long cast-in-place pile temperature based on BOTDR technology. Measurement **179**(8), 109481 (2021)
21. Zhu, Y.Q., Zhu, H.-H., Sun, Y.J.: Model experiment study of pipe driving into soil using FBG-BOTDA sensing monitoring technology. Rock Soil Mech. **35**(S2), 695–702 (2014). (in Chinese)

22. Ding, Y., Wang, P., Yu, S.: A new method for deformation monitoring on H-pile in SMW based on BOTDA. Measurement **70**, 156–168 (2015)
23. Pei, H.F., Yin, J.H., Wang, Z.: Monitoring and analysis of cast-in-place concrete bored piles adjacent to deep excavation by using BOTDA sensing technology. J. Mod. Opt. **66**(7), 703–709 (2019)
24. Gao, L., Ji, B., Kang, G., Xu, H., Li, M., Mahfouz, A.H.: Distributed measurement of temperature for PCC energy pile using BOFDA. J. Sens., 610473 (2015)
25. Zheng, X., Shi, B., Zhu, H.-H., Zhang, C.-C., Wang, X., Sun, M.-Y.: Performance monitoring of offshore PHC pipe pile using BOFDA-based distributed fiber optic sensing system. Geomech. Eng. **24**(4), 337–348 (2021)
26. Gao, L., Cao, Y., Liu, H.L., Kong, G.Q., Zhang, X.L.: Distributed monitoring of deformation of PCC pile under horizontal load using OFDR technology. IOP Conf. Ser. Earth Environ. Sci. **570**, 032064 (2020)

Heat and Mass Transfer in the Freezing Soils

Assel Sarsembayeva[1](✉), Askar Zhussupbekov[1], and Philip Collins[2]

[1] The L.N. Gumilyov Eurasian National University, Nur-Sultan 010008, Kazakhstan
[2] Brunel University London, London UB8 3PN, UK

Abstract. The moisture mass transfer in a gaseous state was previously widely neglected in the exploration of frost heave. The widespread model of coupled heat and water transfer allowed for detailed calculation and estimation of moisture migration by using the numerical methods. However, it could not explain the driving forces and physical nature. In this research, the significantly modified testing method, including the slow freezing techniques and longer samples lengths, was presented, as is an effective way to evaluate the impact of freeze-thaw cycles on soils. The moisture mass transfer in a state of vapour flow was considered during unidirectional freezing on the example of the conducted laboratory testing. The obtained results have improved understanding of the heat and mass transfer phenomenon during the unidirectional freezing of soils. The conceptual model for frost heave in soils was developed based on the vapour mass transfer. The algorithm of vapour flow calculation in unsaturated soils was presented using fundamental thermodynamic equations.

Keywords: Freeze-thaw test · Laboratory testing · Vapour mass transfer

1 Introduction

The subbases and subsoils of the structures in cold countries are subject to the frost heave due to increased thermal conductivity, disruption of natural moisture circulation as well as dynamic loading and application of de-icing chemicals in the winter months. The freezing rate in the highway structure is more rapid, and the temperature distribution is lower than that of adjacent roadside soils [1–3]. According to [4], the convectional heat fluxes can lead to a significant change of the temperature field in highway subsoils and must be accounted for in road design and in the prognosis of subsoils' temperature distribution. Junwei et al. found that the maximum influencing depth for temperature fields under the embankments was over 8 m [5]. Observation of the temperature distribution during the freeze-thaw period on the local highways in Kazakhstan was performed by Teltayev and other researchers [6, 7].

Formation of frost heave within freezing soil is necessarily accompanied by moisture redistribution and heat transfer. Most of the studies were performed in the laboratory condition. Laboratory approaches benefit from better technical control and the ability to set the initial parameters of the soil sample, such as the temperature control plates and freezing rate, as well as being able to adjust these parameters during monitoring. The length of the samples mostly ranged from 15 cm to 20 cm or less [8] and rarely

H.-H. Zhu et al. (Eds.): BRWSG 2021, LNCE 230, pp. 12–32, 2022.
https://doi.org/10.1007/978-981-16-9963-4_2

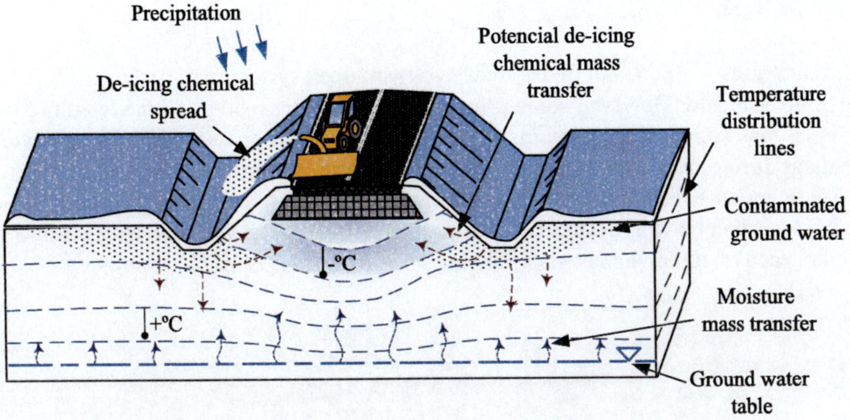

Fig. 1. Mass transfer under the highway pavement in seasonally freezing soils.

longer [9, 10]. The number of cycles is usually two, although in studies for mechanical properties, there might be more [11, 12]. Ming and Li (2015) presented an inverse relationship between the water influx and freezing rate was investigated for different water content samples, and developed a coupled water and heat mass transfer model based on this relationship [13]. Individual testing models have been developed for exclusively saturated soils and eliminating the vapour phase [14].

Some of the experiments were conducted in an adapted triaxial cell, which allows measuring pore pressure, the major principal stress σ_1 in the vertical axis and radial minor principal stress σ_3 [15, 16] tested frozen saline silt under triaxial compression and at close to freezing temperatures, −3.9 °C and −2.4 °C [16]. Conducting the freeze-thaw cycles in the triaxial cell allowed them to observe a relationship between the strain rate and freezing temperature. However, triaxial cell tests enable a contemporaneous mechanical loading of only one soil sample at a time. The size of the tested sample is also restricted by the size of the test cell.

The classical view of frost heave involved mainly considering water-ice interaction and their migration, while the vapour component in soils was widely ignored. Moreover, the widespread model of coupled heat and water transfer allowed for detailed calculation and estimation of moisture migration by using the numerical methods. However, it could not explain the driving forces and physical nature. Nevertheless, the developed methods and approaches formed the basis of the breakthrough of numerical methods and, moreover, achieved tolerable results in frost heave calculations and prognosis.

Vapour convection has been widely underestimated in geotechnics up to the present day, and its calculation has been generally neglected in moisture transfer modeling [17]. To fill this gap in the knowledge it was decided to conduct a cyclical freeze-thaw experiment with a modified testing method, which was adapted to allow for the close examination of mass transfer in the freezing subsoils of highways.

2 Method

The laboratory method with freeze-thaw cycles in 'open system' was based on ASTM D 5918-06 Standard. However, some principle changes were made, and the length of the samples was increased to 0.5 m in order to improve the observation of the temperature gradient during freeze-thaw cycles (see Fig. 2). The freezing rate was conducted with 2 °C temperature reduction of surface temperature per day. The increased capacity environmental chamber was designed and built to enable the testing of nine soil columns at a time, each of 10 cm diameter and of lengths 50 cm and up 1 m.

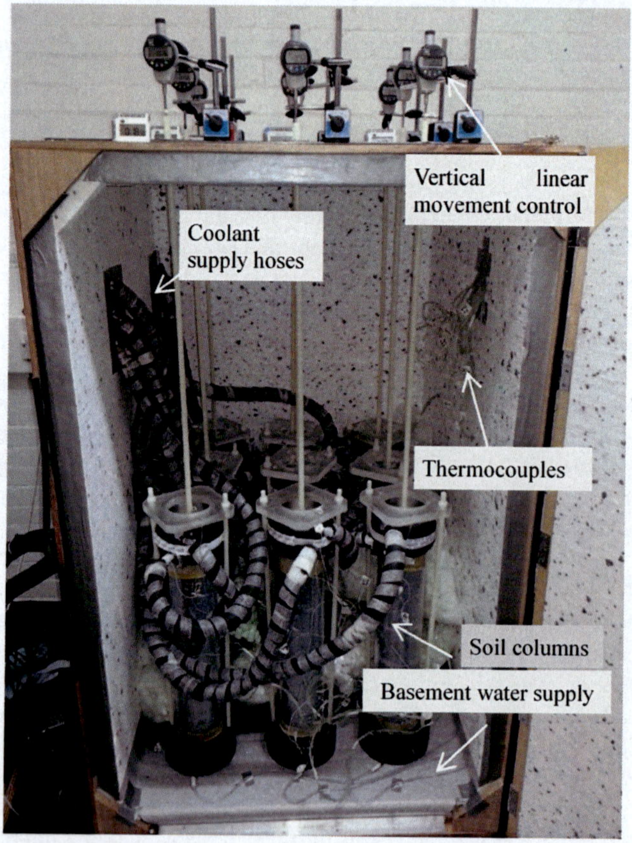

Fig. 2. Environmental chamber.

2.1 Sample Preparation

Each soil column was equipped with a basement water supply from the reservoir and an air tap fitted at 10 cm above the base, allowing the excess air pressure to be released as the column filled with water. The bottom 5 cm of the test soil sample was kept saturated during the test, while the other 45 cm remained in an unsaturated condition. To provide the uniform distribution of water supply over the sample cross section, the feed was provided via a 5 cm thick fine sand filter layer at the base of the column (see Fig. 3). The soil samples were prepared to simulate the engineering properties of soil from Astana, Kazakhstan. Local soils of the area presented the ancient sedimentary rocks, which comprise irregular thicknesses of residual layers and alluvium depositions, according to borehole data from Karaganda GIIZ [18].

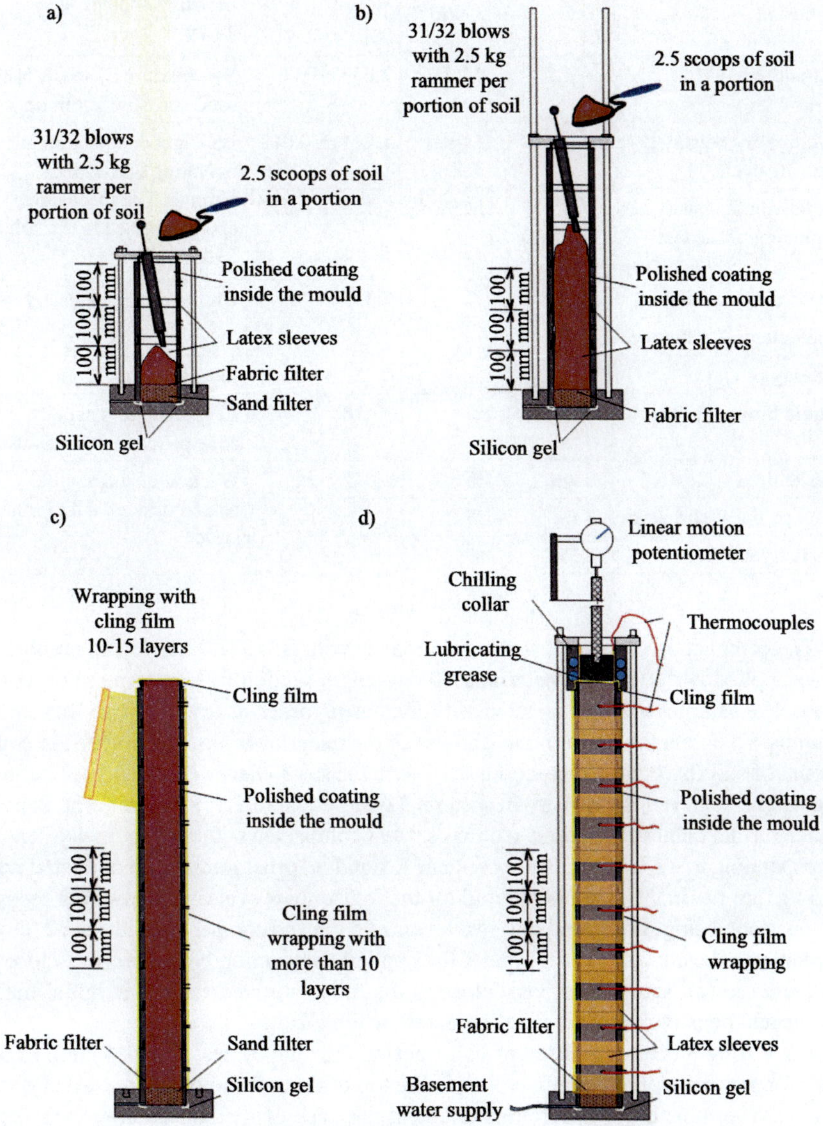

Fig. 3. Sample preparation.

Table 1. Initial characteristics of the soil.

Characteristic	Symbol	Unit	Value	Annotation
Initial moisture content	W	%	17.2	According to 95% max. dry density – moisture content relationship
Angle of internal friction	φ	°	24.1°	CD direct shear test, moisture content $W = 17.2\%$
Cohesion	C	kN/m^2	10	
Particle density	ρ_s	Mg/m^3	2.615	Soil mixture by mass: 50% sand and 50% kaolinite
Average dry density before freezing cycle	ρ_d	Mg/m^3	1.814 ± 0.012	BS Light compaction test operating with 2.5 kg rammer. The mechanical energy applied to the soil is 596 kJ/m^3
Initially bulk density at the beginning of the test	ρ	Mg/m^3	2.128 ± 0.015	
Uniformity coefficient	*Cu*	–	2.4	Uniformly-graded sand
Coefficient of curvature	*Cc*	–	3.65	
Activity of clays	*A*	–	0.25	Inactive clays
Liquid limit	w_L	%	37.18	CI – Medium plasticity cone penetrometer test used
Plastic limit	w_P	%	23.77	Fraction of soil sample passed through 0.425 mm sieve
Average linear shrinkage	L_S	%	5	
Plasticity index	*PI*	%	13	

The compaction of the soil was in accordance with BS1377-1:1990. The number of blows applied per 10 cm diameter and 10 cm length section was corresponding to the effort. The samples were compacted with increasing order of dry density between the columns. To obtain the uniform distribution of the material within the volume, an equal amount of soil (by scoops) was compacted with the same energy effort over the sample length. The achieved dry density in column #1 was equal to 1.178 Mg/m^3. The density of each further column increased. Columns 9 was compacted with maximum dry density corresponding to a 17.2% moisture content. A standard proctor test was conducted with a 2.5 kg rammer in the assembled mold for the freeze-thaw cycles. 63 blows per section volume, each being subdivided into two layers of 5 cm and compacted with 31/32 blows by rotation. The air void content under the applied compaction procedure was close to zero, and the soil sample was very close to the fully saturated state. The initial mass, volume and density parameters are presented in Table 2.

The testing process consisted of two freezing-thawing cycles. After 24–36 h of the conditioning period, the chilling collars at the top of the soil sample were cooled down to 3 °C. The moisture supply temperature at the base of the soil columns was kept stable at +4 °C. After the conditioning period, the circulated coolant temperature at the temperature control machine was reduced to −3 °C and subsequently was dropped daily

Table 2. Density distribution within the soil samples

Column number	Average volume in a mould, cm^3	Weight of assembled mould with sand filter layer, g	Weight of the compacted sample in the mould before test, g	Weight of soil before test, g	Dry density before the test, Mg/m^3
#1	3,966	4,121	9,602	1.382	1.178
#2	3,966	4,105	10,160	1.527	1.302
#3	3,966	4,117	10,372	1.577	1.345
#4	3,966	4,120	10,961	1.725	1.471
#5	3,966	4,095	11,201	1.792	1.527
#6	3,966	4,115	11,591	1.885	1.607
#7	3,966	4,110	11,792	1.937	1.651
#8	3,966	4,114	12,466	2.106	1.795
#9	3,966	4,122	12,485	2.109	1.798

by a further 2 °C. It should be noted that the heat loss between the temperature control unit and the cooling ring on the top of the soil sample was approximately 2–3 °C. The red stepped line in Fig. 4 represents the temperature at the temperature control unit, while the temperature reaction of the soil surface is shown as a flowing line. The freezing rate of the soil sample was dropped by 2 °C every 24 h for 12 days and reduced down to −23 °C at the temperature control unit. When the coolant supply temperature had decreased to −23 °C, it was held there for the next 24 h. Afterward, all the cooling system was switched off, and the environmental chamber was slowly returned to room temperature through natural thawing.

2.2 Post-freeze-Thaw Sampling and Testing

After the end of the second freezing cycle, the soil columns were unplugged from all monitoring and fixing elements, removed from the environmental chamber and weighed immediately in the frozen state. After that, each column was positioned horizontally to avoid the possible draining of melting moisture in a longitudinal direction. The soil columns were disassembled in-ring sections and sampled for moisture content determination every 1 cm in the top and the bottom ring sections and every 10 cm in the middle lengths of the soils. The sampling procedure was generally carried out the same day before the soil started to thaw in order to prevent moisture redistribution or structure change during melting.

The moisture content was found according to standard BS 1377-4:1990 by weight difference of the wet and oven-dried sample at 105 °C for a 24 h period. The total moisture content included that attributed to the mass of ice lenses and the mass of water in a liquid condition. The top 10 cm section was sampled every 1 cm due to the increased

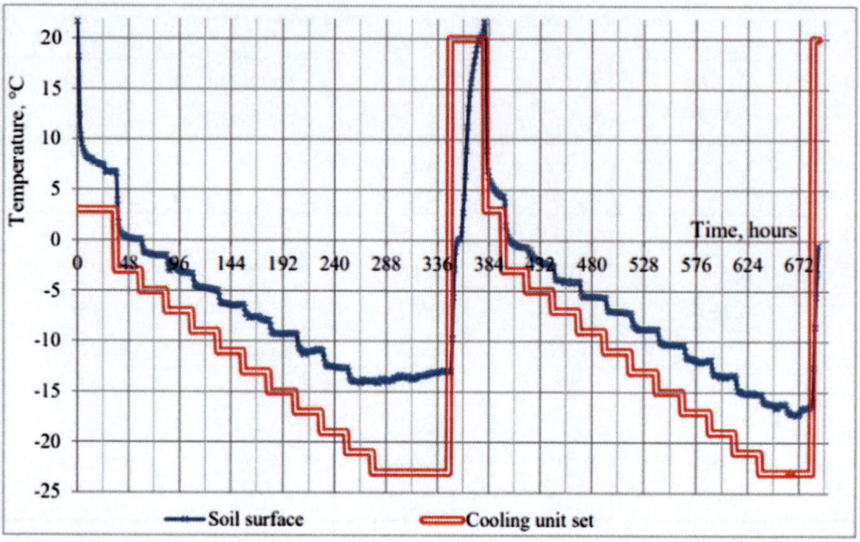

Fig. 4. The cooling unit sets the temperature and the corresponding temperature on the soil surface.

ice lens formation and the anisotropic structure, while the lower part of the soil column was only sampled every 10 cm, mainly at the junctions of the mold sections.

2.3 Calculation of the Vapour Mass Transfer in Unsaturated Freezing Soils

The algorithm for calculation of vapour mass transfer includes determination of volumetric and mass constituents of soils such as water, soil particles, vapour. Porosity and the volume of air are the crucial characteristics for the vapour mass transfer in freezing soils (Fig. 5).

The heat change Q in a mold section over the time t is found as the sum of the cooling heat and the latent heat during the phase transfer:

$$Q \cdot t = Q_1 + Q_2 \tag{1}$$

where, Q_1 is heat energy used for cooling the vapour mass to the temperature ΔT; and Q_2 is heat energy is used for phase transfer.

$$Q_1 = m_{vapour}, t_i \cdot C \cdot \Delta T \tag{2}$$

where m_{vapour}, t_1 a mass of vapour at the starting time t_1, g; C – specific heat of vapour passing through the cumulative air voids cross-section, *J/kg · °C*; ΔT – temperature change, °C; and t – time interval, h.

If part gravitational water remains in the unfrozen condition, the mass of vapour is found by Eq. (2). The density of the vapour is calculated for each period of time, corresponding to the temperature and saturated vapour pressure:

$$\rho_{vapour,t_i} = \frac{m_{vapour,t_i}}{V_{air,t_i}} \tag{3}$$

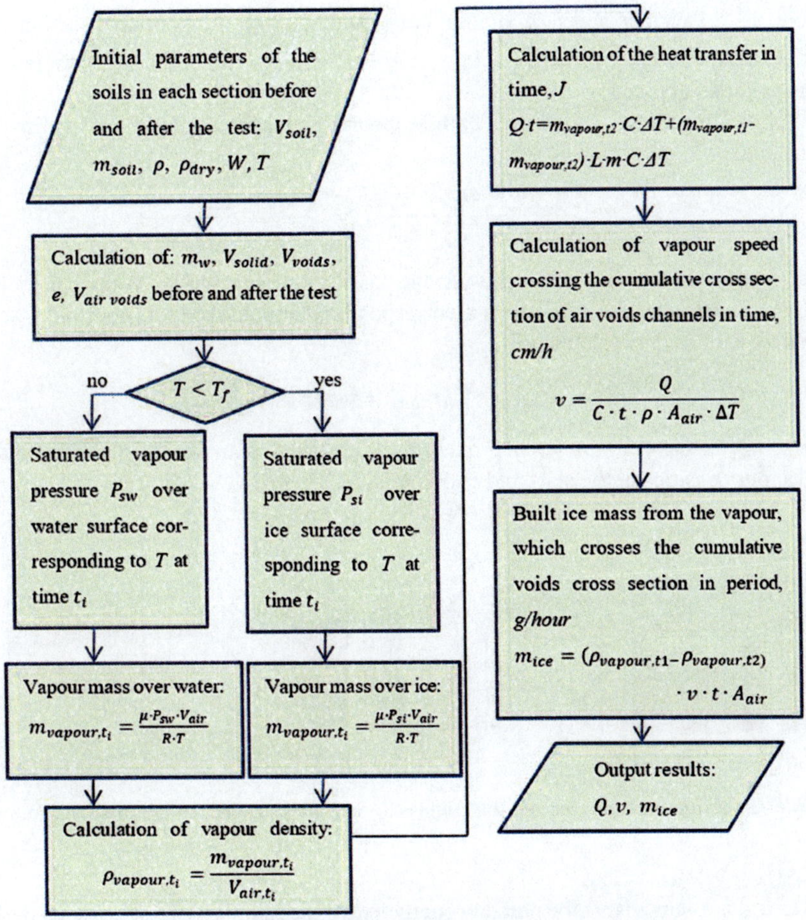

Fig. 5. Algorithm for the vapour mass transfer calculation.

where, ρ_{vapour,t_i} - vapour density for period of time, g/cm^3.

Heat energy for the phase transfer includes the latent heat for the condensation and solidification of the vapour mass difference at the beginning τ_1 and end time τ_2 of the calculation period.

$$Q_2 = \left(m_{vapour,1} - m_{vapour,2}\right) \cdot L \tag{4}$$

where, $m_{vapour,2}$ – mass of the vapour at the end period t_2; L – is a total latent heat $L = L_1 + L_2$, where L_1 - specific latent heat for condensation $L_1 = 2.3 \cdot 10^6$ J/kg and L_2 - specific latent heat for solidification $L_2 = 0.335 \cdot 10^6$ J/kg of 1 kg of water.

The volume of vapour V_{vapour} is equal to the speed of vapour passing through the air voids' cross-section A over the time t:

$$V_{vapour} = v \cdot t \cdot A_{air\ voids} \tag{5}$$

where, v – an average speed of vapour, cm/h; and $A_{air\ voids}$ – cumulative section cross of the air voids' $A_{air\ voids} = \frac{\pi \cdot d_a^2}{4}$, cm^2, corresponding to the porosity coefficient and moisture content (see Fig. 6).

Substituting the V_{vapour} in Eq. (2) the vapour speed is found at the starting time and at the end:

$$v_{vapour} = \frac{4Q}{C \cdot \rho \cdot \pi \cdot d_a^2 \cdot \Delta T \cdot t} = \frac{Q}{C \cdot \rho \cdot A_{air} \cdot \Delta T \cdot t} \tag{6}$$

The mass of ice built from the vapour passing through the air voids channels in a 10 cm length mould section with a correspondent cumulative cross-section A_{air} and speed v over time t is calculated:

$$m_{ice} = \rho_{vapour} \cdot V_{air\ voids} = \rho_{vapour} \cdot v \cdot t \cdot A_{air} \tag{7}$$

where, m_{ice} is mass of built ice in grams; ρ_{vapour} is taken as an average density value of the vapour densities at the start and end time point, g/cm^3.

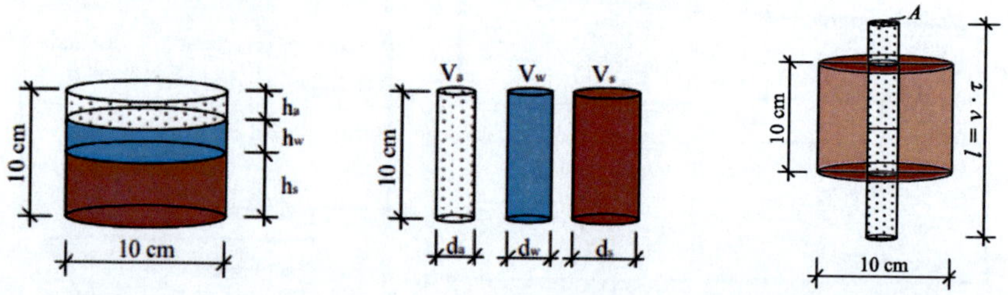

Fig. 6. Vapour rate passing through the cumulative air voids channel in the mold section over time t.

Here, it is assumed that the porosity coefficient of the sample remains constant during the calculation period. Consequently, the volume of the air voids and the cumulative cross-section of the air voids also remain constant. It should be noted that only heat consumed for the gas phase energy exchange was considered in this problem. The solid and liquid parts have also been cooled down with the heat withdrawn by the cooling machine. However, they are not counted because all the necessary energy exchange has been done by an automatic set of temperature controls.

The model is suitable for numerical solutions, like finite element analysis or a model similar to a coupled heat-water transfer, in terms of considering the vapour flow and taking into account the cross-section forces. The latent heat for the phase transitions and the dynamic change of the coefficient of porosity and air void volume also needs to be taken into account.

3 Results

The temperature distribution in the freeze-thaw test with a range of soil density did not obtain a steady pattern of freezing rate in terms of initial density. This is possible because the freezing rate of 2 °C per day provided sufficient time for uniform temperature distribution and stabilization of the temperature field across the range of soil densities. However, a deceleration of the freezing rate was noted in the second freezing cycle.

Figure 7 shows the time intervals when the temperature reached 0, −5 and −10 °C for the part 15 cm from the top of each soil sample, during both the first and second freezing cycles. In column #1, −5 °C was registered at 15 cm depth from the soil surface after 147 h for the first freezing cycle, and this took 518 h in the second freezing cycle (Fig. 8).

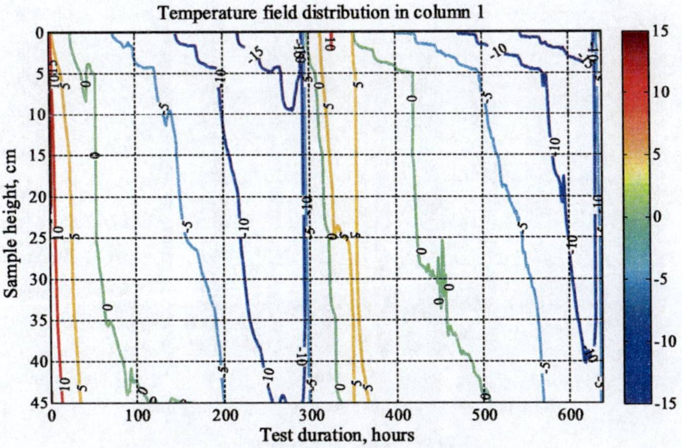

Fig. 7. Temperature field contours.

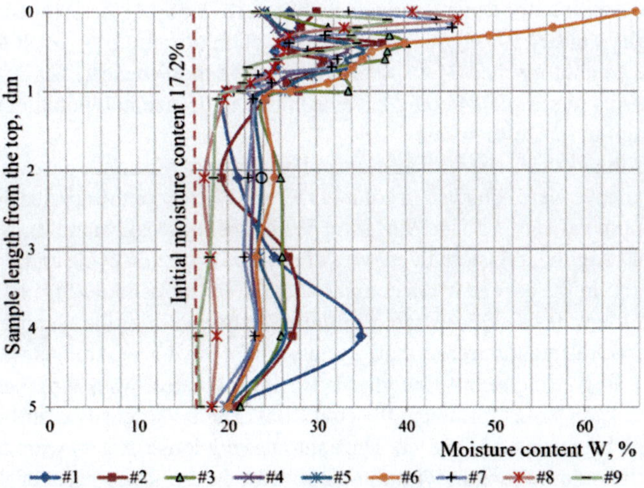

Fig. 8. Moisture intake during the freeze-thaw cycles of the soils in test with a deionized water supply and various densities.

In the top 10 cm layer, ice lens formation caused a highly irregular distribution of moisture, which was justified by the centimeter sampling. Since the groundwater supply table was quite shallow, at 45 cm depth, the moisture intake was higher than for Columns 8 and 9, where the water supply was located at 95 cm depth, and the soil samples were made with maximum dry density. For this reason, the moisture content between 15 and 45 cm from the soil surface was relatively stable and depended just on the density of the soil samples (Fig. 9).

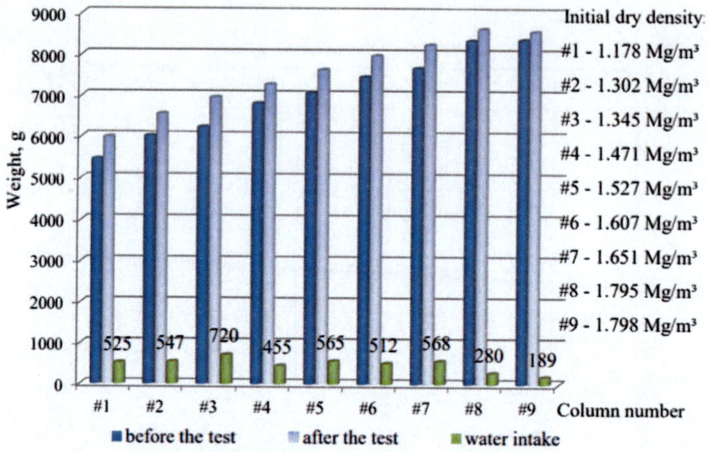

Fig. 9. Sample preparation.

According to the results, the moisture content represents advanced water intake in the loose soil, with a dry density range between 1.18–1.65 Mg/m^3, compared to dense soils with a dry density 1.8 Mg/m^3. With the exception of sample #4, in all the columns, the moisture content reached 24.5% or above by the end of the test. The reason for low moisture intakes in sample #4 could have been due to occasional violation of the water supply during the freeze-thaw test.

In freeze-thaw cycles test with a deionized water supply from the base, the maximum rates of frost heave were achieved in columns #7 and #8, where the dry density was close to the maximum value of 1.65–1.79 Mg/m^3. While the loose soil samples, with dry density 1.18–1.47 Mg/m^3, registered very weak heaving in the first cycle and consolidation or compression in the second cycle compared to the initial volume (Fig. 10).

In freeze-thaw tests with variable initial density packing, the sample with loose soils condensed during the freeze-thaw cycles, whilst the dense samples were loosened (see Fig. 11). Regarding sample length, the top layers were always loosened due to the high content of ice lenses formed at the cold front. The bulk density distribution by the column length is presented in Fig. 12. The columns with loose density were consolidated throughout the column length, with the most value in the lower parts of the samples. Whilst in the dense soil samples, only the 10 cm layer on top was loosened and had attracted a large amount of water to form ice lenses.

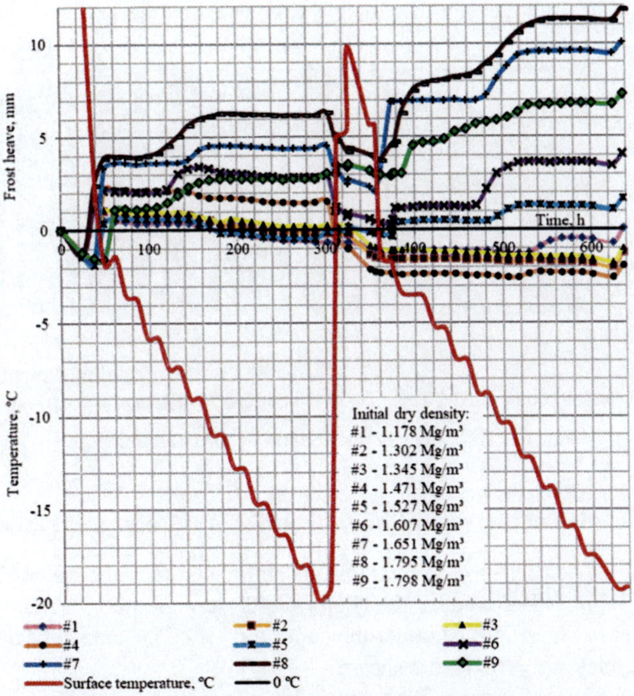

Fig. 10. Frost heave in the samples with varied densities.

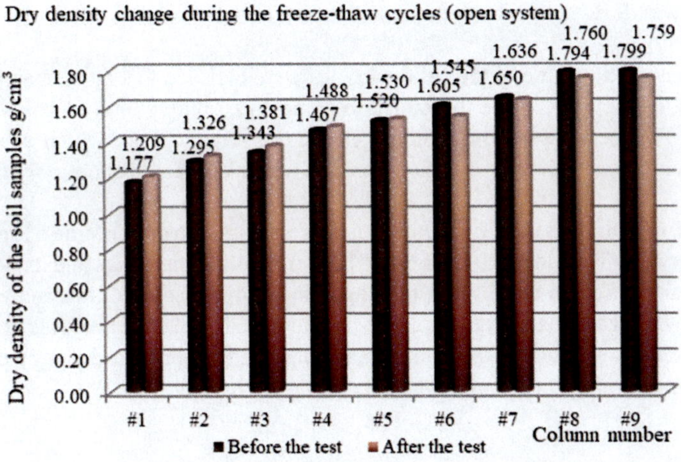

Fig. 11. General density change in the samples with variable dry density.

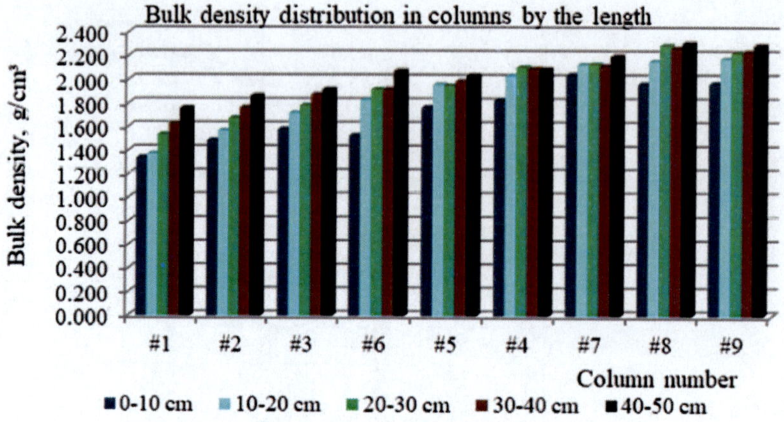

Fig. 12. Bulk density distribution by the column length

3.1 Determination of the Vapour Mass Transfer in Unsaturated Freezing Soils

The results of the density-moisture parameters before and after the freeze-thaw test are presented in Table 3. Here, under the "before test" title, the initial setting parameters at $t = 0$ h are presented. Whilst under the "after test" title, the data collected after two freeze-thaw cycles at $t = 636$ h are shown.

The period of time between 590 h and 614 h, which corresponded to the end of the second freezing cycle period, was considered for analyzing the mass transfer within the frozen soils. Sub-zero temperature distribution is presented in Tables 5a and 5b. The length of the soils sample was 50 cm, which was composed of five assembled mold sections, each 10 cm in length. Moreover, each of the sections had an overlay joint of 5 mm. Because the density varied from the loosest one in column #1 to the densest #8 and #9, it was decided to consider the mass moisture transfer in the unsaturated samples #1 and #7.

The mass and volume of the solid particles were believed to stay unchanged during the freeze-thaw cycles, while the moisture content and the volume of voids had been modified. The ice lenses formed mostly in the top section layer, which caused some linear deformation in the vertical direction and was taken into account only for the top section.

Further calculations were performed on top section #11 in soil column #1 for the time interval between 590 and 614 h (Table 4). The moisture content intake and its distribution over the sample length were set at the beginning and measured at the end of the two freeze-thaw cycles. Because the time slot was considered during the final freezing stage of the test, the moisture content distribution and correlated volumetric-density calculations have been performed according to the data presented in Table 1.

Table 3. Soil sample parameters before and after the freeze-thaw cycles.

Column	Weight of soil, g		Bulk density, average by sample length ρ, Mg/m^3		Moisture content, average by sample length *W*, %		Dry density, average by sample length ρ_{dry}, Mg/m^3		Void ratio, average by sample length *e*	
	Before	After test	Before test	After test	Before test	After test	Before	After test	Before	After test
#1	**5,463**	**5,988**	**1.380**	**1.512**	**17.20**	**25.01**	**1.177**	**1.209**	**1.223**	**1.162**
#2	6,012	6,559	1.519	1.660	17.20	25.23	1.295	1.326	1.020	0.972
#3	6,236	6,956	1.575	1.760	17.20	27.39	1.343	1.381	0.947	0.893
#4	7,452	7,964	1.883	1.842	17.20	23.78	1.605	1.488	0.629	0.604
#5	7,057	7,622	1.783	1.918	17.20	25.35	1.520	1.530	0.721	0.709
#6	6,813	7,068	1.721	1.994	17.20	29.07	1.467	1.545	0.782	0.907
#7	7,662	8,230	1.936	2.037	17.20	24.50	1.650	1.636	0.585	0.598
#8	8,332	8,612	2.105	2.124	17.20	20.70	1.794	1.760	0.457	0.486
#9	8,355	8,544	2.111	2.126	17.20	20.85	1.799	1.759	0.453	0.487

Table 4. Calculation of soil parameters for the 590–614 h time interval on the example of sample #1.

Sample section	Distance from the top surface, cm	Volume in a section, cm^3	Mass of solids, g	Mass of water, g	Mass of the soil in a section, g	Bulk density, Mg/m^3	Average moisture content W, %	Dry density, Mg/m^3
#11	0–10	819	969.2	245.59	1214.76	1.483	**25.34**	1.183
#12	10–20	785.4	923.0	187.28	1110.30	1.414	20.29	1.175
#13	20–30	785.0	923.0	214.14	1137.16	1.448	23.20	1.175
#14	30–40	785.4	923.0	277.37	1200.39	1.528	30.05	1.175
#15	40–50	785.4	923.0	254.75	1177.78	1.500	27.60	1.175

The results of moisture mass transfer calculations in column #5 are presented in Tables 6, 7a and 7b. The average bulk density of the soil at the end of freeze-thaw cycles was 1.89 g/cm^3. The moisture content comprised 30% at the top section and 23–24% in the rest of the sample. The obtained ice mass formation rate was in the range of 6.18 ·

Table 5a. Calculation of the mass moisture transfer in the soils for sample #1

Sample section	Volume of solids, cm^3	Volume of voids, cm^3	Voids ratio e	Volume of air, cm^3	Temperature at 590 h	Saturated vapour pressure over ice, P_{si}, Pa	Mass of vapour at 590, g	Density of the saturated vapour at 590 h, Mg/m^3
#11	**370.62**	**448.46**	**1.21**	**180.55**	**−10.99**	**237.93**	**$3.55 \cdot 10^{-4}$**	**$1.97 \cdot 10^{-6}$**
#12	352.97	432.43	1.23	228.13	−10.30	253.24	$4.76 \cdot 10^{-4}$	$2.09 \cdot 10^{-6}$
#13	352.97	432.43	1.23	198.83	−9.49	272.16	$4.45 \cdot 10^{-4}$	$2.24 \cdot 10^{-6}$
#14	352.97	432.43	1.23	129.85	−8.36	300.63	$3.19 \cdot 10^{-4}$	$2.46 \cdot 10^{-6}$
#15	352.97	432.43	1.23	154.52	−6.46	354.77	$4.45 \cdot 10^{-4}$	$2.88 \cdot 10^{-6}$

Table 5b. Calculation of the mass moisture transfer in the soils for sample #1 (continuation of Table 5a)

Sample section	Temperature at 614 h, °C	Saturated vapour pressure over ice, P_{si}, Pa	Mass of vapour at 614 h, g	The heat realised in 24 hours $Q \cdot t = m \cdot C \cdot \Delta T$, J	Vapour rate $v = 4 \cdot N/(C \cdot \rho \cdot \pi \cdot d^2 \cdot \Delta T)$, cm per 24 h	Vapour rate, cm/h	Build-up of ice mass between the period 590/614 h, g/hour
#11	**−13.30**	**193.26**	**$2.91 \cdot 10^{-4}$**	**0.1702**	**9.008**	**0.375**	**$2.67 \cdot 10^{-6}$**
#12	−12.60	205.98	$3.91 \cdot 10^{-4}$	0.2269	9.014	0.376	$3.56 \cdot 10^{-6}$
#13	−11.68	223.81	$3.69 \cdot 10^{-4}$	0.2016	9.067	0.378	$3.16 \cdot 10^{-6}$
#14	−10.43	250.35	$2.68 \cdot 10^{-4}$	0.1363	9.126	0.380	$2.14 \cdot 10^{-6}$
#15	−8.96	284.98	$3.61 \cdot 10^{-4}$	0.2237	8.956	0.373	$3.51 \cdot 10^{-6}$

Table 6. Calculation of soil parameters for the 590–614 h time interval for sample #5

Sample section	Distance from the top surface, cm	Volume in a section, cm^3	Mass of solids, g	Mass of water, g	Mass of the soil in a section, g	Bulk density, Mg/m^3	Average moisture content W per section, %	Dry density, Mg/m^3
#11	0–10	854.73	1251.96	374.086	1626.05	1.902	29.88	1.465
#12	10–20	785.40	1192.34	280.797	1473.14	1.876	23.55	1.518
#13	20–30	785.40	1192.34	282.347	1474.69	1.878	23.68	1.518
#14	30–40	785.40	1192.34	297.609	1489.95	1.897	24.96	1.518
#15	40–50	785.40	1192.34	276.147	1468.49	1.870	23.16	1.518

Table 7a. Calculation of the moisture mass transfer in the soils for sample #5

Sample section	Volume of solids, cm^3	Volume of voids, cm^3	Voids ratio *e*	Volume of air, cm^3	Temperature at 590 h, °C	Saturated vapour pressure over ice, P_{si}, Pa	Mass of vapour at 590, g	Density of saturated vapour at 590 h, Mg/m^3
#11	478.76	375.97	0.79	0	−13.48	190.14	–	–
#12	455.96	329.44	0.72	23.12	−11.45	228.51	$4.37 \cdot 10^{-5}$	$1.89 \cdot 10^{-6}$
#13	455.96	329.44	0.72	21.43	−10.69	244.59	$4.33 \cdot 10^{-5}$	$2.02 \cdot 10^{-6}$
#14	455.96	329.44	0.72	4.78	−10.29	253.46	$9.99 \cdot 10^{-5}$	$2.09 \cdot 10^{-6}$
#15	455.96	329.44	0.72	28.20	−8.98	284.46	$6.58 \cdot 10^{-5}$	$2.33 \cdot 10^{-6}$

10^{-8}–$3.82 \cdot 10^{-7}$ g/h (Table 7b). In the top 10 cm layer by the initial time of the testing period 590 h, the soil had reached the fully saturated condition, and further mass transfer was implemented via hygroscopic water in a liquid state. Consequently, the vapour mass transfer calculation in the top section was not applicable from that moment onwards.

Table 7b. Calculation of the moisture mass transfer in the soils for sample #5 (continuation of Table 7a)

Sample section	Temperature at 614 h, °C	Saturated vapour pressure over ice, P_{si}, Pa	Mass of vapour at 614 h, g	The heat realised in 24 h$Q \cdot t = m \cdot C \cdot \Delta T$, J	Vapour rate$v = 4 \cdot N/(C \cdot \rho \cdot \pi \cdot d^2 \cdot \Delta T)$, cm per 24 h	Vapour rate, cm/h	Build of ice mass between 590/614 h period, g/h
#11	−15.46	158.49	–	–	–	–	–
#12	−13.61	187.89	$3.63 \cdot 10^{-5}$	0.0199	9.066	0.378	$2.82 \cdot 10^{-7}$
#13	−12.83	201.65	$3.60 \cdot 10^{-5}$	0.0194	9.078	0.378	$2.76 \cdot 10^{-7}$
#14	−12.36	210.49	$8.36 \cdot 10^{-6}$	0.0043	9.113	0.380	$6.18 \cdot 10^{-8}$
#15	−10.92	239.48	$5.58 \cdot 10^{-5}$	0.0265	9.178	0.382	$3.82 \cdot 10^{-7}$

4 Discussion

Against the backdrop of the ongoing global warming and enlargement of the areas with periglacial soil formation [19], the issues of seasonal transitions through zero degrees Celsius and associated frost heaving modify and, in cases of an increased volume of air voids, worsen the conditions of highway subsoils in the wintertime. Because of the potential complexity of soil conditions on-site and the practical issues when implementing field research, a laboratory-based approach using non-saline soil samples in an "open system" was developed. The significantly modified experimental design not only benefited from clear observation of temperature distributions and moisture transport over the sample length with time but also delivered a better understanding of mass moisture transfer in a vapour state at sub-zero temperatures. The maximum frost heave was observed at the freezing front of a 2 cm/day moving rate. Moreover, increased sample length signified the reduction of temperature drop near the transition zone next to the 0 °C isotherm. Such retardation of the freezing rate was necessary for the phase transition to occur and the heat balance to equilibrate or endeavor to do so. The moisture intake and its redistribution within the soil samples were also found to depend on the density and the chemical content of pore water. The total moisture intake in ice and unfrozen water, which was pulled up to the top 10 cm section, prevailed in a deionized water supply test.

There are not many works that have considered the gaseous component of mass transfer. The vapour flux has widely been omitted or underestimated by scientists in the calculation of mass transfer. One of the few works that have been considered was presented by [17].

The main distinctions and similarities of the current proposal with what they put forward are considered here:

1. In the current work, it is accepted that moisture mass transport is implemented primarily in the gas phase, i.e., vapour in the freezing fringe transport zone. Otherwise, it is supplied in the capillary zone from the water table source in a liquid state. Zhang et al. (2016) [17] accept the intake moisture flow as combined from liquid and water fluxes, conforming to thermal and isothermal hydraulic conductivities. The liquid water flow was presented by Richard's equation and explained as driven by water potential or temperature gradient, which is proportional to hydraulic conductivity.
2. In this work, it is assumed the vapour is fully saturated and filling all the volume of air voids in the soil. At the same time, Zhang et al. (2016) calculated the relative humidity Hr and vapour diffusivity Da, including the empirical parameter of an enhancement factor η [17]. The vapour density was also found by the empirical formula of saturated vapour density multiplied by its relative humidity.
3. Zhang et al. (2016) determined the unfrozen water content in equation [17] by temperature, and the coefficients a and b were estimated by regression analysis of the measured data of temperature and unfrozen water content. However, it could be the case that "a" and "b" might vary for different types of soils. In Zhang et al.'s model, the presence of unfrozen water content in the frozen part of the soil was admitted. However, its mobility was considered doubtful and, hence, such water mass transfer was not accounted for unless it had evaporated and was transported in a gas state.
4. Zhang et al. (2016) simulated the natural field impacts of precipitation, evaporation and heat fluxes from the net radiation, which was omitted in the current study [17].
5. In this work, detailed positioning of the soil structure was considered, based on the experimental data of the measured temperature, vertical linear volumetric change registered by testing time and the obtained moisture-density relation.

The engineering properties of the samples after freeze-thaw cycles signified a clear reduction in strength, mainly because of increased moisture content and lowered dry density. The samples with loose density were compacted during the freeze-thaw cycles, while dense soils became loosened. It was observed that a shallow water table induced greater moisture intake to the soil samples and, consequently, their reduction in strength.

As many road subsoils are designed to be exploited in unsaturated conditions or undergo an unsaturated transition stage while being frozen (see Fig. 1), it is crucial to consider the moisture mass transfer in a gaseous state, accompanied by phase transitions. Building on the experimental results, the conceptual model for soil freezing was considered from the perspectives of moisture content, phase transfer and volume changes with temperature within the freezing soil. To determine the moisture mass transfer at different freezing stages, the degree of saturation and the voids ratio with time were set as the input parameters. Moisture mass transfer in the unsaturated freezing soils with a significant voids ratio was implemented predominantly with vapour flow. The concept of the moisture flows in gaseous state benefits from having a clear calculation algorithm, absence of empirical relations and compliance with the laws of physics. Vapour convection was naturally induced by the temperature gradient, where the driving force was presented by saturated vapour pressure difference with further energy release by the phase transfer in the frozen zone. This mode of moisture transfer is applicable for both saline and non-saline soils. Notably, the voids ratio and the moisture content have a significant impact on the mass transfer rate and ice growth. After the pore volume is filled with ice, primarily deposited from the vapour, the further mass transfer has to be considered with hygroscopic water, which stays in a liquid state even at very low temperatures. The migration of the hygroscopic water is 175, explained as being driven to the frozen side by cryosection forces owing to a difference in surface tension.

Considered phenomenon of the vapour mass transfer and the interphase equilibrium in the freezing soils replicate with the many broader issues in the periglacial areas and help to explain many hitherto problems of moisture mass transfer occurring in the freezing soils.

5 Conclusions

Further outcomes have been concluded from the conducted work:

1. The significantly modified testing method, including the slow freezing techniques and longer samples lengths, is an effective way to evaluate the impact of freeze-thaw cycles on soils.
2. The obtained results have improved understanding of the heat and mass transfer phenomenon during the unidirectional freezing of soils.
3. The conceptual model for frost heave in soils was developed based on the vapour mass transfer. The algorithm of vapour flow calculation in unsaturated soils was presented using fundamental thermodynamic equations.
4. Temperature redistribution during the freeze-thaw tests of the samples with variable density between the columns did not have an obvious pattern regarding the temperature field. Such a feature can be explained by the slow freezing technique, which provided sufficient time for uniform freezing regardless of the soil density.
5. Observation of the temperature distribution revealed a reduction of temperature change in the transition zone next to the 0 °C isotherm. While the cooling rate was set to be stable, at 2 °C per day, the temperature drop over the soil length and the resulting frost heave were varied. The maximum frost heave was observed when the freezing rate over the sample length was 2 cm/day. A slow freezing rate was observed in the top 10–15 cm layer, and the boosted freezing was registered in the lower part of the soil samples.
6. Low-density soil samples contributed moisture mass transfer, while increased bulk density, in contrast, decelerated the passage of water. In addition, ice lens formation in the freezing front was prevalent in dense soils free of chemical content, while moisture absorption from the water table predominated in loose soils.

5.1 Key Limitations

A potential limitation in the data is due to the design of the experiment, particularly not being able to extract geotechnical samples below 0 °C. However, as the focus was on the overall impact of freeze-thaw on moisture transport, the overall conclusions are considered to be sound. The total moisture distribution in ice and unfrozen water could not be monitored over the sample lengths with respect to the testing time. As a result, the conclusions are based on the total net moisture transfer measured at the end of each test. It should be noted that these limitations do not detract from the scientific contribution of the work, where vapour flow was identified as a key source for moisture mass transfer.

5.2 Further Research Perspectives of Mass Transfer in the Freezing Soils

According to the findings of the research, the integrated moisture mass transfer in sub-zero temperatures includes moisture transport in a gaseous flow and hygroscopic water transfer in a liquid state, which in turn, depends on the surface tension and the cryosection forces. Consequently, both vapour mass transfer and associated phase transitions need to be taken into account. That is, only after taking into consideration both components of moisture flow and understanding its driving forces will it be possible to move to a new stage of the moisture mass transfer theory. Further research could include vapour mass transfer in quantitative analysis for different types of soils, taking into account the freezing rates, volumetric-moisture characteristics and particle size scenarios. The available techniques need to be reviewed and expressed with formulae explaining the driving force of hygroscopic water transport and vapour transport capacity. Special attention should be given to vapour mass transfer under highways. Application of the dynamic load, causing repeated short-term pressure increases and accompanying phase transition, would have an effect on the volumetric variations, and this needs to be considered in conjunction with site observations. Regarding laboratory experiments, installation of pore pressure probes and equipment sets for determining the voids ratio and moisture content, measured by sample length and recorded hourly, would benefit the calculations of the total flow of moisture with time. It is also important 177 to consider the measurement and monitoring of hygroscopic moisture transport in relation to the testing time. The knowledge of the displacement of this component could entirely solve the problem of integrated moisture mass transfer in soils under sub-freezing temperatures. The presented vapour flow model is suitable for implementation with numerical analysis, including the finite element method, finite volume method, among others, where the mass of built ice can be calculated from the deposited vapour over time, according to the temperature in each mesh, separately and further integrated into the continuous flow. Determination of the hygroscopic moisture is feasible as the difference of the total mass transfer and vapour flow. Application of the presented method would benefit the accuracy of frost heave prognosis by fostering understanding of soil degradation, in particular, in highways subsoils.

References

1. Simonsen, E., Isacsson, U.: Soil behavior during freezing and thawing using variable and constant confining pressure triaxial tests. Can. Geotech. J. **38**(4), 863–875 (2001)
2. Han, C., Jia, Y., Cheng, P., Ren, G., He, D.: Automatic measurement of highway subgrade temperature fields in cold areas. In: 2010 International Conference on Intelligent System Design and Engineering Application (ISDEA), pp. 409–412 (2010)
3. Sarsembayeva, A., Zhussupbekov, A.: Experimental study of deicing chemical redistribution and moisture mass transfer in highway subsoils during the unidirectional freezing. Transp. Geotech. **26**, 100426 (2021)
4. Arenson, L.U., Sego, D.C., Newman, G.: The use of a convective heat flow model in road designs for Northern regions, pp. 1–8. IEEE (2006)
5. Zhang, J., Li, J., Quan, X.: Thermal stability analysis under embankment with asphalt pavement and cement pavement in permafrost regions. Sci. World J. **2013** (2013). Article ID: 549623

6. Sakanov, D.K.: Regional specific features of temperature mode of the road constructions Cand. in techn. sci. Kazakh academy of transport and communication named after M. Tynyshbayev (2007)
7. Teltayev, B.B., Loprencipe, G., Bonin, G., Suppes, E.A., Tileu, K.: Temperature and moisture in highways in different climatic regions. Magazine of Civil Engineering **100**(8), 10011 (2020)
8. Lai, Y., Pei, W., Zhang, M., Zhou, J.: Study on theory model of hydro-thermal–mechanical interaction process in saturated freezing silty soil. Int. J. Heat Mass Transf. **78**, 805–819 (2014)
9. Hermansson, Å.: Laboratory and field testing on rate of frost heave versus heat extraction. Cold Reg. Sci. Technol. **38**(2–3), 137–151 (2004)
10. Nagare, R., Schincariol, R., Quinton, W., Hayashi, M.: Effects of freezing on soil temperature, freezing front propagation and moisture redistribution in peat: laboratory investigations. Hydrol. Earth Syst. Sci. **16**(2), 501–515 (2012)
11. Bi, G.: Study on influence of freeze-thaw cycles on the physical-mechanical properties of loess. In: Smart Materials and Intelligent Systems, vol. 442, pp. 286–290 (2012)
12. Bing, H., He, P.: Experimental investigations on the influence of cyclical freezing and thawing on physical and mechanical properties of saline soil. Environ. Earth Sci. **64**(2), 431–436 (2011). https://doi.org/10.1007/s12665-010-0858-y
13. Ming, F., Li, D.: Experimental and theoretical investigations on frost heave in porous media. Math. Probl. Eng. (2015). Article ID: 198986. https://doi.org/10.1155/2015/19898
14. Wu, D., Lai, Y., Zhang, M.: Heat and mass transfer effects of ice growth mechanisms in a fully saturated soil. Int. J. Heat Mass Transf. **86**, 699–709 (2015)
15. Cui, Z., Zhang, Z.: Comparison of dynamic characteristics of the silty clay before and after freezing and thawing under the subway vibration loading. Cold Reg. Sci. Technol. **119**, 29–36 (2015)
16. Sinitsyn, A.O., Løset, S.: Strength of frozen saline silt under triaxial compression with high strain rate. Soil Mech. Found. Eng. **48**(5), 196–202 (2011). https://doi.org/10.1007/s11204-011-9148-2
17. Zhang, M., Wen, Z., Xue, K., Chen, L., Li, D.: A coupled model for liquid water, water vapor and heat transport of saturated–unsaturated soil in cold regions: model formulation and verification. Environ. Earth Sci. **75**(8), 1–19 (2016). https://doi.org/10.1007/s12665-016-5499-3
18. Zhussupbekov, A., Alibekova, N., Akhazhanov, S., Sarsembayeva, A.: Development of a unified geotechnical database and data processing on the example of Nur-Sultan City. Appl. Sci. **11**(1), 1–20, 306 (2021)
19. Harris, C., et al.: Permafrost and climate in Europe: monitoring and modelling thermal, geomorphological and geotechnical responses. Earth Sci. Rev. **92**, 117–171 (2009)

Characterization Properties of Three In-House Produced Biochars from Different Feedstock

Boneng Chen, Lina Xiao, Xiaoli Huang, Xia Bao, and Ankit Garg(✉)

Shantou University, Shantou 515063, China
{21bnchen,21lnxiao1,19xlhuang1,20xbao1,ankit}@stu.edu.cn

Abstract. Biochar is a carbon product produced from the pyrolysis of animal and plant wastes and is mainly used as a soil conditioner. However, biochar's ability to improve soil varies depending on its microscopic properties. Therefore, it is necessary to study the characterization properties of biochar before large-scale application. In this study, three common plant wastes (Wedelia trilobata, Shaddock peel, Bagasse) were selected to produce biochar and carry on the characterization analysis on them. Biochar produced at different pyrolysis temperatures (300 °C, 600 °C) were characterized using a suite of analytical techniques, including SEM, EDS, FTIR, XRD. When the pyrolysis temperature of biochar increased from 300 °C to 600 °C, the oxygen (O) ratio in SPB and BB decreased from 16.46% and 13.18% to 9.19% and 3.37% respectively, and the degree of carbonization and specific surface area increased correspondingly. In addition, high temperatures will decompose some functional groups of biochar, and the PH values of biochar prepared at different pyrolysis temperatures are different. Such properties of biochar will be useful in selecting suitable soil conditioners for usage in geo-engineering infrastructure.

Keywords: Biochar · Characterization · Pyrolysis · Soil amendment

1 Introductions

Biochar has received increasing attention over the past few decades. It is a kind of black high-porosity material with high carbon content produced by thermal chemical decomposition of organic matter in low oxygen or no oxygen state [1–3]. Biomass molecules are turned into charcoal under thermal pressure, a process called thermal decomposition. The thermal decomposition process of biomass can be divided into carbonization, slow pyrolysis, intermediate pyrolysis, among which slow pyrolysis is the best.

The properties of biochar mainly depend on the feedstocks and pyrolysis temperature [4–6]. It has already been widely used in daily life. At present, biochar is not only widely used in water purification as a catalyst, active agent, electrode material and electrode modifier, but also used as a modified base material to improve the performance of green infrastructure. In this study, modern analytical techniques are used to identify the characterization of different biochar. A lot of research has been conducted on the characterization of biochar, but due to the shortage of available experimental resources and

H.-H. Zhu et al. (Eds.): BRWSG 2021, LNCE 230, pp. 33–43, 2022.
https://doi.org/10.1007/978-981-16-9963-4_3

instruments and other facilities, there is still insufficient practical applications. Biochar characterization is intended to distinguish it from soil organic matter and other forms of black carbon produced by various biomass. Its usefulness depends largely on its physic-ochemical properties. For instance, biochar with a rich microporous structure and strong absorbability can improve soil fertility. The microporous structure of biochar can reduce the bulk density of soil and remove the heavy metals in the passivated soil. The unique stability of biochar can capture and sequester carbon dioxide in the soil, which plays a good role in carbon sequestration.

More research is recently conducted to analyze the practical application of biochar in geo-engineering infrastructures such as slopes, landfill cover and including green infrastructure (green roofs, biofiltration units). Although the mechanism of action of biochar in applications related to agriculture is widely studied, there is still a lack of understanding of the mechanism of application of different types of biochars in engineered soil. As a first step, it will be useful to interpret the physio-chemical properties of biochar produced from different types of feedstock and pyrolysis conditions (i.e., temperature). Biochar characterization can be performed using a variety of modern techniques, such as Scanning Electron Microscopy (SEM), Energy Dispersive Spectrometer (EDS), Fourier infrared spectroscopy (FTIR), and X-ray diffraction (XRD) etc., which were generally used to differentiate biochar from other organic matter in soil [7]. The above characterization techniques can be used to identify different biochar properties, for instance, SEM for biochar morphology, EDS for the content of elements, FTIR for determining functional groups, XRD for Structure and size of crystalline materials etc.

The objective of this study is to compare the characterization of various biochars produced from three different feedstocks (Wedelia trilobata, Shaddock peel and Bagasse) at different pyrolysis temperatures. In this study, the characterization was conducted using SEM, EDS, FTIR and XRD.

2 Materials and Methods

2.1 Preparation of Biochar Samples

The feedstocks used in the experiment were Bagasse, Wedelia trilobata and Shaddock peel. Bagasse and Shaddock peel were collected from the local market, while Wedelia trilobata were collected from the roadside around Shantou University campus. First, the collected feedstocks are screened and cleaned, air-dried and crushed. Feedstocks were kept in a sealed bag before converting them into biochar using the pyrolysis process.

The biochar used in this experiment was produced under nitrogen and controlled temperature conditions. The instrument used for biochar production is a programmable muffle furnace (STM-10-13). The three materials were pyrolyzed at 300 °C and 600 °C (heating rate of 100 $°C \cdot min^{-1}$), respectively, with a residence time of 180 min.

After drying and weighing, biochar was mashed into powder. Most of the biochar particles were able to pass through a 0.9 mm sieve. The biochar was put into a sealed bag. The container used for sample preparation is the petri dishes which measures the diameter is 1.3 cm, and the height is 1.7 cm. In order to distinguish each type of sample clearly, the samples and petri dishes were marked with names in the format of material name abbreviation and pyrolysis temperature. In total, six in-house biochars were produced.

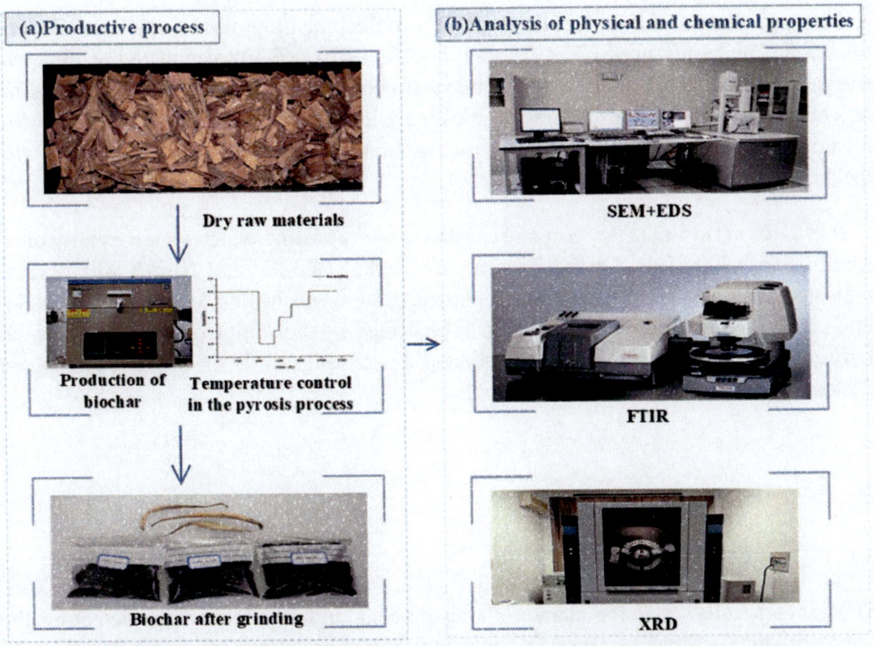

Fig. 1. Schematic diagram of the experimental process and parameter setting

They are named as: Wedelia trilobata 300 °C (WTB300), Wedelia trilobata 600 °C (WTB600); Shaddock peel 300 °C (SPB300), Shaddock peel 600 °C (SPB600); Bagasse 300 °C (BB300), Bagasse 600 °C (BB600) (see Fig. 1).

2.2 Biochar Characterization

The physical structure of biochars prepared at different feedstocks and temperatures was analyzed using a scanning electron microscope (SEM). Energy Dispersive Spectrometer (EDS) is used to analyze the type and content of elements in the material microregion, in conjunction with the use of SEM and transmission electron microscope. Different elements release different energy during the process of energy level transition under X-ray irradiation, thus producing their own characteristic wavelengths. EDS is equipped with probes that can detect the characteristic X-ray wavelengths of various elements in biochar to determine the content of each element in the sample. 5–6 sampling points were selected for each biochar sample, and the element content measured at each target point was averaged.

Fourier Transform Infrared spectroscopy (FTIR) is a powerful technique that can be used to obtain infrared spectra for absorption/emission of solids, liquids or gases. When the infrared radiation passes through the sample, part of the infrared radiation is absorbed by the specific covalent bond of the functional group, while the other part of the infrared radiation directly penetrates the collected spectrum representing the absorption and transmission of the molecule, forming a molecular fingerprint for chemical identification. This study used FTIR (Model: Nicolet is 50, USA) to analyze the chemical structure of biochar. To analyse the chemical structure of biochar, 1–2 mg of dry biochar was mixed with dry KBr (1:80 weight) and then grind it in a bowl for further compression for infrared scanning. FTIR scans 15 times in the range of 4000–400 cm^{-1} waves with the 4 cm^{-1} resolution.

X-Ray Diffraction (XRD). Structure and size of crystalline materials are usually measured using XRD. The X-ray diffractometer works at 40 kV and 20 mA with a scanning step size of 2° min^{-1}, using a graphite crystal monochromator and a CuKα radiation source. Compared with the standards compiled by the Joint Committee on Powder Diffraction and Standards, the data generated at a 2θ range of 0–50° identified the phase peaks.

3 Results

3.1 SEM + EDS

SEM is generally used for characterizing biochar and is effective for detecting the microstructure biochar. Microstructure images of biochar prepared from different materials at different temperatures were captured and compared at a magnification of two thousand times (see Fig. 2).

Under the same pyrolysis temperature, the surface microstructure of various biochars produced by different biomass has a significant discrepancy. Three kinds of biochar microstructures were generated at 300 °C: Wedelia trilobata biochar mainly had wavy sheet structure (a), Shaddock peel biochar mainly with a rough clumpy structure (c), bagasse biochar mainly had smooth sheet structure (e). Simultaneously the pyrolysis temperature also influences the morphology and surface structural changes in biochar [8]. In general, the surface area is found to increase with an increase in pyrolysis temperature. Compared with the biochar samples pyrolyzed at 300 °C, the samples at 600 °C showed more broken, rough, or pitted surface textures and structures (see Fig. 2b, d, f), indicating a higher degree of carbonization of biochars. This is likely due to the decomposition of organic matter, which leads to the formation of more micropores [9].

In an elemental analysis of biochar, the content of C, H, N, O and S in organic components of biochar is essential. In most cases, the amount of S in biochar is negligible [10]. Table 1 summarizes the relative content of elements of various biochars. The carbon (C) content of biochar at different pyrolysis temperatures was basically steady, while the oxygen (O) content was significantly different. When the temperature increases from 300 °C to 600 °C, the oxygen (O) content in SPB and BB declines sharply. The decline in O is possibly due to the reduction in impure volatile products and unstable structures (oxygen (O)) along with the conversion from Alkyl and O-alkyl C group convert to Aryl [11].

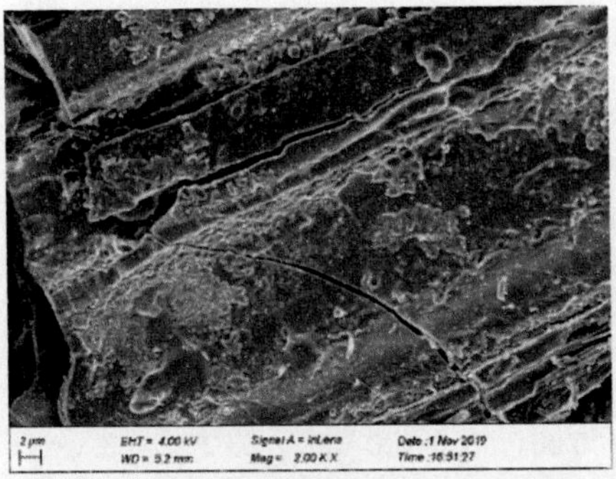

(a) WTB300

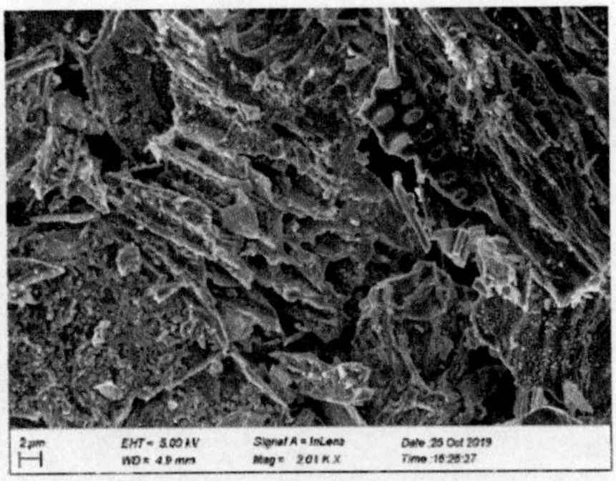

(b) WTB600

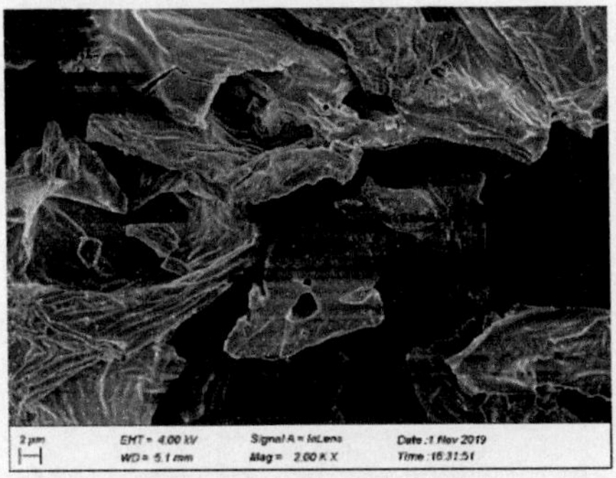

(c) SP300

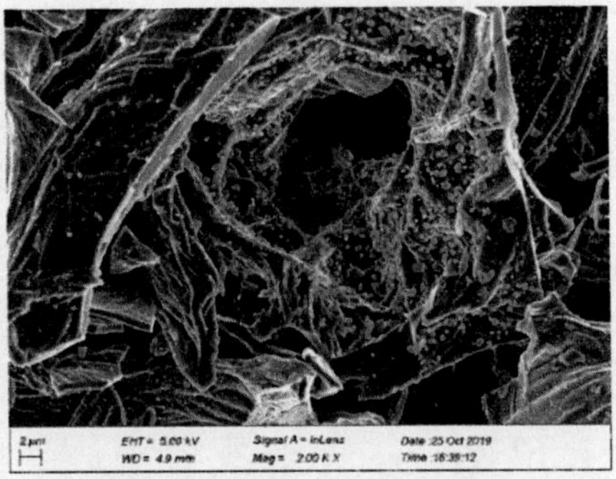

(d) SPB600

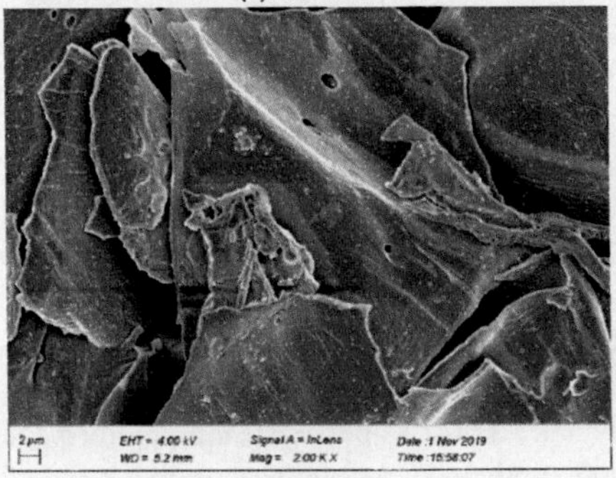

(e) BB300

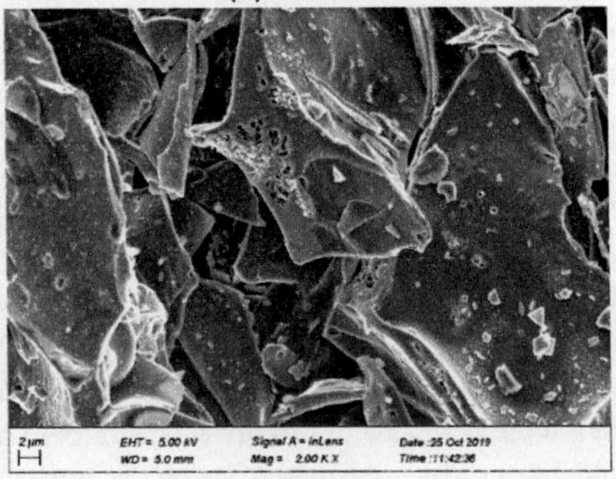

(f) BB600

Fig. 2. SEM images (under same magnification) of different materials that were pyrolyzed at 300 °C (a, c, e) and 600 °C (b, d, f)

Table 1. The average percentages of elements of various biochars at different temperatures

Element (wt.%)	WTB300	WTB600	SPB300	SPB600	BB300	BB600
C	66.85	62.18	77.51	76.10	79.70	81.79
O	18.57	17.68	16.46	9.19	13.18	3.37
Na	0.46	0.27	0.00	0.00	0.06	0.02
Mg	0.22	0.31	0.36	0.44	0.00	0.12
Al	0.00	0.15	0.00	0.00	0.00	0.00
Si	0.67	3.73	0.00	0.00	0.02	0.02
P	0.58	0.67	0.16	0.08	0.06	0.32
S	0.36	0.18	0.09	0.11	0.35	0.36
Cl	1.70	0.82	0.27	0.00	1.62	5.49
K	7.06	10.88	4.21	10.26	4.97	8.51
Ca	3.53	3.13	0.95	3.75	0.00	0.00
Ir	0.00	0.00	0.00	0.00	0.06	0.00
Mo	0.00	0.00	0.00	0.07	0.00	0.00
Total	100.01	100.00	100.01	100.00	100.00	100.00

3.2 FTIR

FTIR was used to detect the characteristics of functional groups on the surface of biochar. Many researchers reveal changes in biochar and the degree of carbonization using FTIR methods with different modes and methods [12]. Figures 3(a) and (b) exhibit the spectra of various biochars produced at pyrolysis temperatures of 300 °C and 600°, respectively. The spectral pattern of the three types of biochar at the two temperatures selected is very similar. Spectrograms (in Fig. 3(a) and (b)) show the different substitution reactions of aromatic C = C and aromatic ring at 2000–500 cm [13]. At 300 °C, the peaks of various functional groups are very obvious, and the contents of some functional groups are higher than those obtained from pyrolysis at 600 °C. At the same time, we can see that many functional groups tend to disappear at 600 °C. At 300 °C, some components of cellulose, hemicellulose and lignin tend to remain in biochar. At 600 °C pyrolysis temperature, the biochar consists mainly of aromatic compounds with an increased carbon and reduced oxygen content [14].

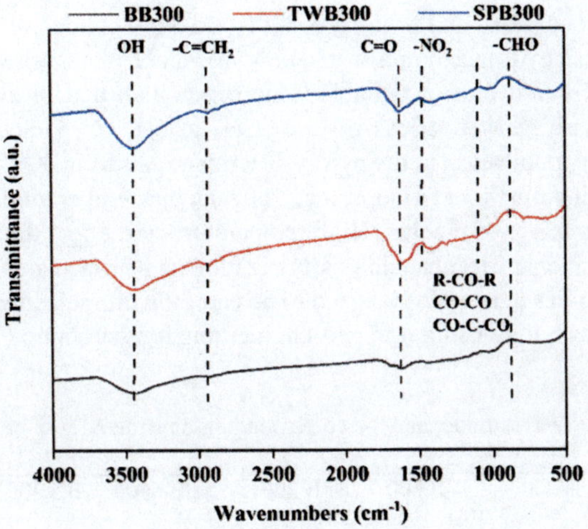

(a)

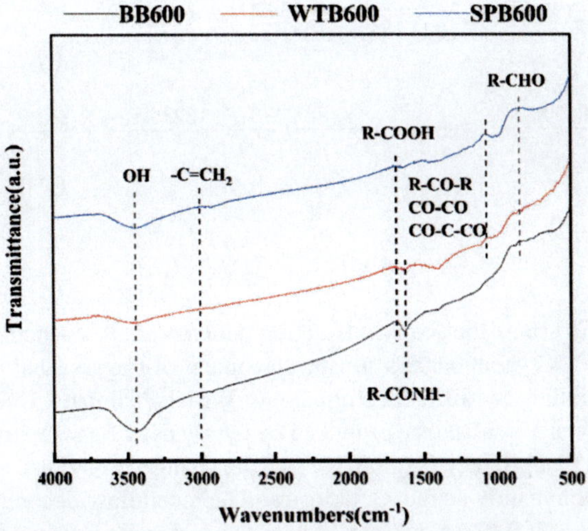

(b)

Fig. 3. (a) FTIR spectrum of biochar prepared at (a) 300 °C and (b) 600 °C, respectively

There are obvious peaks and troughs at 3700 and 3200 cm due to the stretching of O-H. This is because hydrogen bonds of phenols are weaker than those of alcohols, and hence, the general bandwidth of phenols is narrower than that of alcohols. Wedelia trilobata and bagasse showed a loss of $C = CH_2$ at 600 °C, indicating that it was decomposed during an increase in the pyrolysis temperature from 300 °C to 600 °C. By analyzing the missing functional groups, it can be seen that temperature has a significant influence on the structure of biochar. High temperature will affect the characterization characteristics of biochar by enhancing carbonization and surface area but reducing its functionality. It implies that engineers need to be careful while selecting biochar and its pyrolysis conditions before using it in geo-engineering infrastructure (Table 2).

Table 2. Wavenumber and the corresponding transmittance of biochar

Functional Group	TWB 300 (%)	TWB 600 (%)	SPB 300 (%)	SPB 600 (%)	BB 300 (%)	BB 600 (%)
O–H	3706.513	3727.726	3727.726	3743.154	3698.799	3698.799
-C=CH_2	2994.909		2991.052	2981.409	2992.980	
R–COOH	1793.474		1785.760	1700.908	1785.76	1685.48
Cl-C-H	1482.991	1560.13	1482.991	1492.633	1482.991	
R–CO–R		1074.156	1081.87	1079.941		
CO–CO						
CO–C–CO						
–CHO	865.882		883.238	823.455	863.953	908.308

3.3 XRD

Figure 4(a) and (b) show the X-ray diffraction patterns of all biochars at different temperatures. At 300 °C, the biochar's substance content of Bagasse and Shaddock peel is similar. However, they are different from that of Wedelia trilobata. Under the condition of 600 °C, the biochar's substance produced by pyrolysis of Bagasse and Shaddock peel is different from that produced at 300 °C, and the change is obvious. Biochar prepared at 300 °C is predominantly acidic. Cellulose and hemicellulose decompose into organic acids and phenols at 300 °C, so they can be used to adjust soil pH value. Biochar prepared at 600 °C is mainly alkaline, which is very different from biochar prepared at 300 °C. Changes in pH under the influence of temperature will have a corresponding impact on plant growth and survival in green infrastructure. Biochar from decomposition at different temperatures can be used to neutralize and adjust the alkalinity or acidity of existing soils for maximum benefit.

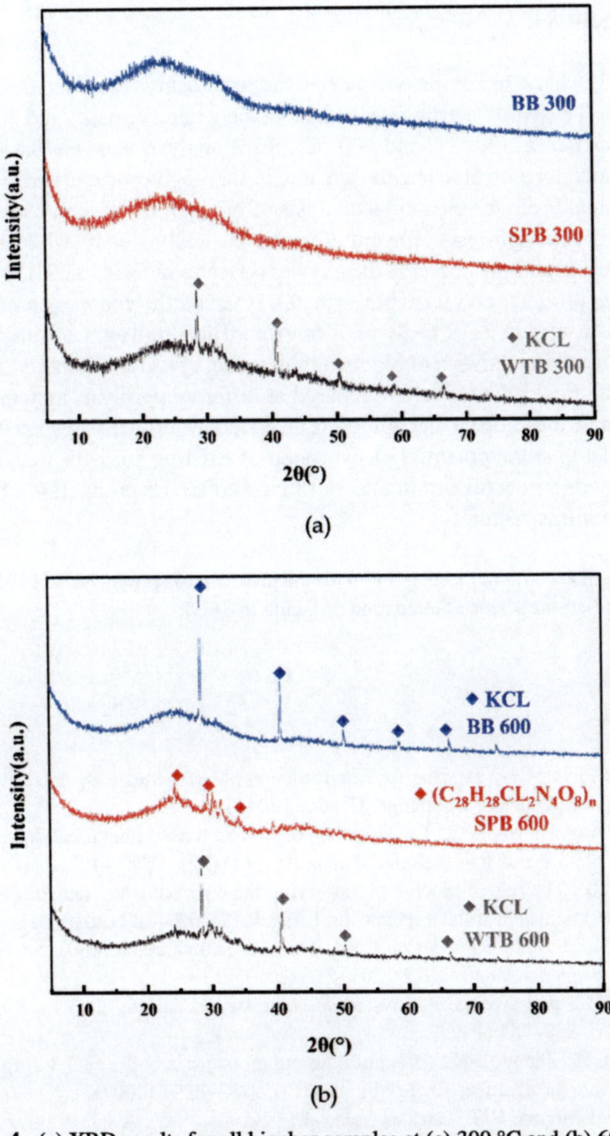

Fig. 4. (a) XRD results for all biochar samples at (a) 300 °C and (b) 600 °C

4 Conclusions

In this study, the characterization was carried out for biochars prepared from three different feedstocks (i.e., Wedelia trilobata, Shaddock peel and Bagasse) and at two different pyrolysis temperatures (300 °C and 600 °C). SEM analysis showed that the higher the pyrolysis temperature of biochar, the greater is the degree of carbonization and specific surface area. From the elemental analysis of biochar, the carbon content of various biochars at different pyrolysis temperatures was basically steady, while an increase in temperature would sharply decrease the oxygen (O) ratio in SPB and BB, possibly due to impure volatile products and unstable structure (O), and the conversion of alkyl and O-alkyl C to aryl groups. In FTIR analysis, it was found that high temperature will affect the characterization characteristics of biochar, thus reducing its functionality. XRD analysis showed that the final pH of biochar prepared at different pyrolysis temperatures would be different, and the biochar decomposed at different temperatures could be used to neutralize and adjust the potential of hydrogen of existing soils for maximum benefit. Such a study will be useful for engineers to customize the production of biochars for usage in green infrastructure.

Acknowledgment. The authors are grateful for the project grant (Grant no. 41907252) supported by the National Natural Science Foundation of China (NSFC).

References

1. Lehmann, J., Joseph, S.: Biochar for Environmental Management: Science, Technology and Implementation, 2nd edn. Routledge, London (2015)
2. Liu, W.J., Jiang, H., Yu, H.Q.: Development of biochar-based functional materials: toward a sustainable platform carbon material. Chem. Rev. **115**(22), 12251–12285 (2015)
3. Kookana, R.S.: The role of biochar in modifying the environmental fate, bioavailability, and efficacy of pesticides in soils: a review. Soil Res. **48**(7), 627–637 (2010)
4. Alexis, M.A., et al.: Fire impact on C and N losses and charcoal production in a scrub oak ecosystem. Biogeochemistry **82**(2), 201–216 (2007)
5. Boehm, H.P.: Some aspects of the surface chemistry of carbon blacks and other carbons. Carbon **32**(5), 759–769 (1994)
6. Yip, K., Wu, H., Zhang, D.K.: Effect of inherent moisture in collie coal during pyrolysis due to in-situ steam gasification. Energy Fuels **21**(5), 2883–2891 (2007)
7. Jeffery, S., Verheijen, F.G., van der Velde, M., Bastos, A.C.: A quantitative review of the effects of biochar application to soils on crop productivity using meta-analysis. Agric. Ecosyst. Environ. **144**(1), 175–187 (2011)
8. Uchimiya, M., Wartelle, L.H., Klasson, K.T., Fortier, C.A., Lima, I.M.: Influence of pyrolysis temperature on biochar property and function as a heavy metal sorbent in soil. J. Agric. Food Chem. **59**(6), 2501–2510 (2011)
9. Katyal, S., Thambimuthu, K., Valix, M.: Carbonisation of bagasse in a fixed bed reactor: influence of process variables on char yield and characteristics. Renew. Energy **28**(5), 713–725 (2003)
10. Cheah, S., Malone, S.C., Feik, C.J.: Speciation of sulfur in biochar produced from pyrolysis and gasification of oak and corn stover. Environ. Sci. Technol. **48**(15), 8474–8480 (2014)

11. Verheijen, F., Jeffery, S., Bastos, A.C., Van der Velde, M., Diafas, I.: Biochar application to soils. A critical scientific review of effects on soil properties, processes, and functions. EUR24099, 162 (2010)
12. Chia, C.H., Gong, B., Joseph, S.D., Marjo, C.E., Munroe, P., Rich, A.M.: Vibrational spectroscopy imaging of mineral-enriched biochar by FTIR. Raman and SEM—EDX62, 248-57(2012)
13. Demopoulos, G.P., Distin, P.A.: On the structure and composition of Kelex 100. Hydrometallurgy **11**(3), 389–396 (1983)
14. Zhang, H., Voroney, R.P., Price, G.W.: Effects of temperature and processing conditions on biochar chemical properties and their influence on soil C and N transformations. Soil Biol. Biochem. **83**, 19–28 (2015)

Effect of Matric Suction on Permeability and Shear Strength of Vegetated Soils

Qi Liu[1,3], Lijun Su[1,2,3](✉), Chonglei Zhang[1], Bingli Hu[1,3], and Siyou Xiao[4]

[1] Key Laboratory of Mountain Hazards and Earth Surface Process, Institute of Mountain Hazards and Environment, CAS, Chengdu 610041, China
[2] China-Pakistan Joint Research Center on Earth Sciences, Islamabad 45320, Pakistan
[3] University of Chinese Academy of Sciences, Beijing 100049, China
[4] Liupanshui Normal University, Liupanshui 553000, China

Abstract. Matric suction is an important parameter to designate the mechanical properties of unsaturated soil. To explore the influence of matric suction on the properties of unsaturated soil covered by vegetation, a series of laboratory tests were conducted to investigate the effects of matric suction on the permeability and shear strength indexes in vegetated soils. 8 test boxes compacted with sandy silt were prepared, six of which were covered with *Cynodon dactylon* and *Indigofera amblyantha* while two boxes were left bare as control. The results showed that the water retention capacity of vegetated soil was significantly improved compared with the bare soil. The variations in water infiltration amount presented a similar trend with time in bare and vegetated soils, and the water infiltration amount of vegetated soil (Herbs) and vegetated soil (Shrubs) was about 1/2 and 2/3 of bare soil when infiltration time was 2 h, respectively. The permeability coefficient of bare and vegetated soils gradually decreased as matric suction increased, and the permeability coefficient of vegetated soils was obviously lower than bare soil. Permeability coefficient of vegetated soils decreased to the magnitude of 1.00E-08 when soil matric suction was about 25 kPa. According to the soil-water characteristic curve of sandy silt, shear strength indexes of bare soil and vegetated soils increased as matric suction augmented in the boundary effect region and the transition region. The rate of increase in soil cohesion slowly diminished, and internal friction angle increased gradually as the matric suction augmented. The consequences are of great significance for using vegetation to prevent slope instability and other geological disasters, meanwhile furtherly enriching the hydraulic mechanism of vegetation in slope stability.

Keywords: Vegetated soil · Matric suction · Soil-water characteristic curve · Permeability coefficient · Shear strength

1 Introduction

The interaction between soil and water is a major topic in the research field regarding slope engineering [1]. Rainfall infiltration can increase pore-water pressure of slope soil

H.-H. Zhu et al. (Eds.): BRWSG 2021, LNCE 230, pp. 44–59, 2022.
https://doi.org/10.1007/978-981-16-9963-4_4

mass, reduce the shear strength of the potential sliding surface of rock-soil mass, and then induce slope instability [2, 3].

Vegetative protection is an effective way to increase slope stability. In particular, for the shallow slopes with a depth of 1–2 m, plants can not only improve slope stability by root reinforcing but also dissipate the excess pore-water pressure in soil via root water uptake [4–6]. Moreover, plant transpiration induces soil matric suction can increase the shear strength of slope superficial soil [7, 8]. For cohesive slopes, vegetated soil can maintain the matric suction within 5–20 kPa in the root zone under one 10-year return period rainfall [9, 10] and the matric suction more than 9 kPa under one 100-year return period rainfall [10, 11].

As one of two independent variables which control the stress-strain characteristic of unsaturated soil [12], the variation of matric suction will induce the change of soil properties, such as permeability and shear strength [3, 13–15]. The influence of vegetation on the matric suction distribution in superficial soil reduces when rainfall intensity is high [16]. Plant roots could effectively limit the development of soil fissures, whereas straight roots also provided infiltration channels for rainfall and the dominant flow, which would increase soil permeability [17, 18]. However, plant roots occupied the soil pores, which could improve soil-water retention capacity and then reduce soil permeability coefficient [11, 19–22].

Matric suction also makes a significant contribution to the shear strength of unsaturated soil [1]. For unsaturated loess specimens, when matric suction is weak, the volumetric strain of specimens changes gradually from shear shrinkage to shear dilation. Soil cohesion increases with the advance of matric suction, and the change in internal friction angle is not obvious [23]. The cohesion of unsaturated reticulate red clay is more sensitive to matric suction change than the internal friction angle. As matric suction increases, the cohesion has an increasing trend followed by a reduction [24].

At present, researchers regarding unsaturated soil mainly adopt the methods of remodeling the samples to control soil water content, which does not take soil pore-water pressure into consideration, thus ignoring the influence of matric suction on soil shear strength in practice. At the same time, the impact of matric suction of vegetated soil on soil permeability characteristics is absent. Based on this, in this paper, sandy silt was selected as the carrier, and *Cynodon dactylon* and *Indigofera amblyantha* were chosen as tested plants. Soil homogeneity and the consistency of plants' growing environment were controlled. As a result, the influences of matric suction on permeability and shear strength of vegetated soils were studied so as to investigate the hydraulic mechanism of soil reinforcement and slope protection via vegetation.

2 Materials and Methods

2.1 Materials and Equipment

The soil samples used for testing were obtained from a cutting slope on the first phase urban expressway in Xiazhou avenue in Yichang, China, and were representative of sandy silt in this region. Tested soil was chosen below the surface of 0.3 m, and impurities in the soil were eliminated. The soil was air-dried, crushed and sieved through a 2.0 mm sieve. After that, soil particle size distribution was obtained from vibrating screens and

Malvern spray particle size analysis (Fig. 1). Table 1 presents several basic properties of the tested soil, such as specific gravity, optimum water content, pH value, plastic limit and liquid limit (Table 1).

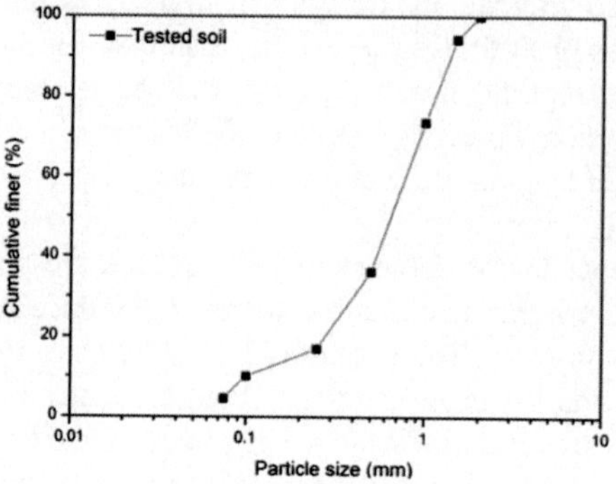

Fig. 1. Grain size distribution curve of tested soil

Table 1. Basic physical properties of experimental soil

SG	OMC (%)	MDD (%)	pH	LI (%)	PI (%)	SPC/m·s^{-1}
2.69	18.6	1.61	6.5	27.5	22.3	5.75×10^{-6}

Note: SG is specific gravity, OMC is optimum moisture content, MDD is maximum dry density, *LI* is a liquid limit, *PI* is the plastic limit, and *SPC* is saturated permeability coefficient.

Cynodon dactylon and *Indigofera amblyantha,* used widely in slope greening projects, were selected as test plants. *Cynodon dactylon* is a gramineae perennial herb, possess rhizomes and stolon. It can grow well in light sand and light acid-alkali land, as well as strong ability in drought resistance. Indigofera amblyantha belongs to a perennial deciduous shrub, and its growing period is 6 months approximately, possesses a strong ability in drought resistance and barren resistance. Both the two plants being mentioned above are the most common soil-water conservation plants in tropical and subtropical regions.

Soil permeation instrument: the WS-55 type permeation instrument produced by Nanjing Soil Instrument Co., Ltd in China was used in this paper. The sample size is Ø61.8 mm × 40 mm, and specimens were prepared according to the instructions in the Geotechnical Test Method Standard (GB/T50123–1999) in order to measure the soil permeability coefficient.

Direct shear apparatus: the ZJ strain-controlled four-type direct shear instrument was employed in this paper, produced by Nanjing Soil Instrument Co., Ltd in China. The instrument with a maximum vertical load of 400 kPa, and the sample size is Ø61.8 mm × 20 mm. Shear velocities including 2.4, 0.8, 0.1 and 0.02 mm/min, respectively.

Mini-lysimeter: the instrument is a SoilTron automatic weighing lysimeter, which consists of the multi-layer profile detectors and the data collector. The plants were cultivated in a woody cubic box. In the front surface of the box, several holes were set orderly from top to bottom so that detectors could be inserted to measure soil matric suction. Moreover, the data collector was utilized to automatically collect and record the data of each detector regularly.

2.2 Test Design and Methods

The size of the woody cubic box is 300 mm × 300 mm × 300 mm. Homogeneous soil through a 2 mm sieve was placed in the box, soil height and compactness were 250 mm and 90%, respectively [10, 25]. For both *Indigofera amblyantha* and *Cynodon dactylon*, three replicates each were set. In addition, four bare replicates were prepared as comparisons. In the planting experiment, the planting density of *Cynodon dactylon* was 15 g/m^2, while which of *Indigofera amblyantha* was 80–100 plants per square meter [25, 26].

The planting experiment was conducted in an open space behind a laboratory in China, Three Gorges University. After spreading plant seeds, planting samples were watered completely. All planting samples were subjected to the same conditions of rainfall, temperature and light. The planting period of *Cynodon dactylon* was half a year, while which of *Indigofera amblyantha* was 2 years (Fig. 2).

Fig. 2. Planting experiment

Transpiration Test. The transpiration test was conducted in a laboratory in China, Three Gorges University. The laboratory was set at a temperature of 20 °C ± 2 °C, a humidity of 50%–60%, an illumination height of 1m, cold light illuminator (plant growth lamp) is 36 W, and thermal light illuminator is 275 W [10, 27].

A pair of 5 mm diameter holes were set from top to bottom every 50 mm on the planting box. Then detectors of lysimeter were inserted into the holes to measure soil matric suction. The planting samples were uniformly watered before conducting the test until accumulated water occurred on the surface of the samples. Subsequently, a drying test (soil evaporation or plant transpiration) was performed, and the variations of matric suction at 5 cm, 10 cm, 15 cm and 20 cm depth were monitored automatically once every 10 min through the lysimeter (Fig. 3).

Fig. 3. Transpiration test

Direct Shear Test. After the drying test was completed, the above-ground biomasses of planting samples were eliminated, and the planting box was disassembled. Specimens were taken horizontally at different depths of the center of the samples by using the cutting ring, which matched with the direct shear instrument. Twenty cutting-ring specimens were taken at the same depth, and consistent root biomass was ensured. Five groups of specimens were set, with four cutting-ring specimens in each group. Each set of specimens was humidified or dried, and the matric suction was controlled by 5 kPa, 10 kPa, 15 kPa, 20 kPa and 25 kPa, respectively.

Considering the confining pressure on vegetated soil in the actual practice, the cutting-ring specimens were loaded under the normal stress of 25 kPa, 50 kPa, 75 kPa and 100 kPa and sheared at the rate of 0.02 mm/min until the readings of the dynamometer do not change with the increase of shear displacement (Fig. 4). In the shear test, the maximum shear stress under each vertical load was obtained [28].

Fig. 4. Direct shear test

The test results were plotted in a τ-P curve, and then the cohesion c and the internal friction angle φ of specimens were obtained. The relationship between shear stress and normal stress can be expressed by the followed formula:

$$\tau = c + p\tan\varphi \quad (1)$$

where τ is the shear stress of specimens, kPa; c is the cohesion, kPa; φ is the internal friction angle, °; p is the normal stress, kPa.

Fig. 5. Penetration test

Penetration Test. The sampling procedures were the same as the direct shear test. Matric suction of cutting-ring specimens within the same group was kept the same, and the ones with a large difference in matric suction were humidified or dried. The permeability coefficient of bare soil, vegetated soil of *Indigofera amblyantha* and *Cynodon dactylon* were measured by the constant head penetration test, with a duration time of 2 h.

Porous stone and sealing ring was set into the base of the permeation instrument, and then Vaseline was applied on the inner wall of the sleeve. The in-situ cutting-ring specimens were put into the permeation instrument and arranged on the base. The sealing ring, porous stone, upper cap, and tighten screw were arranged on the permeameter in a row to avoid leaking of air and water. The inlet pipe was connected to the water supply device, and then the permeameter was laid flat to ensure water and air venting. After water overflowed from the outlet pipe, the measurement was started. During the test, the water head tube was filled with water to the required height, then the tube clamp was closed, and the stopwatch was started. Meanwhile, the initial water head h_1 at the moment t_1 was measured. After the seepage was stable, the final water head h_2 at the moment t_2 was also recorded. Water temperatures at the beginning and end of the test were measured (Fig. 5).

$$k = 2.3\frac{aL}{A(t_2 - t_1)}log\frac{h_1}{h_2} \tag{2}$$

where a- cross-sectional area of the head pipe, cm^2; L-sample height, cm; A-cross-sectional area of the sample, cm^2; h_1-initial water head (t_1), cm; h_2-final water head (t_2), cm. In this test, the cutting-ring sample has a cross-sectional area of 30 cm^2 and a height of 40 mm.

3 Results and Discussion

3.1 Soil-Water Characteristic Curve

In recent years, a large number of empirical models have been proposed to describe the properties of soil-water characteristic curves. For example, Brooks-Corey, Fledlund-Xing and Van-Genuchten models of which fitting parameters were obtained based on experimental observations [29–31]. According to the soil-water characteristic curve, as matric suction of unsaturated soil increased, the soil saturation decreased with several recognizable phrases. The matric suction can be divided into three regions according to the tangent slope of the soil-water characteristic curve, namely, the boundary effect region, the transition region and the unsaturated residual region, where the transition region can be further subdivided into primary and secondary transition regions [32].

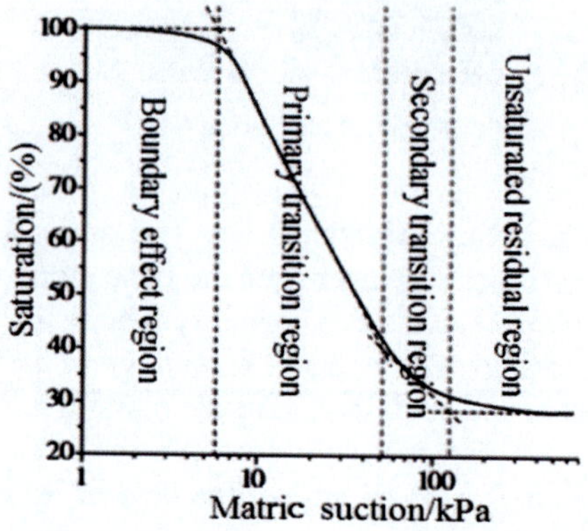

Fig. 6. The typical soil-water holding capacity curve [33]

Through pressure plate test, the water retention capacity curve of unsaturated remolded loess can be divided distinctly into four sections. And turning point values of the boundary effect region, the primary transition region, the second transition region and the unsaturated residual region were around 5 kPa, 60 kPa and 170 kPa, respectively (Fig. 6) [33]. Based on the viewpoint that plant roots occupy soil particle gaps and reduce soil porosity, the void ratio of vegetated soil was deduced via the three-phase diagram of soil [8]:

$$e = \frac{e_0 - R_v(1 + e_0)}{1 + R_v(1 + e_0)} \tag{3}$$

where e_0-the void ratio of bare soil, R_v-the volume ratio of plant roots.

A numerical model regarding soil saturation and soil matric suction was established so as to simulate the effect that plant roots reduce soil porosity and increase soil-water retention capacity [20]:

$$S_r = \left[1 + \left[\frac{se^{m_4}}{m_3}\right]^{m_2}\right]^{-m_1} \tag{4}$$

where S_r-the soil saturation; s-soil matric suction; m_1, m_2 and m_4 are dimensionless parameters; m_1 and m_2 control the basic shape of the soil-water characteristic curve; m_3 and m_4 are related to the intake value of bare soil.

In this paper, computed conclusions showed that the average void ratio (e_0) of bare soil is 0.558, whereas which of vegetated soil (*Indigofera amblyantha*) is 0.536. Through the compacted kaoline test, the recommended values of $m_1 = 0.03586$ and $m_2 = 3.746$ were offered [34, 35]. Based on the soil-water characteristic curve of sandy silt, $m_3 =$ 37.161 kPa and $m_4 =$ 8.433 kPa were employed to carry out a numerical simulation for bare soil [36–38]. In addition, 5 groups date of vegetated soil were verified, and the results were demonstrated in Fig. 7.

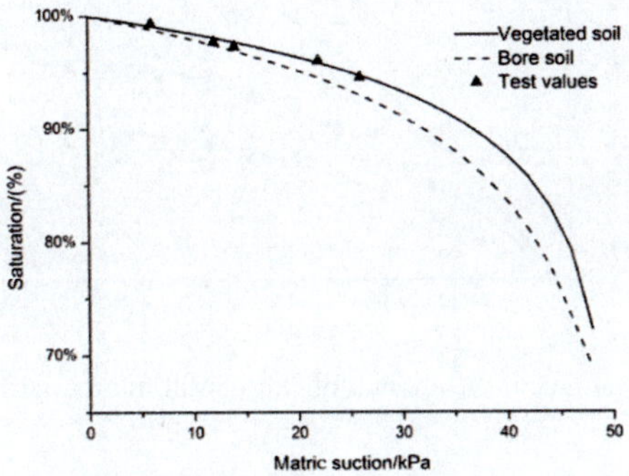

Fig. 7. Water holding capacity of bare soil and vegetated soil (Sandy silt)

Figure 7 showed that the equation of soil matric suction and soil saturation established by Gallipoli et al. [20] could effectively simulate the experimental data with high accuracy. The water retention capacity of vegetated soil is higher than that of bare soil, but the water retention capacity curves of those two are similar. This indicates that plant roots in vegetated soil can effectively increase the air intake value of soil but do not have much influence on the rate of change in soil water content.

During the plant growth period, soil particles were arranged in a definite direction and order under gravity stress, and a certain pore structure was formed [39–41]. During the drying process of soil, gas initially entered the pores with a large size and drained the pore water in soil pores. This part of pore water accounts for a large proportion of the total pore water. As soil matric suction increase, the water in soil pores decrease gradually, non-pore water is drained slowly, and soil presents a high water retention capacity. Plant roots in the vegetated soil can effectively fill the pores among soil particles, and soil porosity is lowered. During the drying process of vegetated soil, it is difficult for gas to enter the gaps of soil particles. Thus air intake value of the vegetated soil is higher than which of the bare soil.

3.2 Permeability Coefficient

As shown in Fig. 8, the amounts of water permeation of both vegetated soil and bare soil increased gradually with a similar variation tendency during the process of the penetration test. When the duration time of penetration test was 2 h, the amount of water infiltration of vegetated soil of *Cynodon dactylon* was about 1/2 of that of the bare soil, whereas which of *Indigofera amblyantha* was around 2/3, which indicated that vegetated soil could effectively reduce soil penetration rate. The permeation rate of both vegetated soil and bare soil presented a gradual decline tendency and reached a stable seepage state. However, the permeation rate of vegetated soil was always lower than that of bare soil.

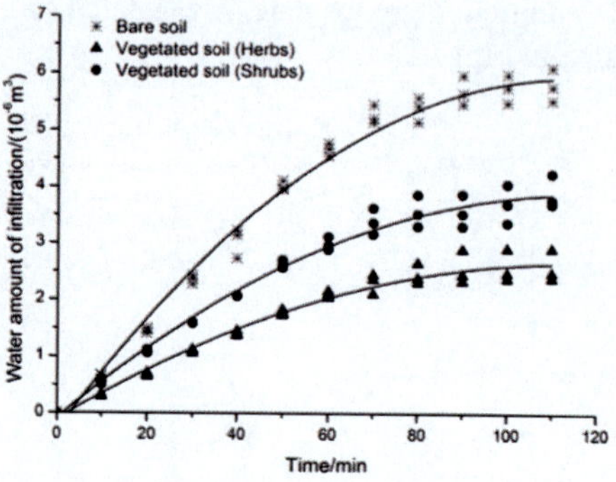

Fig. 8. Variations of the amount of water infiltration in different soils

Results could be explained by plant roots exerting growing pressure on the surrounding soil during the plant growing process, soil relative density increase and void ratio decrease [8, 41]. Lots of bifurcations and fibrous roots can promote soil consolidation and improve soil structure by network connecting and twining, and then the cohesion and friction force between soil particles and plant roots is increased. Meanwhile, the cementing materials produced by biochemical actions between plant roots and soil particles, the organic and inorganic colloids such as cationic bridges and Van der Waals forces formed through multivalent cations in the soil, all those mentioned above can combine soil and root system into an organic complex. Therefore, the integrity of soil is improved, and the penetration rate of soil decreases.

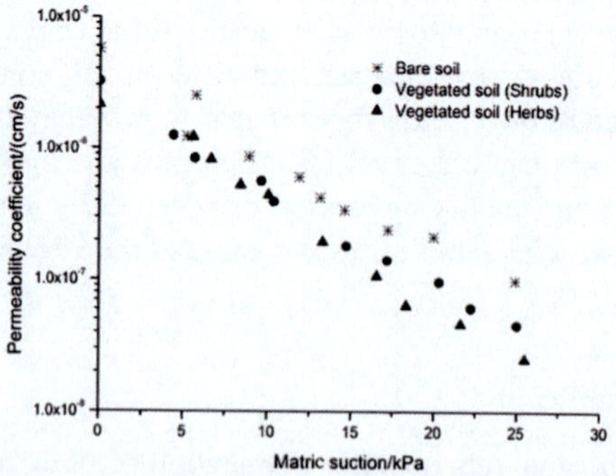

Fig. 9. Relation between soil matric suction and permeability coefficient

The relationship between permeability coefficient and matric suction of vegetated soil and bare soil can be observed in Fig. 9. The permeability coefficient of specimens decreased gradually with the advance of matric suction, and the permeability coefficient of vegetated soil is significantly lower than that of bare soil. In unsaturated soil, pore water is a hydrophilic medium, and air is a hydrophobic medium, so water can only flow through the space occupied by pore water. When matric suction is augmented, the soil moisture content is decreased. Furtherly, the soil permeability coefficient is reduced correspondingly. When the matric suction of vegetated soil is 25 kPa, the permeability coefficient is reduced to the magnitude of 1.00E-08. That indicated the increase in soil pore air pressure impeded the paths of water flow, which reduced the permeability of vegetated soil.

The permeability coefficient of bare soil is higher than that of vegetated soil. In the process of rainfall, the variation of matric suction in vegetated soil is mild, resulting in a longer response time and thereby presumably to a greater soil shear strength in the corresponding period.

3.3 Shear Strength

Figure 10 illustrated that under different normal stresses, the shear strength of bare soil and vegetated soil increase with the advance of matric suction, and the various characteristics of shear strength are almost the same. This manifests that matric suction can be regarded as an independent variable in the shear strength of unsaturated soil. Moreover, two relationship curves show a correlation and both all have significant phases. One is about soil shear strength and matric suction, and the other is about soil saturation and matric suction.

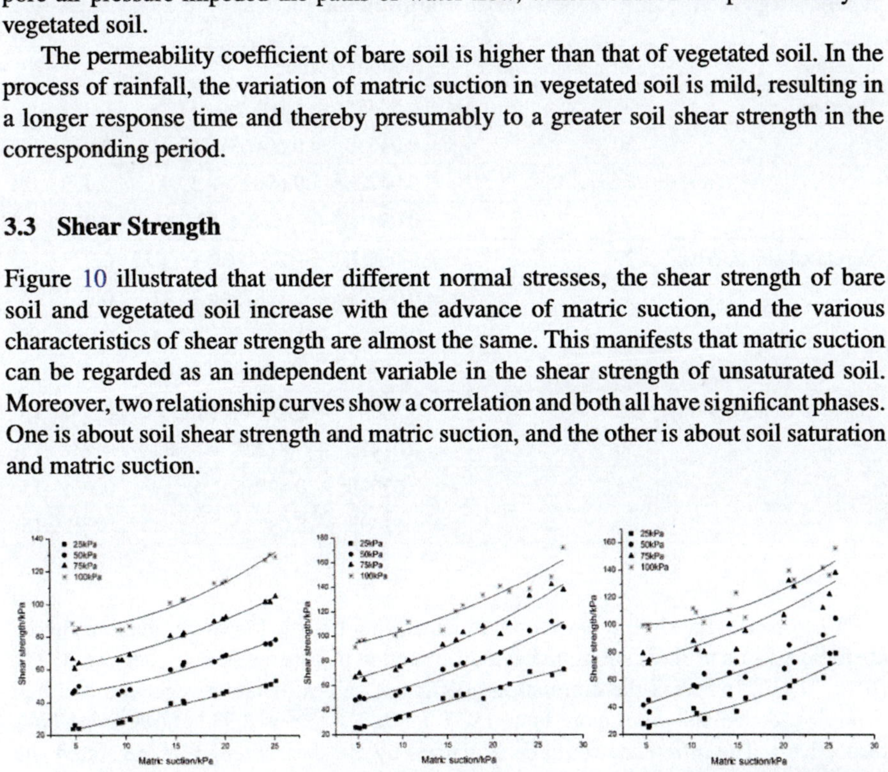

a) Bare soil b) Vegetated soil (Herbs) c) Vegetated soil (Shrubs)

Fig. 10. Relationship between shear strength and matric suction of unsaturated sandy silt

According to Fig. 6, there are four significant phases in the soil-water characteristic curve of unsaturated soil. The matric suction of specimens in this paper is mainly concentrated in the boundary effect zone and the main transition zone. The slope turning point of the soil-water characteristic curve regarding sandy silt in Fig. 7 and the slope turning point of the relationship curve between shear strength and matric suction in Fig. 10a are almost the same, which is the air intake value of saturated sandy silt. In Fig. 10b and Fig. 10c, when the normal stresses are low, the slope variation of the relationship curve between shear strength and matric suction of vegetated soils (Herbs and Shrubs) is small, even remaining stable until matric suction is 25 kPa. According to the soil-water characteristic curve (Fig. 7) of vegetated soil and the shear strength variation characteristics of bare soil, it is straightforward to conclude that compared with bare soil, the air intake value of vegetated soil is higher.

It can be obtained from Fig. 10 that soil shear strength increase due to an increase in matric suction. The Least Square method was adopted to fit the data, and the results were summarized in Table 2.

Table 2. Regression equation of shear strength and matric suction under different normal stresses

Soil type	Normal stress/kPa	Regression equation	R2	n
Bare soil	25	$\tau = 0.0047S^2 + 1.3012S + 17.48$	0.981	15
	50	$\tau = 0.0353S^2 + 0.6046S + 41.21$	0.936	15
	75	$\tau = 0.0472S^2 + 0.6902S + 57.92$	0.979	15
	100	$\tau = 0.0841S^2–0.1156S + 82.01$	0.972	15
Vegetated soil (Herbs)	25	$\tau = -0.0073S^2 + 2.3454S + 13.55$	0.973	15
	50	$\tau = 0.0432S^2 + 1.3682S + 39.93$	0.953	15
	75	$\tau = 0.0582S^2 + 1.4022S + 57.94$	0.963	15
	100	$\tau = 0.0562S^2 + 1.0845S + 88.09$	0.937	15
Vegetated soil (Shrubs)	25	$\tau = 0.1009S^2–0.8960S + 102.59$	0.774	15
	50	$\tau = 0.0141S^2 + 1.8230S + 64.82$	0.924	15
	75	$\tau = 0.0307S^2 + 0.4975S + 52.19$	0.874	15
	100	$\tau = 1.6743S + 18.09$	0.971	15

It can be observed in Table 2 that under different normal stresses, the correlation coefficient value of shear strength and matric suction of bare soil was higher than 0.936 ($R^2 > 0.936$), whereas the correlation coefficient values of herbs vegetated soil and shrubs vegetated soil were more than 0.937 and 0.774 ($R^2 > 0.937$ and $R^2 > 0.774$), respectively. The difference could be explained by the shear strength of vegetated soil being related not only to matric suction but also to mechanical reinforcement induced by plant roots. Due to the difference in root amounts in each specimen varying from one to one when the specimens of vegetated soil were obtained through cutting-ring, the deviation may have occurred in the results. Based on this, on the same matric suction, three specimens were taken by cutting-ring for data statistics and fitting, which could reduce the bias of the results.

From Fig. 10, the cohesion and the internal friction angle of bare soil and vegetated soil both increased as matric suction increased. However, the rate of increase in cohesion decreased, and the internal friction angle increased. The soil-water characteristic curve can be divided into four phases according to Fig. 6, so the matric suction of specimens in this paper is mainly concentrated in the boundary effect region and the primary transition region.

The cohesion of sandy silt is mainly induced by the capillary action of pore water in the soil, and capillary action mainly occurs in the transition region of the soil-water characteristic curve [15]. When soil matric suction value locates in the boundary effect zone of the soil-water characteristic curve, the soil is almost saturated, and few gases within the soil are surrounded by pore water. In this situation, the water-gas system belongs to the gas seal system. The surface tension of the water-air interface shrink membrane in the gas seal system didn't apply to soil particles directly, nor did any obvious cohesion is induced [13, 14]. In addition, lubrication among soil particles is significant due to soil particles are surrounded by pore water. As a result, the variation magnitude of the internal friction angle at this phase is small.

The soil condition begins transforming from saturated to unsaturated in the transition region, air starts entering the soil pores, and the water-air interface shrinkage membrane begins to enlarge. The surface tension of the shrinkage membrane acts inversely on soil particles, and compressive stress is generated, and then soil cohesion increase. At the same time, as the soil was under an incompletely saturated condition, few gases occurred in soil micro-pores. Matric suction generated in soil micro-pores can effectively combine soil particles together, and lubrication among soil particles drops comparatively. In addition, both the soil cohesion and internal friction angle began to increase [42].

When matric suction continues to increase, the pore water in the soil decrease gradually, and a large quantity of gas enter the soil pores. The shrinkage membrane of the water-air interface in the soil pores declines correspondingly, and the total compressive stress act on soil particles decreases slightly, which results in the rate of increase in cohesion slowly dropping. The increased amplitude of internal friction angle increases as the lubrication diminishes and frictional resistance augments among soil particles [43].

As shown in Table 3, as the matric suction of bare soil advanced, the cohesion increased from 5.708 kPa to 25.779 kPa and the internal friction angle developed from 37.013° to 45.905°. By contrast, the variation amplitude of the internal friction angle was relatively small. The cohesion increment of herbs vegetated soil was 34.995 kPa, and the variation range of the internal friction angle was 44.872° ± 3.005°. Moreover, the cohesion increment and the internal friction angle variation range of shrubs vegetated soil were 24.678 kPa and 46.119° ± 2.964°, respectively. Based on the mentioned above, this might lead to the conclusion that compared with bare soil, herbs have a greater influence on the soil cohesion, while shrubs are weighed more on soil internal friction angle. The specimens were obtained in situ at side-by-side locations, but the results may have been influenced by in situ heterogeneity and possible differences in root amounts.

The results may be explained in another way. The root system of *Cynodon dactylon* can be characterized by luxuriant fibrous roots (d < 1.0 mm), with no strong taproot [44]. Indigofera amblyantha with a horizontal developed root system, including a lot of the branches and fibrous roots, and the root diameter is mainly concentrated in 1.0–2.5 mm [45].

Jackson believed that the roots with a diameter <2.0 mm are the principal organ for plants to uptake water [46]. Owing to the large surface area, the increase in connect area between fibrous roots (d < 2.0 mm) and soil particles intensified the permeability of root epidermis and water transmission tunnels. Whereas the roots of which diameter is greater than 2.0 mm, liquid water inside the roots is easily vaporized under high negative pressure to generate cavitation. The decline in the channels in which liquid

Table 3. Cohesion and internal friction angle of soil under different matric suction

Soil type	Shear strength indexes	5 kPa	10 kPa	15 kPa	20 kPa	25 kPa
Bare soil	Cohesion/kPa	5.708	10.081	19.799	21.838	25.779
	Internal friction angle/°	37.013	38.365	39.485	42.642	45.905
Herbs vegetated soil	Cohesion/kPa	3.330	8.543	26.824	28.912	38.325
	Internal friction angle/°	41.867	42.704	43.495	46.873	47.876
Shrubs vegetated soil	Cohesion/kPa	4.915	11.931	20.898	28.854	29.593
	Internal friction angle/°	43.155	43.392	43.852	45.102	49.082

water flows results in lower soil permeability, and plant roots play a role in mechanical reinforcing soil. The root system of *Cynodon dactylon* can improve soil structure by cementing, consolidating soil, and dissipating pore water pressure. Whereas the root system of Indigofera amblyantha presents work in friction effect of soil-root interface and mechanical reinforcement effect. As a result, *Cynodon dactylon* (Herbs) may impact soil cohesion, and *Indigofera amblyantha* (Shrubs) has a significant influence on soil internal friction angle.

4 Conclusions

Through transpiration, penetration and direct shear tests, in this paper, the effects of matric suction on the permeability and shear strength of vegetated soil were investigated. The main findings can be summarized as follows:

(1) According to the results of the vegetative test and the water retention capacity curve of sandy silt, the wilting coefficients of *Cynodon dactylon* and *Indigofera amblyantha* corresponding to the matric suction of sandy silt is 35–40 kPa. To ensure the tests were conducted normally, the matric suction of specimens was maintained within 30 kPa, which is located in the boundary effect region and the initial stage of the transition region.
(2) Compared with bare soil, the water retention capacity of vegetated soil significantly improved. When the duration time of the penetration test is 2 h, the amount of water infiltration of vegetated soil of *Cynodon dactylon* was about 1/2 of that of the bare soil, whereas which of *Indigofera amblyantha* was around 2/3. Permeability coefficient of vegetated soil decreased to the magnitude of 1.00E-08, when soil matric suction was about 25 kPa.
(3) The rate of increase in soil cohesion slowly diminished and internal friction angle increased gradually as the matric suction augmented. Compared with bare soil, *Cynodon dactylon* (Herbs) impact more on soil cohesion and *Indigofera amblyantha* (Shrubs) has a significant influence on soil internal friction angle.

Acknowledgments. This study was financially supported by the National Natural Science Foundation of China (Grant No. 41790432), NSFC Key Project for International Cooperation (Grant

No. 41761144077), the International Partnership Program of Chinese Academy of Sciences (131551KYSB20180042) and the Youth Innovation Promotion Association CAS (Grant No. 2019364). Great appreciation also goes to the editorial board and the reviewers of this paper.

References

1. Wu, J.J., Wang, C.H., Li, G.X.: Influence of matric suction in unsaturated soils on slope stability. Rock Soil Mech. **25**, 732–743 (2004)
2. Ng, C.W.W., Shi, Q.: A numerical investigation of the stability of unsaturated soil slopes subjected to transient seepage. Comput. Geotech. **22**, 1–28 (1998)
3. Zhan, T.L.T., Ng, C.W.W., Fredlund, D.G.: Field study of wetting infiltration into a grassed unsaturated expansive soil slope. Can. Geotech. J. **44**, 392–408 (2007)
4. Barker, D.H.: Vegetation and Slopes: Stabilization, Protection, and Ecology. Thomas Telford Press, London (1995)
5. Simon, A., Andrew, A., Collison, J.C.: Quantifying the mechanical and hydrologic effects of riparian vegetation on streambank stability. Earth Surf. Proc. Land. **27**, 527–546 (2002)
6. Normaniza, O., Barakbah, S.S.: Parameters to predict slope stability-soil water and root profiles. Ecol. Eng. **28**, 90–95 (2006)
7. Pollen-Bankhead, N., Simon, A.: Hydrologic and hydraulic effects of riparian root networks on streambank stability: is mechanical root-reinforcement the whole story? Geomorphology **116**, 353–362 (2010)
8. Ng, C.W.W.: A tmosphere-plant-soil interactions: theories and mechanisms. Chinese J. Geotech. Eng. **39**, 1–47 (2017)
9. Macneil, D.J., Steele, D.P., McMahon, W., Carder, D.R.: Vegetation for slope stability-prepared for quality services, civil engineering, highways agency. TRL Report Press, Berkshire (2001)
10. Ng, C.W.W., Woon, K.X., Leung, A.K., Chu, L.M.: Experimental investigation of induced suction distribution in a grass-covered soil. Ecol. Eng. **52**, 219–223 (2013)
11. Ng, C.W.W., Leung, A.K., Woon, K.X.: Effects of soil density on grass-induced suction distributions in compacted soil subjected to rainfall. Can. Geotech. J. **51**, 311–321 (2014)
12. Woon, K.X., Leung, A.K., Ng, C.W.W., Chu, L.M.: Effects of grass transpiration on suction induced in near-saturated soil condition. In: Mancuso, C., Jommi, C., D'Onza, F. (eds) Unsaturated Soils: Research and Applications. Springer, Heidelberg (2012). https://doi.org/10.1007/978-3-642-31343-1_30
13. Bishop, A.W., Blight, G.E.: Some aspects of effective stress in saturated and partly saturated soils. Géotechnique **13**, 177–197 (1963)
14. Fredlund, D.G., Morgenstern, N.R.: Stress state variables for unsaturated soils. J. Geotech. Eng. Division **103**(5), 447–466 (1977)
15. Lin, H.Z., Li, G.X., Yu, Y.Z., Lv, H.: Influence of matric suction on shear strength behavior of unsaturated soils. Rock Soil Mech. **28**, 1931–1936 (2007)
16. Lim, T.T., Rahardjo, H., Chang, M.F., Fredlund, D.G.: Effect of rainfall on matric suction in a residual soil slope. Can. Geotech. J. **33**, 618–628 (1996)
17. Li, X.W., Kong, L.W., Guo, A.G.: Permeability and mechanical characteristics of expansive and cut slope protection mechanism under vegetation action. Rock Soil Mech. **34**, 85–91 (2013)
18. Chen, J.L., Li, J.H., Cheng, P., Song, L., Zhou, T.: Field test on seepage performance of soil cover with different type of vegetation. Rock Soil Mech. **39**, 222–228 (2018)
19. Gabrm, A., Akran, M., Taylor, H.M.: Effect of simulated roots on the permeability of silty soil. Geotech. Test. J. **18**(1), 112–115 (1995)

20. Gallipoli, D., Wheeler, S.J., Karstunen, M.: Modelling the variation of degree of saturation in a deformable unsaturated soil. Géotechnique **53**, 105–112 (2003)
21. Huat, B.B.K., Ali, F.H.J., Low, T.H.: Water infiltration characteristics of unsaturated soil slope and its effect on suction and stability. Geotech. Geol. Eng. **24**, 1293–1306 (2006)
22. Aravena, J.E., Berli, M., Ghezzehei, T.A., Tyler, S.W.: Effects of root-induced compaction on rhizosphere hydraulic properties – X-ray microtomography imaging and numerical simulations. Environ. Sci. Technol. **45**, 425–431 (2011)
23. Guo, N., Yang, X.H., Chen, Z.H., Gao, D.H., Lai, C.J.: Influence of metrical suction on strength and deformation characteristics of unsaturated loess in intact form. J. Lanzhou Univ. Technol. **43**, 120–125 (2017)
24. Song, H.J., Li, J.Z., Ge, Y.M.: Research on the effects of matric suction on shear strength of unsaturated reticulate red clay. Ind. Constr. **47**, 7–11 (2017)
25. Ng, C.W.W., Ni, J.J., Leung, A.K., Zhou, C., Wang, Z.J.: Effects of planting density on tree growth and induced soil suction. Géotechnique **66**, 711–724 (2016)
26. Xu, W.N., et al.: Technical research of vegetation restoration in disturbance area of hydropower project. Science Press, Beijing (2017)
27. Garg, A., Leung, A.K., Ng, C.W.W.: Comparisons of soil suction induced by evapo-transpiration and transpiration of S heptaphylla. Canadian Geotech. J. **52**, 2149–2155 (2015)
28. Xia, Z.Y., Liu, Q., Xu, W.N., Rao, Y.K., Zhang, H.: Characteristics of interface friction between Indigofera amblyantha root system and soil. J. Soil Water Conserv. **32**, 128–134 (2018)
29. Brooks, R.H., Corey, A.T.: Hydraulic properties of porous media. Hydrological Paper **3**, 27 (1964)
30. Van-Genuchten, M.T.: A closed-from equation for predicting the hydraulic conductivity of unsaturated soils. Soil Sci. Soc. Am. J. **44**, 892–898 (1980)
31. Fredlund, D.G., Xing, A.: Equations for the soil-water characteristic curve. Can. Geotech. J. **31**, 521–532 (1994)
32. White, N.F., Duke, H.R., Sunada, D.K., Corey, A.T.: Physics of desaturation in porous materials. J. Irrig. Drainage Div. **96**, 165–191 (1970)
33. Yan, Y.J., Wen, B.P., Ji, B.X.: Contribution of matric suction to shear strength of un-saturated remoulded loess soils. J. Eng. Geol. **19**, 865–874 (2011)
34. Sivakumar, V.: A critical state framework for unsaturated soil. University of Sheffield, UK (1993)
35. Thu, T.M., Rahardjo, H., Leong, E.C.: Effects of hysteresis on shear strength envelopes from constant water content and consolidated drained triaxial tests. In: 4th International Conference on Unsaturated Soils (2006)
36. Xiao, J.H., Liu, J.K., Peng, L.Y., Chen, L.H.: Effects of compactness and water Yellow River alluvial silt content on its mechanical behaviors. Rock Soil Mech. **29**, 409–414 (2008)
37. Peng, L.Y., Liu, J.K., Chen, L.H.: Strength and yielding characteristics of unsaturated compacted silts. Rock Soil Mech. **29**, 2241–2245 (2008)
38. Wang, G.Y., Li, B., Luo, J., Fu, H.Y.: Study of soil-water characteristics and matric suction measurement device for unsaturated silty soil. Rock Soil Mechan. **31**, 3678–3682 (2010)
39. Bengough, A.G., Mullins, C.E.: Mechanical impedance to root growth: a review of experimental techniques and root growth responses. Soil Sci. **41**, 341–358 (1990)
40. Lipiec, J., Hakansson, I.: Influences of degree of compactness and matric water tension on some important plant growth factors. Soil Tillage Res. **53**, 87–94 (2000)
41. Leung, A.K., Garg, A., Ng, C.W.W.: Effects of plant roots on soil-water retention and induced suction in vegetated soil. Eng. Geol. **193**, 183–197 (2015)
42. Fredlund, D.G., Morgenstem, N.R., Widger, R.A.: The shear strength of unsaturated soils. Can. Geotech. J. **15**, 313–321 (2002)

43. Baets, S.D., Poesen, J., Gyssels, G., Knapen, A.: Effects of grass roots on the erodibility of topsoils during concentrated flow. Geomorphology **76**, 54–67 (2006)
44. Wei, H.W., Luo, H.B., Zhang, Y.H.: Root distribution characteristics and tensile strength of Cynodon dactylon. Bullet. Soil Water Conserv. **31**(4), 185–189 (2011)
45. Yang, Y.S., Xia, Z.Y., Xiao, H., Jia, F.G., Hu, M.Z., Deng, L.: Experimental research on shear strength of Indigofera amblyantha root-soil composite. J. Yangtze River Sci. Res. Inst. **31**, 72–76 (2014)
46. Jackson, R.B., Mooney, H.A., Schulze, E.D.: A global budget for fine root biomass surface area and nutrient contents. Proc. Natl. Acad. Sci. USA **94**, 7362–7366 (1997)

On the Use of Optical Fiber Sensors for Debris Flow Monitoring: A Review of Recent Achievements

Luca Schenato[1,2,3](✉) and Alessandro Pasuto[1,2,3]

[1] Research Institute for Geo-Hydrological Protection, National Research Council, Corso Stati Uniti 4, 35127 Padova, Italy
luca.schenato@cnr.it
[2] Sino-Italian Joint Laboratory on Geological and Hydrological Hazards, No. 9, Block 4, South Renmin Road, Chengdu, China
[3] National Inter-University Consortium for Telecommunications - CNIT, Pisa, Italy

Abstract. Debris flows are among the most common and dangerous types of landslides on the planet, often occurring in mountainous areas. Because of their tremendous speeds and the sheer destructive power of their flow, they are extremely hazardous to life and property. Geotechnical instrumentation is critical for monitoring and warning debris flows. However, traditional monitoring systems are hampered by limitations, such as limited spatial resolution and sensitivity, reduced robustness, and powering and cabling issues. These limitations can be surpassed using optical fiber sensors, which are gaining increasing interest by the scientific and stakeholders community. In this paper, we have selected and review some of the most significant achievements and trends on this topic.

Keywords: Debris flow · Fiber optic sensor · Distributed fiber optic sensor

1 Introduction

Climatic changes are dramatically causing a significant increment of high-intensity precipitations worldwide in terms of total amounts of rainfall and the frequency of extreme events [1]. In that regard, all-natural events induced by water circulation (e.g., soil erosion, floods, landslides) are predicted to have a growing environmental incidence on natural and human structures with serious collateral damage. Debris flows are undoubtfully among the most hazardous natural phenomena, affecting the life and security of people and facilities. They are dangerous to life and property because of the flow's sheer destructive momentum caused by their high density and speed.

People and property are becoming more exposed to debris flow due to rapid and enormous socio-economic development in mountain areas. Adaptation and protection strategies can include different actions, from strengthening civil infrastructures to careful land use planning, devising suitable countermeasures, and defining alerting plans. Furthermore, the last UN Conference on Disaster Risk Reduction [2] has emphasized

H.-H. Zhu et al. (Eds.): BRWSG 2021, LNCE 230, pp. 60–70, 2022.
https://doi.org/10.1007/978-981-16-9963-4_5

the need for disaster preventive investments to increase the resilience of vulnerable populations through non-structural mitigation methods like monitoring and early warning systems.

Debris flows, which are produced by heavy rain in mountainous areas, or rapid glacier melting, are among the most destructive water-related occurrences. The decades-long effort to monitor these occurrences and the corresponding protective structures (e.g., debris flow trap basins) has investigated a wide range of technologies [3, 4] and has lately begun to include fiber optic sensors (FOSs) as well. FOSs has many benefits when it comes to electromechanical technology for geo-hydrological monitoring applications [5]. The key feature is that the fiber serves as both a sensing element and a communication channel. As a result, only the fiber cable, which is passive, intrinsically robust, and relatively inexpensive, should be exposed to the monitored hostile environment. At the same time, the interrogation unit can be placed in a more secure location. Several single-point optical fiber sensors and specialty cables are currently available and may be customized for specific applications.

Furthermore, FOSs can be connected and controlled over large distances (up to several kilometers). They have the unique capacity to enable distributed sensing, allowing continuous space-time monitoring of structures or sites. Because of these characteristics, FOSs are well-suited to monitoring dikes, channels, and even ravines where debris flows may occur.

This paper will review and discuss some of the most recent and relevant works on fiber optic sensors and systems for debris flows monitoring. Please note that the description of the working principles of fiber optic sensors is beyond the scope of this paper (see [6, 7] for an overview).

2 Fiber Optic Sensors for Debris Flow Monitoring

In the context of hazard assessment, land-use and civil-protection planning, mitigation measures design, and emergency response, debris flow monitoring is critical. Ultrasonic sensors, tripwires, geophones, and other technical solutions have been used thus far. In particular, most of the FOSs proposed so far aim to replace geophones or arrays of geophones, but FOS-based strain and pressure sensors have also been proposed. Some recent examples of sensors belonging to these categories will be presented and commented on in the following subsections.

2.1 FOS-Based Seismic Sensors

One of the most promising techniques for detecting debris flow surges, velocity, and rheological behavior is to monitor ground vibrations caused by the propagation of the solid-fluid mixture [8–10].

Geophones and ultrasonic sensors, in particular, have been the most extensively employed equipments for debris flow monitoring up to this point. The ease of installation and data recording allowed them to be used in mountainous areas. However, given the difficulties in deploying a large number of such sensors, the accuracy of these measurements is generally poor. Consequently, their employment is usually limited to surge detection and average velocity estimation, with significant uncertainties in wave shape recognition as recorded by various sensors placed along the channel.

According to the literature [11], the frequency of ground vibration induced by debris flows was estimated to be limited to a range between 30 and 80 Hz, then enlarged to 10 and 100 Hz with the components associated with the surge front containing large boulders falling in the range of 10–30 Hz [12, 13]. The flowing tail, mainly containing sand and gravel mud, generates a signal whose frequency is primarily in the range of 60–80 Hz. More recently, the bandwidth of interest has been slightly extended: individual rocks within debris flows should generate ground vibrations ranging from 10 to 150 Hz, with bigger stones causing ground vibrations with lower peak frequencies [14]. The data collected in Ai-Yu-Zi Creek, Nan-Tou, Taiwan, in 2004 revealed that the debris flow's front generated vibrations lower than 50 Hz, which varied between 50 and 100 Hz after the front passed. Overall, the frequency spectrum encompassed a range from 10 to 250 Hz when the main front was closest to the sensor. Thus, individual rock vibration frequencies were within the frequency range of ground vibrations collected during the debris flow, indicating that rock or boulder contact with the channel bed is one of the major sources of ground vibration generated by debris flows. Furthermore, Kogelnig et al. [15] reported that additional low-frequency infrasonic signals (4–15 Hz) are generated and correlated with seismic signals.

About the energy involved, the debris flows energy discharge is proportional to the amplitude of the ground vibration. As a result, debris flow should be recognized if the amplitude of ground vibrations surpasses a given threshold across a certain frequency range for an extended interval. Compared to earthquakes, debris flows generate weaker responses in the low-frequency range (i.e., less than 5 Hz), with more energy at the higher frequencies, where the attenuation of the ground vibration waves is more significant [16]. This means that only the geophones closed to the debris flow path can effectively detect the corresponding signals. Although this issue may be solved by placing sensors near the source or path of debris flows, the need for long cabling causes signal attenuation and transmission problems in mountain areas. FOSs are easier to deploy over long distances and incorporate into communication networks with respect to geophones. Moreover, they are intrinsically insensitive to electromagnetic interferences.

The acoustic bandwidth of seismic signals generated by debris flows nicely matches the bandwidth of standard geophones. Still, as shown by very early works on FOSs, this requirement is also easily satisfied by optical fiber sensors. For example, Lin and Liang [17, 18], compared a traditional geophone with a fiber optic sensor composed of a mandril wrapped with fiber and a Sagnac interferometer (Fig. 1), showing that the two sensors have a very similar sensitivity and frequency response. In particular, the frequency range of the FOS fully covered the bandwidth (from 10 to 250–300 Hz). A very similar sensor, yet implemented in a dual-loop configuration, was then proposed for debris flow monitoring by the same group [19]. The main drawback of this interferometric sensor is the intrinsic sensitivity of the entire fiber loop to the external perturbations (not only the fiber wrapped around the mandrel), making the practical use challenging to implement on the field. Also, complex multiplexing schemes may be needed to interrogate multiple sensors [20].

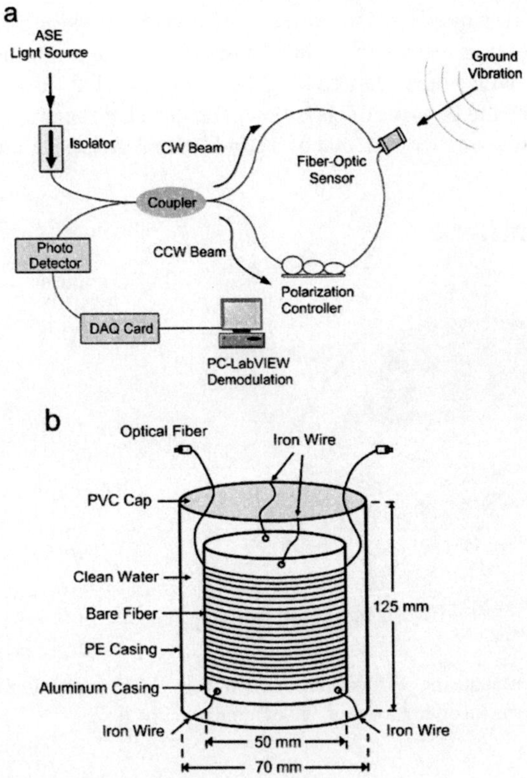

Fig. 1. Interferometric sensor for debris flows seismic monitoring proposed in [17, 18].

Despite the high sensitivity usually achieved by interferometric sensors for acoustic measurement, over the years, researchers mainly looked at the more mature fiber Bragg gratings (FBGs) technology to implement accelerometers and vibration sensors feasible for debris flow monitoring [21]. FOSs implemented by FBG add to the advantage of low transmission losses, the easiness of multiplexing. Therefore, initial efforts on the front of FBG-based debris flow systems focused on verifying the feasibility of the technology, even by exploiting commercially available sensors, as done by Huang et al. [22]. The sensors employed in the tests were commercial single-axis FBG accelerometers with a sensitivity of 10 pm/g and large bandwidth of 0–300 Hz. Four of them, interrogated simultaneously by using a standard frequency-division multiplexing (FDM) technique [23], were installed in an artificial flume, where some debris and rocks were let rolling (Fig. 2). In these tests, the fiber optic sensing system performed better than the conventional sensing system that uses geophones in terms of sensitivity and signal-to-noise-ratio (SNR), with a broad response up to higher frequencies.

In most of the later works, authors put their effort into proposing new sensors designs. For example, Liang and Lin [24] implemented an FBG sensor by embedding an FBG in a layer of a compliant material to which a proof mass is hung. Under vertical acceleration, the mass induces a compression or an expansion of the compliant layer, dragging the lengthening or shortening of the fiber Bragg grating via the Poisson's effect (Fig. 3). The sensor performance was estimated by hitting the riverbed at a 3 m distance where the sensor was buried, with a repeated free fall 30 kg-weight rock from a 1.5 m-height place in series to simulate the impact of debris flow. The detected spectrum is mainly located at 10–250 H, slightly smaller than that of a conventional geophone used for reference.

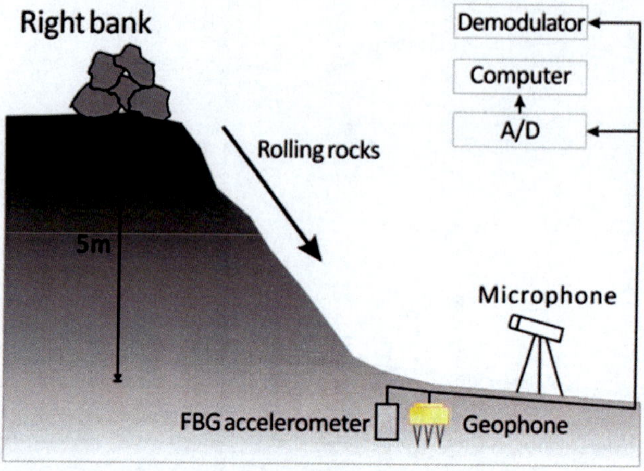

Fig. 2. Field test to compare the performance of a commercial fiber optic vibration sensor to those of conventional systems involving a geophone or a microphone [22].

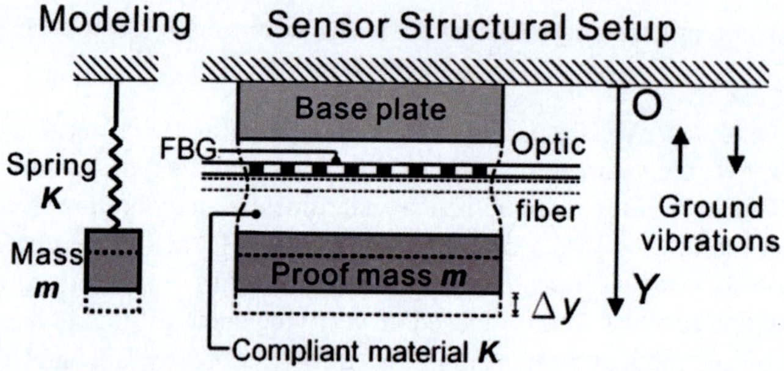

Fig. 3. FBG sensor developed in [24] for debris flow monitoring.

Similarly, Han et al. [25] proposed an inertial sensor but employing a cantilever scheme, with two FBGs undergoing stretching upon ground vibration. The bandwidth of the sensor was limited to 200 Hz, with a marked resonance at 380 Hz. By employing thin-coated FBGs, the sensitivity of this accelerometer was boosted by a factor of 3 with respect to standard-coated FBG.

Distributed acoustic sensing (DAS) is another technology that is having a big impact on geophysics. By exploiting the coherent interference that is generated within the Rayleigh backscattering of a fiber illuminated by a coherent source, the optical fiber can be used as an equivalent array of in-phase accelerometers, monitoring the surrounding sound or vibration field. In that way, DAS can sense vibrations, at a nano strain scale, with a resolution of a few meters across a distance of several kilometers [26, 27]. Unlike a geophone, which captures ground motion (i.e., ground velocity), a DAS generally provides a strain rate that should undergo spatial integration to be compared with geophone's data [28]. Moreover, the sensitivity of geophones is typically better than that of DAS per single sensing point/gauge length. Nonetheless, DAS technology can probe a massive number of coherent sensing points simultaneously, overcoming, to some extent, the reduced local sensitivity [29]. As proof of this, in the last couple of years, DAS has enabled the implementation of distributed earthquake seismometers by exploiting existing telecommunication fibers already installed underground [30–32].

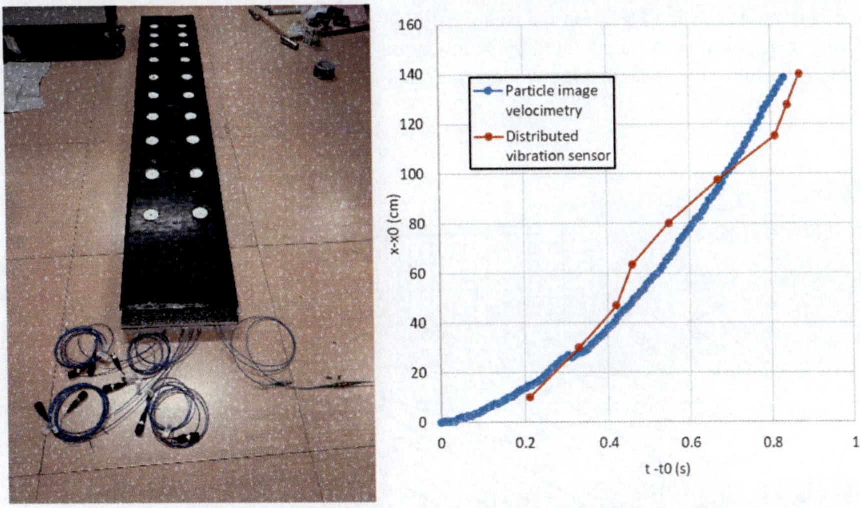

Fig. 4. On the left: the mattress with the 20 fiber-mandrels, where 800 m of fiber was coiled and probed with a DAS system for debris flow monitoring. On the right: the comparison of the front surge time of arrival, calculated from the DAS measurement and particle image velocimetry [33].

Based on these premises, DAS has the potential to be used as a replacement for geophones also in debris flows monitoring. The feasibility of this approach has been recently validated in a physical model of debris flows [33, 34], whose 2 m long flume was instrumented with an engineered mattress encasing 20 fiber-mandrels, each coiling 40 m of fiber, up to 800 m of fiber length (Fig. 4, on the left). The DAS system was capable of precisely following the runoff, providing the distributed characterization of the spectral response. Furthermore, from the analysis of the vibration waves generated by the runoff, it was possible to determine the front surge velocity, consistent with the one obtained by particle image velocimetry. Despite being limited to a small-scale physical model and many practical issues related to the installation, which are still unsolved, this study unveils the potential of DAS technology for debris flow monitoring. The same concept has also been proposed for vibration-based landslide monitoring [35], with excellent performances (Fig. 5).

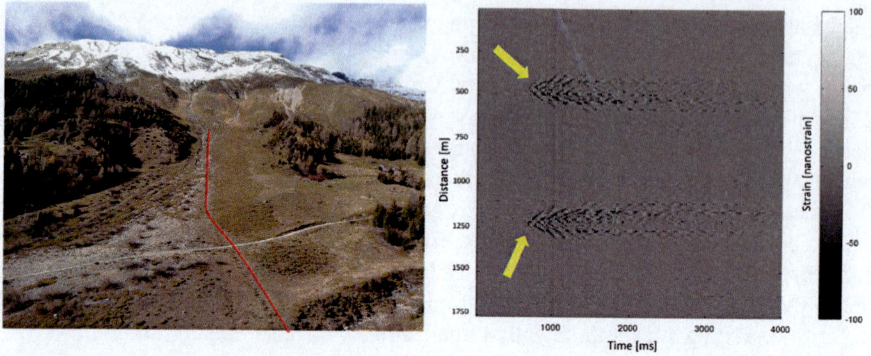

Fig. 5. On the left: test site equipped with a telecom optical cable (in a loop configuration, red line) and probed with a DAS. On the right: waterfall plot of the vibration detected by the DAS during an explosion test (highlighted by yellow arrows). Two positions along the fiber show the explosions due to the deployment in a loop configuration. After [35].

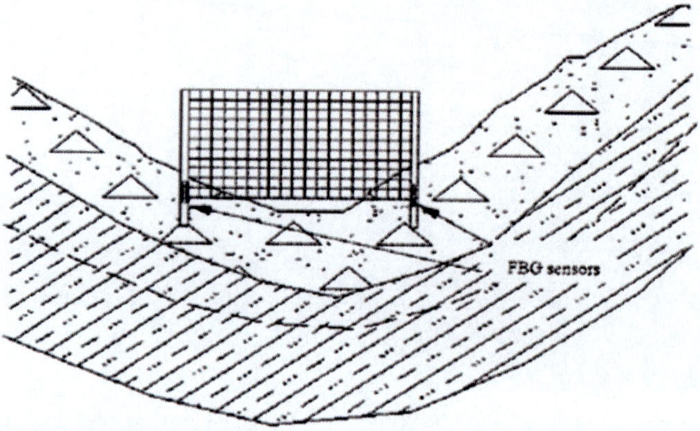

Fig. 6. FBG-based column net for the retention and detection of debris flow. After [36].

2.2 FOS-Based Strain and Pressure Sensors

It is worthy of mentioning that other authors have focused on parameters related to debris flows other than the vibrations. For example, Pei and coauthors [36] developed an FBG-based column-net system for monitoring debris flows. In this device, FBG sensors have been glued on the bottom of two steel pipes, to which a steel wire net was fixed (Fig. 6). In that configuration, the FBG sensors can measure the strain at the bottom of the steel pipe caused by the impact forces occurring when debris flow passes by the cross-section. The idea of instrumenting flex barrier was then proposed a few years later by Yin et al. [37], as shown in Fig. 7.

The FBG technology was also successfully used to investigate the strain distribution information in the slope mass during erosion in China [38]. Starting from the strain measurements, then correlated to the pore water pressure and soil saturation, the authors

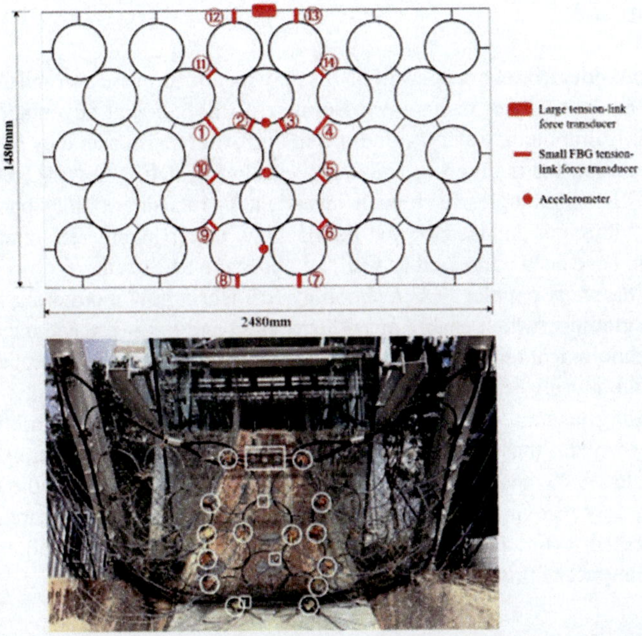

Fig. 7. FBG-based flex-barrier for retaining boulder and granular flow. After [37].

identified four evolution phases during the flow initiation induced by runoff, i.e., the water absorption stage, the deformation stage, the shear zone formation stage, and the fluidization stage. This study, yet performed in an artificial flume, shed light on the triggering of debris flow and pointed to the FBG sensing technology's great potential in developing an effective early warning system for landslides and debris flows initiation.

Another significant debris flow parameter is the pressure fluctuation generated by the flow, normal, and shear, which is often monitored by load cells. They are employed in research activities to describe the flow's water/solid composition and dynamic behavior. However, they have inherent installation problems and is inappropriate for early warning purposes if employed without any other sensors. About the measurements of pressure, several FOSs for pressure sensing in soil have been proposed so far [39–43]. Still, very few have been implemented practically, specifically for debris flows.

In this framework, Xu et al. [44] developed an early-warning system triggered by a given amount of rainfall, which integrated an FBG-based sensor to measure normal earth pressure in gravel material due to the debris flows. The sensor embeds an oil pressure chamber within two diaphragms, one of which is exposed to external pressure. Therefore, any force applied to this diaphragm is transferred to the other one, to which an FBG is attached. An additional FBG, anchored to the external case, is used for temperature compensation. In this design, the geometrical parameters and the materials chosen for the diaphragms provide enough degrees of freedom to tune the sensor performance. The sensor prototypes developed within the work showed an accuracy able to fulfill the requirement of debris flow monitoring.

3 Conclusions

This paper provides a brief overview of debris flow monitoring research and development utilizing fiber optic sensor technology. Because of the inherent advantages of FOSs, including EMI immunity, waterproofness, and corrosion resistance, they have emerged as a viable option for monitoring and early warning of debris flows. Over the years, different FOSs technologies have been investigated, focusing mainly on monitoring the ground vibrations induced by the debris flow. Other parameters related to such phenomenon have been addressed as well, i.e., pressure and strain.

Indeed, the most popular FOS technology for debris flow monitoring is based on fiber Bragg gratings, which enable quasi-distributed and long-distance measurements. Thus, its technological readiness well-fits the need to integrate or even replace standard point-wise monitoring techniques. Nonetheless, very recently, fiber optic distributed acoustic sensing has started to gain attention, enabling unprecedented seismic monitoring capabilities of events that may be too weak to be recorded by a few sparse geophones.

The results on the application of FOS systems are encouraging, but the technology is still not widely recognized nor fully developed. Nonetheless, the research and experiences collected so far, and partially covered by this paper, witness and envisage the tremendous impact of this technology on this topic.

References

1. Fischer, E.M., Knutti, R.: Observed heavy precipitation increase confirms theory and early models. Nature Clim Change. **6**, 986–991 (2016). https://doi.org/10.1038/nclimate3110
2. Aitsi-Selmi, A., et al.: Reflections on a science and technology agenda for 21st century disaster risk reduction. Int. J. Disaster Risk Sci. **7**(1), 1–29 (2016). https://doi.org/10.1007/s13753-016-0081-x
3. Coviello, V., et al.: Combining instrumental monitoring and high-resolution topography for estimating sediment yield in a debris-flow catchment. Environ. Eng. Geosci. **27**, 95–111 (2020). https://doi.org/10.2113/EEG-D-20-00025
4. Marchi, L., Cazorzi, F., Arattano, M., Cucchiaro, S., Cavalli, M., Crema, S.: Debris flows recorded in the Moscardo catchment (Italian Alps) between 1990 and 2019. Nat. Hazards Earth Syst. Sci. **21**, 87–97 (2021). https://doi.org/10.5194/nhess-21-87-2021
5. Schenato, L.: A review of distributed fibre optic sensors for geo-hydrological applications. Appl. Sci. **7**, 896 (2017). https://doi.org/10.3390/app7090896
6. Grattan, L.S., Meggitt, B.T. (eds.): Optical Fiber Sensor TechnologyProducts on AliExpress: Advanced Applications - Bragg Gratings and Distributed Sensors. Springer, New York (2000). https://doi.org/10.1007/978-1-4757-6079-8
7. Hartog, A.: An Introduction to Distributed Optical Fibre Sensors. CRC Press (2017). https://doi.org/10.1201/9781315119014-2
8. Arattano, M., Marchi, L.: Systems and sensors for debris-flow monitoring and warning. Sensors (Basel) **8**, 2436–2452 (2008). https://doi.org/10.3390/s8042436
9. Hübl, J., Zhang, S.C., Kogelnig, A.: Infrasound measurements of debris flow. Presented at the June 10 (2008). https://doi.org/10.2495/DEB080011
10. Hübl, J., Kogelnig, A., Suriñach, E., Vilajosana, I., McArdell, B.W.: A review on acoustic monitoring of debris flow. Presented at the May 29 (2012). https://doi.org/10.2495/DEB120071

11. LaHusen, R.G.: Detecting debris flows using ground vibrations. U.S. Geological Survey (1998). https://doi.org/10.3133/fs23696
12. Huang, C.J., Shieh, C.L., Yin, H.Y.: Laboratory study of the underground sound generated by debris flows. J. Geophys. Res. **109** (2004)
13. Berti, M., Genevois, R., LaHusen, R., Simoni, A., Tecca, P.R.: Debris flow monitoring in the Acquabona watershed on the Dolomites (Italian Alps). Phys. Chem. Earth Part B **25**, 707–715 (2000). https://doi.org/10.1016/S1464-1909(00)00090-3
14. Huang, C.-J., Yin, H.-Y., Chen, C.-Y., Yeh, C.-H., Wang, C.-L.: Ground vibrations produced by rock motions and debris flows. J. Geophys. Res. **112**, F02014 (2007). https://doi.org/10.1029/2005JF000437
15. Kogelnig, A., Hübl, J., Suriñach, E., Vilajosana, I., McArdell, B.W.: Infrasound produced by debris flow: propagation and frequency content evolution. Nat. Hazards **70**(3), 1713–1733 (2011). https://doi.org/10.1007/s11069-011-9741-8
16. Toksöz, M.N., Johnston, D.H.: Society of Exploration Geophysicists: Seismic wave attenuation. Society of Exploration Geophysicists, Tulsa, Okla (1981)
17. Lin, Y.-L., Liang, T.-C.: Design and fabrication of a debris flow sensor using a fiber optic interferometer. In: Proceedings of the 3rd Asia Pacific Optical Sensors Conference (2012). https://doi.org/10.1117/12.914036
18. Liang, T.-C., Lin, Y.-L.: A fiber-optic sensor for the ground vibration detection. Opt. Commun. **306**, 190–197 (2013). https://doi.org/10.1016/j.optcom.2013.05.057
19. Cheng, C.-T., Huang, H.-S., Liang, T.-C.: Improving the performance of Sagnac interferometer for debris flow monitoring system. Presented at the (2017). https://doi.org/10.1109/ICAMSE.2016.7840258
20. Miliou, A.: In-fiber interferometric-based sensors: overview and recent advances. Photonics **8**, 265 (2021). https://doi.org/10.3390/photonics8070265
21. Zhu, H.-H., Shi, B., Zhang, C.-C.: FBG-based monitoring of geohazards: current status and trends. Sensors **17**, 452 (2017). https://doi.org/10.3390/s17030452
22. Huang, C.-J., Chu, C.-R., Tien, T.-M., Yin, H.-Y., Chen, P.-S.: Calibration and deployment of a fiber-optic sensing system for monitoring debris flows. Sensors **12**, 5835–5849 (2012). https://doi.org/10.3390/s120505835
23. Sahota, J.K., Gupta, N., Dhawan, D.: Fiber Bragg grating sensors for monitoring of physical parameters: a comprehensive review. OE **59**, 060901 (2020). https://doi.org/10.1117/1.OE.59.6.060901
24. Liang, T.-C., Lin, Y.-L.: Ground vibrations detection with fiber optic sensor. Opt. Commun. **285**, 2363–2367 (2012). https://doi.org/10.1016/j.optcom.2012.01.037
25. Han, X., Wen, H., Liu, S., Xiong, Y., Luo, Z.: Dual-fiber-Bragg gratings accelerometer for the detection of geosound caused by debris flow. Opt. Eng. **56** (2017). https://doi.org/10.1117/1.OE.56.5.056104
26. Muanenda, Y.: Recent advances in distributed acoustic sensing based on phase-sensitive optical time domain reflectometry. J. Sens. **2018**, 1–16 (2018). https://doi.org/10.1155/2018/3897873
27. Zhan, Z.: Distributed acoustic sensing turns fiber-optic cables into sensitive seismic antennas. Seismol. Res. Lett. **91**, 1–15 (2020). https://doi.org/10.1785/0220190112
28. Lior, I., Sladen, A., Mercerat, D., Ampuero, J.-P., Rivet, D., Sambolian, S.: Strain to Ground Motion Conversion of DAS Data forEarthquake Magnitude and Stress DropDetermination. The evolving Earth surface/Seismics, seismology, geoelectrics, and electromagnetics/Seismology (2021). https://doi.org/10.5194/se-2020-219
29. Correa, J., et al.: Analysis of signal to noise and directivity characteristics of DAS VSP at near and far offsets — A CO2CRC Otway Project data example, 7

30. Lindsey, N.J., et al.: Fiber-optic network observations of earthquake wavefields: fiber-optic earthquake observations. Geophys. Res. Lett. **44**, 11792–11799 (2017). https://doi.org/10.1002/2017GL075722
31. Ajo-Franklin, J.B., et al.: Distributed acoustic sensing using dark fiber for near-surface characterization and broadband seismic event detection. Sci Rep. **9**, 1328 (2019). https://doi.org/10.1038/s41598-018-36675-8
32. Sladen, A., et al.: Distributed sensing of earthquakes and ocean-solid Earth interactions on seafloor telecom cables. Nature Commun. **10**, 5777 (2019). https://doi.org/10.1038/s41467-019-13793-z
33. Schenato, L., et al.: Distributed acoustic sensing of debris flows in a physical model. In: Optical Fiber Sensors Conference 2020 Special Edition, OSA Technical Digest. Optica Publishing Group (2020). https://doi.org/10.1364/OFS.2020.Th4.27. Paper Th4.27
34. Zarattini, F., et al.: Preliminary numerical and experimental test for the study of vibration signals in dry granular flows. In: VI International Conference on Particle-Based Methods, p. 8. International Association for Computational Mechanics (IACM), Barcelona (2019)
35. Ravet, F., Briffod, F., Goy, A., Rochat, E.: Mitigation of geohazard risk along transportation infrastructures with optical fiber distributed sensing. J. Civil Struct. Health Monit. **11**(4), 967–988 (2021). https://doi.org/10.1007/s13349-021-00492-x
36. Pei, H., et al.: Monitoring and warning of landslides and debris flows using an optical fiber sensor technology. J. Mt. Sci. **8**, 728–738 (2011). https://doi.org/10.1007/s11629-011-2038-2
37. Yin, J.-H., Qin, J.-Q., Tan, D.-Y., Zhu, Z.-H.: Monitoring a flexible barrier under the impact of large boulder and granular flow using conventional and optical fibre sensors. In: Wu, W., Yu, H.-S. (eds.) Proceedings of China-Europe Conference on Geotechnical Engineering. SSGG, pp. 755–758. Springer, Cham (2018). https://doi.org/10.1007/978-3-319-97112-4_169
38. Li, H.-J., Zhu, H.-H., Li, Y.-H., Hu, W., Shi, B.: Fiber Bragg grating-based flume test to study the initiation of landslide-debris flows induced by concentrated runoff. Geotech. Test. J. **44** (2021). https://doi.org/10.1520/GTJ20190290
39. Correia, R., et al.: Fibre Bragg grating based effective soil pressure sensor for geotechnical applications. Presented at the October 5 (2009). https://doi.org/10.1117/12.835751
40. Schenato, L., et al.: Highly-sensitive FBG pressure sensor based on a 3D-printed transducer. J. Lightwave Technol. **37**, 1–6 (2019). https://doi.org/10.1109/JLT.2019.2919917
41. Yin, J.-H., Qin, J.-Q., Feng, W.-Q.: Novel FBG-based effective stress cell for direct measurement of effective stress in saturated soil. Int. J. Geomech. **20**, 04020107 (2020). https://doi.org/10.1061/(ASCE)GM.1943-5622.0001724
42. Qin, Y., Wang, Q., Xu, D., Yan, J., Zhang, S.: A fiber Bragg grating based earth and water pressures transducer with three-dimensional fused deposition modeling for soil mass. J. Rock Mech. Geotech. Eng. (2021). https://doi.org/10.1016/j.jrmge.2021.07.009
43. Schenato, L., Pasuto, A., Galtarossa, A., Palmieri, L.: An optical fiber distributed pressure sensing cable with pa-sensitivity and enhanced spatial resolution. IEEE Sensors J. **20** (2020). https://doi.org/10.1109/JSEN.2020.2972057
44. Xu, D.-S., Dong, L.-J., Borana, L., Liu, H.-B.: Early-warning system with quasi-distributed fiber optic sensor networks and cloud computing for soil slopes. IEEE Access. **5**, 25437–25444 (2017). https://doi.org/10.1109/ACCESS.2017.2771494

Development of Optic Fiber Sensing Technology for Geotechnical Application - From Laboratory Measurement to Geotechnical Monitoring

Dao-Yuan Tan(✉) and Jian-Hua Yin

The Hong Kong Polytechnic University, Hong Kong, China

Abstract. Optic fiber sensing (OFS) technology has been attracted much attention from geotechnical engineers considering its advantages including small size, light weight, immunity to electromagnetic interference (EMI) and corrosion. It has been applied to provide potential solutions to geotechnical engineering problems. This study first reviews the state-of-the-art applications of OFS in different geotechnical engineering areas, then focuses on several scientific issues in geotechnical engineering from laboratory measurement (determination of effective stress in saturated soils and measurement of surface strain for cylindrical rock sample) to geotechnical monitoring (monitoring the behavior of tree and retaining wall under extreme weathers), finally introduces the advanced methods and transducers based on the optic fiber sensing technology to make contributions to those scientific issues.

Keywords: Optic fiber sensing · Geotechnical engineering · Laboratory experiments · Geotechnical monitoring

1 Brief Review of OFS Application in Geotechnical Engineering

Optical fiber sensing technology is gaining lots of appreciation and acceptance in geotechnical engineering applications. Compared with traditional electrical resistance strain gauges and vibrating wire transducers, OFS takes the lead in many aspects compared with conventional electronic sensors: capable of fully distributed or quasi-distributed measurement rather than discrete measurement; absolute measurement and high sensitivity with the resolution of 1 $\mu\varepsilon$ for strain measurement and 0.1 °C for temperature measurement; and immunity to EMI and corrosion. During the last decade, the innovative development of optical fiber sensors has led to many new applications in the geotechnical engineering field, from laboratory high-accuracy measurement to geotechnical structure monitoring [1].

OFS could be a very useful tool for high-accuracy measurement of strain or temperature [2]. Chen et al. (2020) developed a new measurement device based on fiber Bragg grating (FBG) to determine the radial strains of a soil specimen under both static and cyclic loadings [3]. By laboratory calibration, this transducer is capable of capturing the full range of radial strain during shearing and accurately measuring the radial strain at

H.-H. Zhu et al. (Eds.): BRWSG 2021, LNCE 230, pp. 71–80, 2022.
https://doi.org/10.1007/978-981-16-9963-4_6

the level of 10^{-5} strain. Qin et al. (2021) proposed a novel fiber Bragg grating (FBG) transducer fabricated by 3D-printing technique [4]. This transducer can measure earth and water pressures simultaneously, and the effective stress can be calculated based on the effective stress principle. Hong et al. (2019) developed an FBG pressure cell using the 3D-printing technique, which provides a simple and quick encapsulation method for the fabrication of OFS [5]. FBG sensors have also been implanted at the center of 3D-printed clay soils manufactured by the Addictive Manufacturing (AM) technique to monitor the horizontal strain distributions at different testing conditions [6].

The application of OFS in underground structure monitoring has been studied widely. Wang et al. (2021) utilized FBG sensors to measure radial strains on a buried pipe section and monitor soil-pipe interaction under surcharge loadings and proposed a feasible method to monitor the radial displacement of pipe section under surcharge loading [7]. Song et al. (2021) applied FBG technology to monitor the entire process of tunnel excavation [8]. FBG-based inclinometer for monitoring the deformation in the surrounding soil and FBG sensors attached to the existing piles for monitoring the deformation of piles induced by tunnel excavation was developed and implemented in a tunnel excavation physical modelling study successfully.

Due to the high sensitivity to temperature, the OFS technology has also been innovatively adopted for field monitoring of soil parameters such as water content of natural soils and ice content of frozen soils based on the thermal conductivity difference [9]. Wu et al. (2021) proposed a new method of ice content measurement for frozen soils by measuring the thermal conductivity using the actively heated fiber Bragg grating (AH-FBG) sensors [10]. With the robustness of OFS in extreme environments, those studies developed a reliable tool for long-term monitoring of frozen soils in cold regions to study soil response to climate change, which is very important for many Belt and Road countries.

Slope monitoring is of great necessity for early warning of landslide hazards [11, 12]. As a new sensing technology, OFS plays an important role in slope displacement monitoring. Pei et al. (2020) proposed a new FBG-based inclinometer [13]. Measurement of inclined angle can be achieved by the measurement of magnetostrictive strain under different magnetic fields, which overcomes the limitation of the brittle FBG sensors, which may be broken under large strains or excessive bending. Li et al. (2021) used FBG sensors to measure the strain distribution within a slope model under erosion in a flume test, which shows that the OFS technique can provide precise and stable performance in monitoring slope deformation and the great potential of establishing an effective early warning system for landslides and debris flows [14]. Xu et al. (2021) developed a new risk assessment and monitoring method based on the measurement of large deformation inside construction solid waste landfills using a newly proposed fiber-optic based transducer, which can measure large deformation of up to 1200 mm with high accuracy [15].

2 Development of OFS for Geotechnical Application: Laboratory Measurement

2.1 FBG-Based Effective Stress Cell

Since Terzaghi first revealed the principle of effective stress in 1923, effective stress has been widely accepted as an important and fundamental variable in saturated soils for geotechnical engineering research [16]. However, direct measurement and accurate determination of effective stress have always been a problem for geotechnical engineers. The normal way relies on the utilization of specific transducers such as pore pressure transducer (PPT) and earth pressure cell (EPC). Effective stress in fully-saturated soil is normally estimated by abstracting the pore-water pressure u of PPT from the total stress σ of EPC, as shown in Eq. 1.

$$\sigma' = \sigma - u \tag{1}$$

This traditional method has the following limitations:

(a) Since the effective stress is determined by the calculation of the measured pore-water pressure and total stress, the accuracy of this method relies on the resolutions of both transducers: PPT and EPC for pore pressure and total stress measurement. The errors of either type of transducer will weaken the reliability of the calculated effective stress.
(b) Normally the PPT and EPC are not installed in the same location, and different locations lead to misalignment measurement of pore-water pressure and total stress. Even for the effective earth pressure sensor invented by Correia et al. (2009) using two diaphragms as sensing elements for independent measurement of the total stress and pore-water pressure separately, the stresses on different diaphragms still will not be exactly the same [17].

Considering the above inherent disadvantages of the traditional method, Yin et al. (2020) developed a novel effective stress cell based on the fiber Bragg grating (FBG) sensing technology for direct measurement of effective stress in saturated soil [18]. The working principle of this transducer is based on a different method from Eq. 1.

The design of this novel transducer is shown in Fig. 1. The effective stress cell consists of (1) a thin plate as a "sensing plate" with an FBG sensor adhered to the center of the back surface and (2) a "perforated disc base" on the back of the cell. For field applications where the temperature may fluctuate, a dummy FBG sensor that is mechanical strain-free can be placed inside the cell to measure temperature change and reach temperature compensation. In practical applications, a filter stone/paper covering the perforated disc was used to prevent the perforated disc from being clogged with soil particles. Instead of measuring the pore water pressure, the pore pressures acting on both front and back surfaces of the sensing plate are balanced and do not induce the deflection of the sensing plate. Thus the measured deflection of the sensing plate is caused by the effective stress only.

Figure 2 plots the measured results of conventional transducers and FBG-based effective stress cells under complex overburden pressures. The good agreement of measure

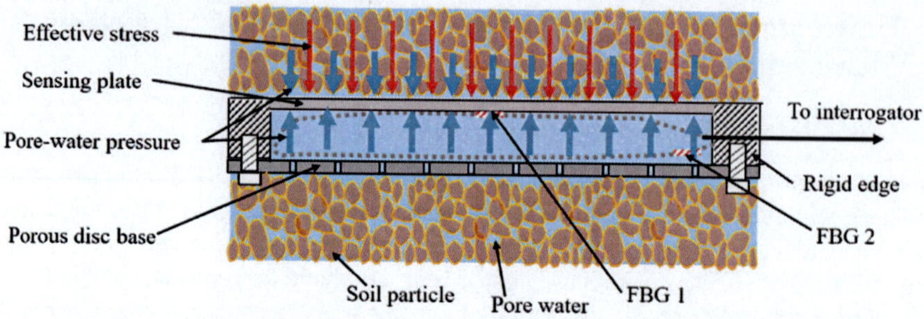

Fig. 1. Illustration of the novel effective stress transducer placed in saturated soil [18]

data by FBG-based effective stress cells and calculated values from the conventional transducers demonstrate that the developed effective stress cell can be utilized in saturated soils subjected to complex loading conditions. Furthermore, the effective stress can be directly obtained with higher accuracy in the measurement and less disturbance on the soil by making use of a single FBG-based effective stress cell, in comparison with embedding two transducers: an earth pressure transducer and a pore-water pressure transducer at different locations for the same purpose.

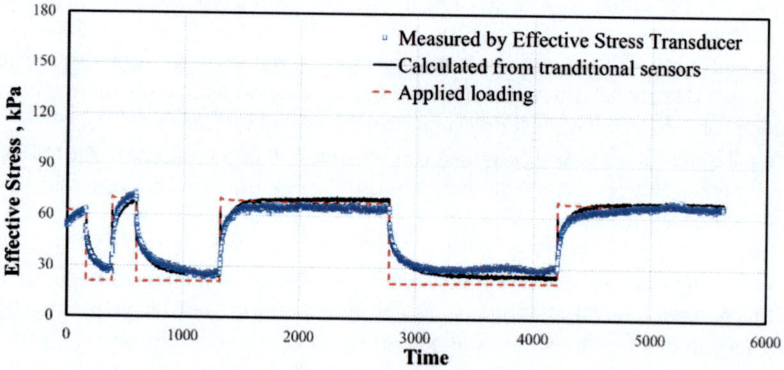

Fig. 2. Comparison of the performance of the novel effective stress transducer with traditional method [19]

2.2 Distributed Fiber Optic Sensor (DFOS) for Surface Strain Monitoring

At present, the uniaxial compression test is one of the basic and widely used laboratory test methods to estimate the mechanical properties and failure mechanisms of rocks through strain measurement [20–22]. Various technologies have been developed for strain measurement in laboratory tests, for example, electrical resistance strain gauges (ESG), linear variable differential transformers (LVDTs), FBG and digital image correlation (DIC) [23–28]. Even though traditional techniques are currently believed to be reliable methods for rock surface strain measurement, they can only provide strain measurement of a single point or at limited locations so that the mechanical properties of rock would be determined by a limited view of the strain field. Digital image correlation (DIC), as a non-intrusive method, has been introduced into strain field measurement because of its capacity for measuring strain distribution within a monitoring window. Currently, the development from two-dimensional (2D) to three-dimensional (3D) DIC solved the measurement errors of 2D DIC caused by the relatively small out-of-plane motions and achieved full-field 3D surface strain measurement [26]. However, there are still some limitations for DIC, for example, dependence on strict external conditions, complicated post-test data processing, and limited strain resolution of 50 $\mu\varepsilon$ [27]. Considering these shortcomings of current techniques for surface strain measurement, Lin et al. (2021) proposed a novel approach to surface strain measurement for cylindrical rock specimens using distributed fiber optic sensing technology [29]. The potential of this method for predicting the possible sequence of crack occurrence and measuring rock crack opening displacement for rock fracturing characteristics research has been further investigated.

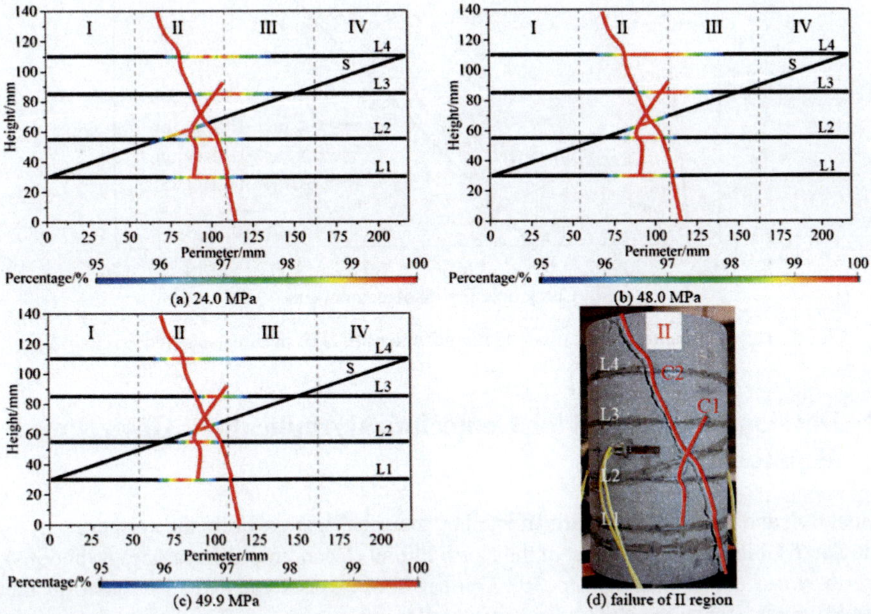

Fig. 3. Strain localization characteristics at different heights in the sandstone under different axial stresses [29]

The OFDR sensing technology adopted in that study can achieve a spatial resolution of 1 mm in a sensing range of 100 m with an outstanding measuring accuracy of $\pm 1\ \mu\varepsilon$, which is fully suitable for strain measurement and micro-crack detection in rocks in laboratory experiments. By gluing a combination of hoop fibers and spiral fibers on the outer surface of a cylindrical specimen, the lateral and axial strains on the specimen surface and the Poisson's ratio of the specimens' material can be measured. Furthermore, the promising performance of this novel approach to surface strain measurement for cylinder rock samples has demonstrated immense potential in local deformation monitoring, especially in those concerning non-uniform deformation. Through a series of UCS tests on rock specimens, this method has been applied to investigate strain localization and possible sequence of crack occurrence, as shown in Fig. 3, and monitor the development of the opening displacements of cracks occurring on the surface of hard rock at a very early stage, as shown in Fig. 4.

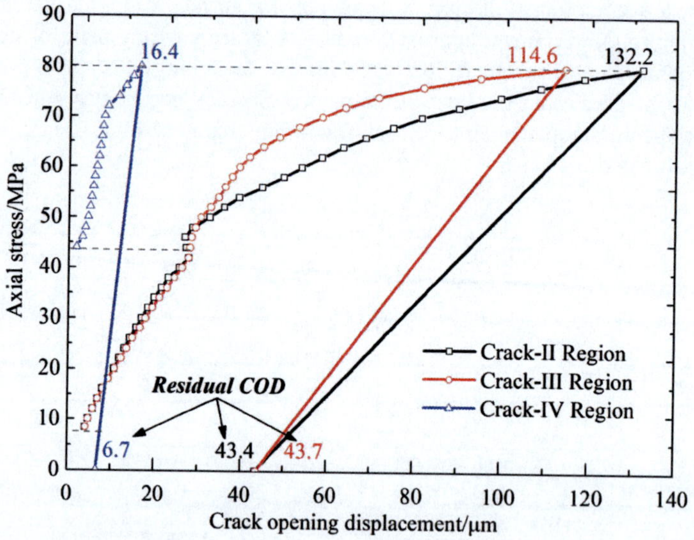

Fig. 4. Development of crack opening displacement (COD) of three selected cracks [29]

3 Development of OFS for Geotechnical Application: Geotechnical Monitoring

Most Belt and Road countries are affected by marine disasters. From the northern part of the South China Sea to the coast of the North Indian Ocean, tropical cyclones (typhoons) are the most common disasters [30]. Optical fiber sensors can directly transform the sensed parameter to the shift of optical wavelength, which is independent of the optical energy, light level, and fiber losses. Apart from the monitoring systems based on conventional electronic sensors, optical fiber sensing techniques, such as fiber Bragg grating (FBG), Brillouin optical-fiber time-domain analysis (BOTDA), and Brillouin optical time-domain reflectometer (BOTDR), have been successfully used for structural health monitoring [31–34].

Since the invention of the first FBG sensor by Hill et al. (1978), FBG sensing technology has shown its advantages in high accuracy, good reliability, immunity to electromagnetic interference, and capacity of multiplexing [35]. Therefore, FBG technology can be a promising technique for monitoring the dynamic behavior of structures. Wu et al. (2021) developed an FBG-based strain gauge specifically for measuring the strain

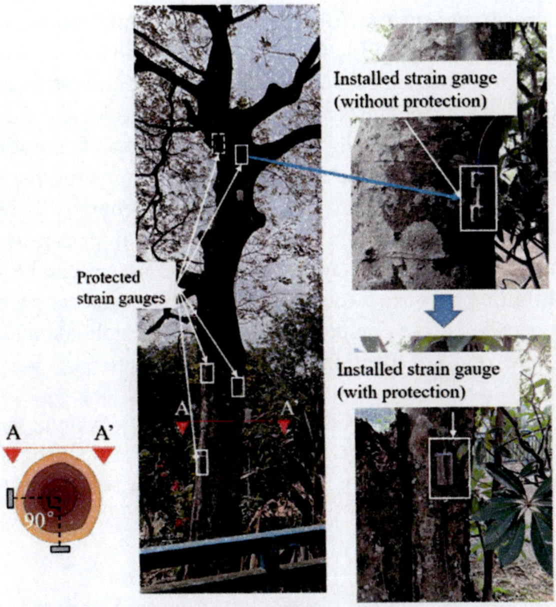

Fig. 5. Arrangement of instrumentation for tree monitoring

distribution of tree trunks under static or dynamic loading and implemented this transducer for the monitoring of tree sway during typhoons [36]. To monitor 2D-strain of the tree trunk under wind loads, a pair of strain gauges were installed perpendicular to each other at the same height of a tree, and the strain at the given height on the tree can be expressed as a combination of two orthogonal components, which are measured by the two strain gauges. A series of FBG sensors can be installed in a row along the tree trunk to measure the strains at different heights, as shown in Fig. 5. From four-month monitoring of dynamic behavior of trees experiencing cyclone disturbance, it is found that this transducer can well capture the dynamic behavior, including the natural frequency of a tree under dynamic wind loads, which is very important for tree stability analysis. If a wind pulse arrives at a frequency near the tree dominant sway frequency, wind force amplifies tree sway as oscillatory motion and may cause tree failure [37]. With the ability to monitor the dynamic response of valuable trees to wind loads, this finding provides a potential application of this transducer in avoiding resonance of trees by proper measures, such as the pruning of tree limbs to prevent possible instability of trees.

Due to the high ecological and cultural values of the stonewall trees and tree-colonized masonry retaining walls in Hong Kong, those structures in service need intelligent health monitoring for evaluating their deformation and safety, further control their possible failures when exposed to severe weathers such as typhoons, heavy rains, increase of groundwater or water pressure. Wu et al. (2021) established a monitoring system for masonry retaining walls, stonewall trees, and large trees. The monitoring system consists of (a) transducers based on FBG sensing technology, (b) data acquisition system, (c) data processing system, (d) remote data transmission module, and (e) solar power system, as shown in Fig. 6 [38]. Optic signals of transducers installed on masonry retaining walls and trees are detected by the data acquisition system and transferred to the data processing system for further analysis. Figures reflecting the dynamic movement of monitored masonry retaining walls and deflection of monitored trees can be plotted and presented on a visualized user interface by analyzing the real-time data of the transducers, which was achieved by MATLAB GUI program. Threshold values for both movements of masonry retaining walls and sway of trees can be pre-set for early warning of possible stability risks. Warning or alarm messages can be distributed to the pre-assigned distribution list once those threshold values are exceeded. Effective and stable coordination between those components enables the establishment of a monitoring system with the following key functions:

(1) Measurement and display of dynamic deflections of trees under wind loads,
(2) Issue of early warnings based on the set thresholds of critical strain and tilt of trees,
(3) Deployment in remote areas with the assistance of a solar power system and 4G data transmission device.

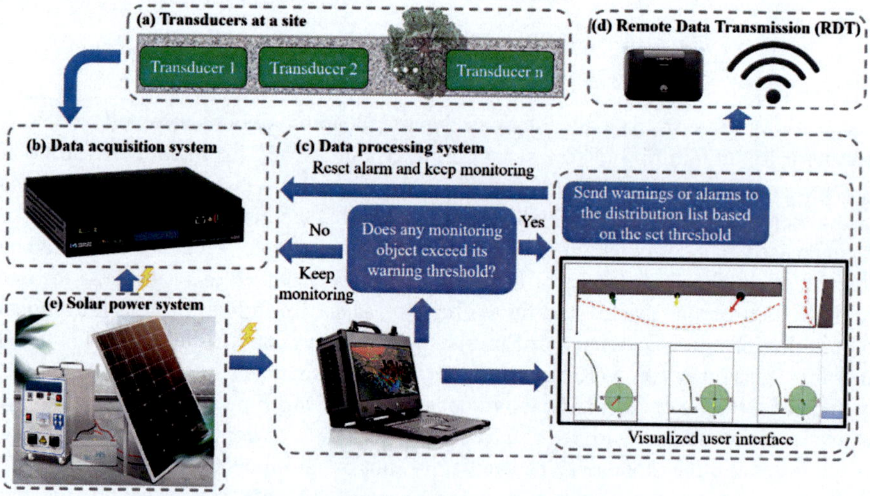

Fig. 6. Illustration of the monitoring system [38]

Four-month successive monitoring has demonstrated the feasibility, accuracy, serviceability, and reliability of the application of optical fiber sensing technology for monitoring dynamic deflections of trees and deformations of masonry retaining walls. Potential of this developed monitoring system includes: (a) study the dynamic behavior of trees under extreme weathers, (b) assess and enhance the resistance of forests under changing climate, (c) investigate efficient measures to manage and prevent possible instability of trees, and (d) provide service life monitoring of important structures and slopes in remote areas.

References

1. Shi, B., et al.: DFOS applications to geo-engineering monitoring. Photonic Sens. **11**(2), 158–186 (2021)
2. Guo, Y., Zhou, W., Xiong, L., Zhou, X., Li, L.: A fiber Bragg grating sensor for positive and negative displacement measurement. IEEE Sens. J. **21**(19), 21564–21571 (2021)
3. Chen, W.B., Feng, W.Q., Yin, J.H., Qin, J.Q.: New fiber Bragg grating (FBG)-based device for measuring small and large radial strains in triaxial apparatus. Can. Geotech. J. **58**(7), 1059–1063 (2021)
4. Qin, Y., Wang, Q., Xu, D., Yan, J., Zhang, S.: A fiber Bragg grating based earth and water pressures transducer with three-dimensional fused deposition modeling for soil mass. J. Rock Mech. Geotech. Eng. (2021)
5. Hong, C.Y., Yuan, Y., Yang, Y., Zhang, Y., Abro, Z.A.: A simple FBG pressure sensor fabricated using fused deposition modelling process. Sens. Actuators A **285**, 269–274 (2019)
6. Hong, C.Y., Wang, X., Han, K., Su, D., Chen, Z.: Performance investigation of 3D printed clay soil using fiber Bragg grating technology. Acta Geotechnica, 1–10 (2021)
7. Wang, D.Y., Zhu, H.H., Wang, B.J., Shi, B.: Performance Evaluation of buried pipe under loading using fiber Bragg grating and particle image velocimetry techniques. Measurement, 110086 (2021)
8. Song, H., Pei, H., Zhu, H.: Monitoring of tunnel excavation based on the fiber Bragg grating sensing technology. Measurement **169**, 108334 (2021)
9. Cao, D.F., Zhu, H.H., Wu, B., Wang, J.C., Shukla, S.K.: Investigating temperature and moisture profiles of seasonally frozen soil under different land covers using actively heated fiber Bragg grating sensors. Eng. Geol. **290**, 106197 (2021)
10. Wu, B., Zhu, H.H., Cao, D., Xu, L., Shi, B.: Feasibility study on ice content measurement of frozen soil using actively heated FBG sensors. Cold Reg. Sci. Technol., 103332 (2021)
11. Zhu, H.H., Shi, B., Yan, J.F., Zhang, J., Wang, J.: Investigation of the evolutionary process of a reinforced model slope using a fiber-optic monitoring network. Eng. Geol. **186**, 34–43 (2015)
12. Shi, X.S., Liu, K., Yin, J.: Effect of initial density, particle shape, and confining stress on the critical state behavior of weathered gap-graded granular soils. J. Geotech. Geoenviron. Eng. **147**(2), 04020160 (2021)
13. Pei, H., Jing, J., Zhang, S.: Experimental study on a new FBG-based and Terfenol-D inclinometer for slope displacement monitoring. Measurement **151**, 107172 (2020)
14. Li, H.J., Zhu, H.H., Li, Y.H., Hu, W., Shi, B.: Fiber Bragg grating–based flume test to study the initiation of landslide-debris flows induced by concentrated runoff. Geotech. Test. J. **44**(4), 986–999 (2021)
15. Xu, D., Zhu, F., Lalit, B., Fan, X., Liu, Q.: Construction solid waste landfills: risk assessment and monitoring by fibre optic sensing technique. Geom. Nat. Haz. Risk **12**(1), 63–83 (2021)
16. Bishop, A.W., Blight, G.E.: Some aspects of effective stress in saturated and partly saturated soils. Géotechnique **13**(3), 177–197 (1963)
17. Correia, R., et al.: Fibre Bragg grating based effective soil pressure sensor for geotechnical applications. In: 20th International Conference on Optical Fibre Sensors, vol. 7503, p. 75030F. International Society for Optics and Photonics (2009)
18. Yin, J.H., Qin, J.Q., Feng, W.Q.: Novel FBG-based effective stress cell for direct measurement of effective stress in saturated soil. Int. J. Geomech. **20**(8), 04020107 (2020)
19. Qin, J.Q., Feng, W.Q., Wu, P.C., Yin, J.H.: Fabrication and performance evaluation of a novel FBG-based effective stress cell for directly measuring effective stress in saturated soils. Measurement **155**, 107491 (2020)

20. Basu, A., Mishra, D., Roychowdhury, K.: Rock failure modes under uniaxial compression, Brazilian, and point load tests. Bull. Eng. Geol. Env. **72**(3), 457–475 (2013)
21. Kuhinek, D., Zorić, I., Hrženjak, P.: Measurement uncertainty in testing of uniaxial compressive strength and deformability of rock samples. Meas. Sci. Rev. **11**(4), 112–117 (2011)
22. Xie, H., Pei, J., Zuo, J., Zhang, R.: Investigation of mechanical properties of fractured marbles by uniaxial compression tests. J. Rock Mech. Geotech. Eng. **3**(4), 302–313 (2011)
23. Clayton, C., Khatrush, S.: A new device for measuring local axial strains on triaxial specimens. Géotechnique **36**(4), 593–597 (1986)
24. Ibraim, E., Di Benedetto, H.: New local system of measurement of axial strains for triaxial apparatus using LVDT. Geotech. Test. J. **28**(5), 436–444 (2005)
25. Sun, Y., Li, Q., Yang, D., Fan, C., Sun, A.: Investigation of the dynamic strain responses of sandstone using multichannel fiber-optic sensor arrays. Eng. Geol. **213**, 1–10 (2016)
26. Zhang, G., Xing, Y., Wang, L.: Comprehensive sandstone fracturing characterization: integration of fiber Bragg grating, digital imaging correlation and acoustic emission measurements. Eng. Geol. **246**, 45–56 (2018)
27. Abdulqader, A., Rizos, D.C.: Advantages of using digital image correlation techniques in uniaxial compression tests. Results Eng. **6**, 100109 (2020)
28. Munoz, H., Taheri, A., Chanda, E.K.: Pre-peak and post-peak rock strain characteristics during uniaxial compression by 3D digital image correlation. Rock Mech. Rock Eng. **49**(7), 2541–2554 (2016)
29. Lin, S.Q., Tan, D.Y., Yin, J.H., Li, H.: A Novel Approach to surface strain measurement for cylindrical rock specimens under uniaxial compression using distributed fibre optic sensor technology. Rock Mech. Rock Eng., 1–15 (2021)
30. Peng, C., Regmi, A., Qiang, Z., Yu, L., Xiaoqing, C., Deqiang, C.: Natural hazards and disaster risk in one belt one road corridors. In: Mikoš, M., Tiwari, B., Yin, Y., Sassa, K. (eds.) WLF 2017, pp. 1155–1164. Springer, Cham (2017). https://doi.org/10.1007/978-3-319-53498-5_131
31. Feng, W.Q., Yin, J.H., Borana, L., Qin, J.Q., Wu, P.C., Yang, J.L.: A network theory for BOTDA measurement of deformations of geotechnical structures and error analysis. Measurement **146**, 618–627 (2019)
32. Pei, H.F., Teng, J., Yin, J.H., Chen, R.: A review of previous studies on the applications of optical fiber sensors in geotechnical health monitoring. Measurement **58**, 207–214 (2014)
33. Hong, C.Y., Zhang, Y.F., Li, G.W., Zhang, M.X., Liu, Z.X.: Recent progress of using Brillouin distributed fiber optic sensors for geotechnical health monitoring. Sens. Actuators A **258**, 131–145 (2017)
34. Xu, D.S., Liu, H.B., Luo, W.L.: Development of a novel settlement monitoring system using fiber-optic liquid-level transducers with automatic temperature compensation. IEEE Trans. Instrum. Meas. **67**(9), 2214–2222 (2018)
35. Hill, K.O., Fujii, F., Johnson, D.C., Kawasaki, B.S.: Photosensitivity on optical fiber waveguides: application to reflection filter fabrication. Appl. Phys. Lett. **32**, 647–649 (1978)
36. Wu, P.C., Tan, D.Y., Chen, W.B., Malik, N., Yin, J.H.: Novel fiber Bragg Grating-based strain gauges for monitoring dynamic responses of Celtis sinensis under typhoon conditions. Measurement **172**, 108966 (2021)
37. Baker, C.J.: Measurements of the natural frequencies of trees. J. Exp. Bot. **48**(5), 1125–1132 (1997)
38. Wu, P.C., et al.: Development of a monitoring and warning system based on optical fiber sensing technology for masonry retaining walls and trees. J. Rock Mech. Geotech. Eng., 1647–7755 (2021)

A Comparative Analysis of the Energy Security Index in the ASEAN Region

Saleh Shadman[1,3](✉), Phahmee Ahanaf Khalid[1,3], Christina Chin May May[1], Novita Sakundarini[1], and Eng Hwa Yap[2]

[1] University of Nottingham Malaysia, Jalan Broga, 43500 Semenyih, Selangor, Malaysia
saleh1shadman@gmail.com

[2] School of Robotics, XJTLU Entrepreneur College (Taicang), Xi'an Jiaotong-Liverpool University, Suzhou 215123, Jiangsu, People's Republic of China

[3] EcoThinkers Consultancy, House 59, Road 1, Block i, 1213 Banani, Dhaka, Bangladesh

Abstract. In this study, a comparative analysis of four ASEAN countries has been done to assess these countries' energy security (ES) levels. With an overall degradation in the climate change situation, it calls for improving the ES level for each country through proper policy implications and implementation. This study aims to provide evidence and a data-driven framework to assess these countries' overall energy security index (ESI) to ensure long-term ES can be achieved. With stakeholder engagement and data curation, normalization is the critical method to be followed in this study. The results suggested an overall improvement in each country's ESI level, with Thailand showing the highest margin of gain of 58.6% and Singapore with the lowest. Indonesia and Malaysia have a similar growth pattern for ESI but not a significant one indicating the need to implement the current energy policies to their potential best. ASEAN as a region has very high potential to achieve higher ES scores with better regional coordination and understanding of the critical challenges.

Keywords: Energy security · Sustainability · Renewable energy · Environment · Technology

1 Introduction

Different groups of distinguished researchers in the field of energy security (ES) and sustainability have defined the concept of ES in the context of a country's or a region's energy demand and security of supply [1, 2], geopolitics, environmental sustainability goals, economic planning, technological advancement, and efficiency [3, 4]. As a result, Chester (2010) [5] stated that ES is "multi-dimensional" in nature, thus allowing researchers to create a framework with a set of dimensions on their own [6, 7]. The ultimate objective of ES, according to Liu et al. (2019) [8], is to preserve energy independence by having better and higher production than consumption, regardless of whether energy is imported or exported to or from other nations. This is a simplified version of a more detailed description, such as the feature in which a connected system works ideally and sustainably in all dimensions, free of any dangers [9]. ES relies on

H.-H. Zhu et al. (Eds.): BRWSG 2021, LNCE 230, pp. 81–94, 2022.
https://doi.org/10.1007/978-981-16-9963-4_7

the global flow of primary and secondary energies and the trade of commodities and services generated using energy [10].

According to Dincer and Acar (2015) [11], meeting escalating energy demands in an environmentally friendly and long-term manner is a daunting job, particularly for developing nations. Safari et al. (2019) [12] discussed how a country's economic expansion has an environmental footprint, resulting in environmental constraints. Environmental issues are often regarded as humanity's most significant issue today. In 2019, it was estimated that 84.3% of energy will come from fossil fuels, which are a non-renewable source of energy, while 11.4% will come from renewable sources of energy such as hydropower, the solar, wind, and biofuels, with nuclear power accounting for the remaining 4.3% [13]. When compared to the year 2000, fossil fuels provided 86.1% of energy. ASEAN should make diversification of energy sources a top priority in order to minimize its reliance on fossil fuels. ASEAN should promote diversification of energy sources as a primary concern in an effort to reduce its dependency on fossil fuels. As the world advances toward more sustainable development, ASEAN should increase awareness of ES vulnerabilities and mitigate them by ensuring that the ES components are addressed. The issue stems from an over-reliance on fossil fuels such as coal and natural gas, which strains these energy sources and energy reserve to output ratio. The issue also arises from a policy framework that relies on fossil fuels to satisfy the majority of its energy needs [14].

There is a lack of studies that quantitatively assess the ES of the ASEAN region. Few studies [15–18] initiated the process and have successfully quantified it through various data collection methods and data curation. However, there is a gap in engaging stakeholders to verify and validate the indicators and hence the quantitative data of these indicators are being studied. Some studies have quantified ES for certain countries, but very few have assessed a region overall or multiple countries within a region. In this study, the gap assessment suggests a need to determine ES for the ASEAN region quantitatively. The current study aims to fill this gap by analyzing 5 dimensions of ES; availability, affordability, environmental sustainability, the applicability of technology and accessibility of energy. These dimensions have been identified through thorough stakeholder engagement in the previous studies [19]. A total of 26 indicators from these 5 dimensions have been quantified. The initial assessment was done for Malaysia as the dimensions, and the indicators were selected for Malaysia. However, these same sets of dimensions and their indicators can be applied to Singapore, Thailand, and Indonesia as well based on a similar energy outlook and pattern of energy demand and consumption.

Upon successful completion of the study's objective, a comparison was made to the four countries within ASEAN to understand the level of ES of each country. However, there are certain data limitations in terms of data unavailability for the listed indicators in Table 1. Firstly, the country-specific analysis of the results is done in Sect. 3.1 followed by an overall comparison of the ES level in Sect. 3.2. The results are followed by discussion and policy implications in Sect. 4, finally conclusive remarks, limitations, and future direction of the framework developed in this study.

1.1 Existing Frameworks for the ASEAN Region

There are multiple existing frameworks and data curation, normalisation, and weighting methods for the quantitative data collected for ES measurement. Table 1 summarises some of the studies with a similar approach in terms of data normalization but different frameworks with different numbers of indicators and dimensions based on the requirements of the study.

In Table 1, the cases within Asia have been selected with the same data normalization approach to ensure consistency is maintained throughout the study. While other literature quantify ES using different approaches are excluded.

Table 1. Existing frameworks for ES assessment in Asia

Source	Framework	Dimensions	Indicators	Key methodology followed
[17]	4A's framework	Availability, applicability, affordability, and acceptability	16	**Normalisation:** Min-Max approach **Scale:** 1–10 **Weighting**: Equal weight
[20]	4 A's framework	Availability of energy, applicability of technology, acceptability by the society, and affordability of energy resources	20	**Normalisation:** Min-Max approach **Scale:** 1–10 **Weighting**: Equal weight
[21]	4 A's framework	Availability, applicability, acceptability, and affordability	16	**Normalisation:** Min-Max approach **Scale:** 1–10 **Weighting**: Equal weight
[22]	Case study basis	Availability, affordability, efficiency, sustainability, and governance	20	**Normalisation:** Min-Max approach **Scale:** 0–100 **Weighting**: Equal weight
[15]	Case study	Energy supply and demand and, economic dimensions	6	**Normalisation:** Min-Max approach **Weighting**: Equal weight

2 Methods and Data

In this study, the primary data sources play the most crucial role in terms of data validation. The entire method is designed based on the previous research by Shadman et al. (2021) [23], which solely described the practices utilized to collect data and engagement with stakeholders. Data collection plays a vital role in addressing the research objectives in this study. Hence it is treated with the highest significance of protocol. A semi-structured interview was used to obtain qualitative data from stakeholders. The framework of ES was then created through qualitative data gathering and based on the main characteristics and the associated indicators for Malaysia ES. This framework provides the structure and foundation for establishing a quantitative system dynamic model and energy security index (ESI).

2.1 Data Collection Tools and Methods

To identify the essential characteristics of ES for Malaysia and other ASEAN nations, the first set of qualitative data was collected through stakeholder engagement [24]. In prior research by Shadman et al. (2021) [19], the data collecting procedure includes sample selection, rationale and with a list of stakeholders described in detail. A total of 16 stakeholders were questioned utilizing a series of semi-structured interview (SSI) questions, and the information gathered was captured and transcribed on audio/video. To construct emergent themes from the transcribed data, researchers used the qualitative data analysis (QDA) program Quirkos. This is an inductive method of producing theories and frameworks based on empirical evidence; in this instance, the data was gathered and recorded from stakeholders via SSI. As explained by Charmaz in the research by Hesse-Biber and Leavy [25], the grounded theory (GT) method is inductive, and its approach is iterative, as depicted in Fig. 1.

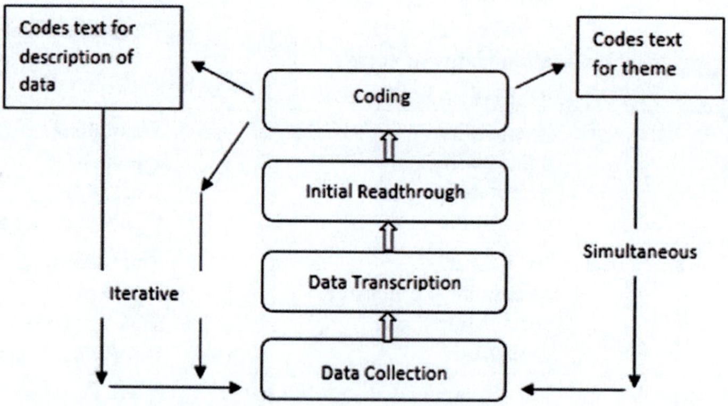

Fig. 1. The grounded theory process by Charmaz [25]

Quantitative data are collected from verified sources and under regulatory and statutory bodies of the respective countries. This is to ensure that secondary data collected from the public domain are of high quality and accuracy. The raw data, in particular, were collected from the following documents and websites within the public domain and are freely available for use [26–34]. The benefits of adopting secondary data from these sources are highly verified because Malaysian and ASEAN official agencies release it and are available in huge quantities at no cost. Additionally, secondary data has extensive background work, including literature reviews, case studies, published texts, and statistics [35].

Table 2 consists of the dimensions and the indicators resulting from the data collection process to be explained in this section. A combination of critical and systematic literature review as shown in Table 1 and stakeholder engagement process leads to the final set of dimensions and the indicators in Table 2.

Table 2. Dimensions and their respective indicators

Dimension	Indicators
Availability of energy (AV)	Total Primary Energy Supply per capita (TPES/POP)
	R/P oil
	R/P gas
	R/P coal
	Energy self-sufficiency ratio (total energy production/TPES)
	Total final energy consumption/POP
Affordability of energy (AF)	GDP per unit energy Unit: PPP per kg of oil equivalent
	Liquid fuel retail price/GDP for diesel and gasoline
	Energy consumption per capita (TFEC/POP)
	Electricity tariff Unit: sen/kWh
	Crude oil price Unit: USD/barrel
	Natural gas price Unit: Henry-hub
	Population with access to electricity (%)
Environmental sustainability (ENV)	CO_2 emission/POP
	CO_2 emission/GDP
	CO_2 emission/TPEC
	Share of RE in electricity generation (%)
	Non-carbon share/TPES
Applicability of technology and efficiency (APE)	Energy supply intensity Unit: toe/GDP in million RM
	Industrial energy intensity
	Overall energy intensity (TPEC/GDP) Unit: toe/GDP in million RM

(*continued*)

Table 2. (*continued*)

Dimension	Indicators
Accessibility (AC)	Access to electricity (%)
	Crude oil market concentration risk
	Electrification level (EPOP/POP) Unit: None
	Access to clean fuel and technology for cooking (%)

2.2 Data Coding, Curation and Normalisation

To ensure consistency across the index, it is necessary to convert all of the indicators to a single standard scale for data standardization. Data normalization is the term for this approach. For this investigation, the min-max method of data normalization was chosen. This approach evaluates the performance by comparing the best and worst results and to validate the min-max technique of data normalization, the research of [17, 18, 21] and [20] were rigorously followed. This technique is easy, accurate, and reasonable for a small or large amount of raw data. The weighting of the five dimensions has been equalized and maintained to avoid bias since there is no clear indication of the weight of each dimension.

The raw values for the indicators were transformed into normalized values with a range of 1–10. The following formula was used for the transformation of the indicators with positive attributes to ES. The high value of X' corresponds with high ES.

$$X' = 1 + \left(\frac{X - MinA}{MaxA - MinA}\right) \times (10 - 1)$$

For inversely related indicators like CO_2 emission per capita, the higher value would indicate to lower ES. Hence, for these indicators, the formula is changed to ensure that the maximum value in the scale is considered Min A and vice versa. The procedure used for the inversely proportional indicators is

$$X' = 1 + \left(\frac{X - MaxA}{MinA - MaxA}\right) \times (10 - 1)$$

X': The transformed value of the indicator
X: Raw value of the indicator
A: Range of the raw value
Max A: Maximum value of the indicator in the scale
Min A: Minimum value of the indicator in the scale.

Finally, the value of the dimension is calculated using the formula below, which is a simple average of each of the indicators for that respective year since the weight of each indicator within the dimension is considered equal.

$$Dimensional\ value = \frac{Indic.1 + Indic.2 + ..Indic.n}{n}$$

3 Results and Discussion

The results for Malaysia, Singapore, Indonesia and Thailand are presented in this section and discussed. All the tables in this section represent the normalized data generated following the methods in Sect. 2.2.

3.1 Malaysia

Table 3 shows the values of the respective dimensions after the indicators have been normalized and the average is taken to calculate the dimension value.

Table 3. Dimensional values for Malaysia

Year	AV	AF	AC	APE	ENV
2010	4.03	7.49	6.07	4.63	3.73
2011	4.54	6.47	8.05	4.68	4.58
2012	5.45	6.78	9.49	6.46	2.98
2013	6.01	5.08	7.99	6.16	3.4
2014	6.18	5.89	6.59	4.05	3.02
2015	5.44	8.82	5.5	1.59	2.78
2016	3.95	8.04	5.5	3.42	2.65
2017	3.15	5.55	10	5.23	5.14
2018	5.63	5.09	10	5.96	6.51

With scores of 5.19 and 6.64, the total ESI has increased from 2010 to 2018 as shown in Fig. 2. This improvement is an indication that Malaysia is moving in the righteous trajectory in terms of ES. However, it is only a 27.9% growth over 8 years, which might have been higher if environmental sustainability, technological applicability, and efficiency dimension performance had been better. This highlights the need for more significant research into energy-efficient technology, as well as the necessity to protect the environment from carbon-based emissions and raise the percentage of renewable energy to enhance the economic state, as documented in [36, 37] and [38]. Policy implications are required to ensure strict measures are taken against emission levels [39]. Danish et al. (2020) [39] have also highlighted that economic policy uncertainty may negatively impact environmental quality. Thus a nation's policy framework must be consistent. Expanded research and development funding and infrastructure can improve

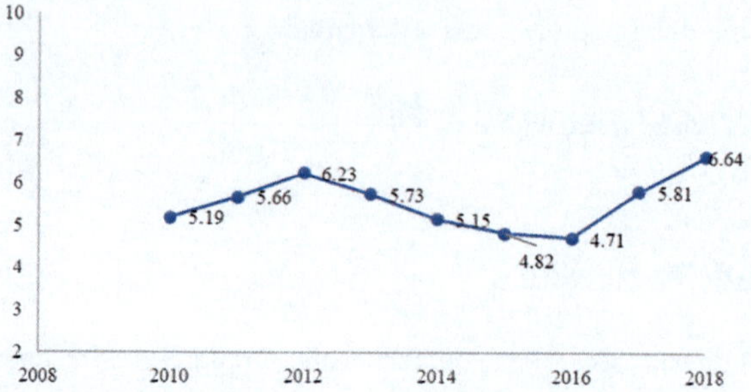

Fig. 2. Overall ESI level of performance for Malaysia from the year 2008–2018

the application of technology to enable more effective use of available assets, therefore protecting the current dimension of energy availability. The availability of energy has decreased dramatically from 2014 to 2017, affecting overall performance; otherwise, the ESI by the end of 2018 would have been higher.

3.2 Indonesia

Table 4 outlined the dimensional values for Indonesia's ESI from the year 2010–2018.

Table 4. Dimensional values for Indonesia

Year	AV	AF	ENV	APE	AC
2010	2.76	2.89	6.41	5.91	1.00
2011	4.17	5.31	2.68	5.42	2.63
2012	5.62	6.28	4.51	6.08	4.78
2013	5.16	5.59	8.97	5.84	6.09
2014	4.47	6.74	5.29	3.56	7.41
2015	4.39	4.89	5.61	2.60	8.53
2016	4.02	5.52	7.36	2.01	9.08
2017	3.88	8.22	5.89	2.62	9.24
2018	5.71	10.00	3.13	5.98	10.00

While Fig. 3 is the overall level of ESI for Indonesia, suggesting a very positive trend of increasing the level of ESI from 3.79 in 2010 to 6.96 in 2018 out of the highest achievable score of 10. This indicates that the ES of Indonesia has grown to a great extent. However, after 2013, the expected ESI, according to the trend, should have been higher for the upcoming years, which would eventually mean that by 2018, the ESI score should have been better than 6.96. There are a few key takeaways from the analysis of Indonesia, namely;

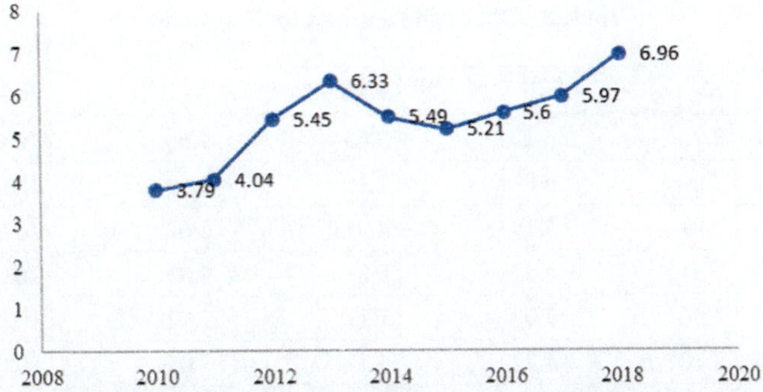

Fig. 3. ESI level for Indonesia from the year 2010–2018

- Energy availability has increased from the year 2010–2018, but it does not indicate an increase in the share of renewables or alternative fuel sources to a great extent. The rise in availability represents more fossil fuels over alternative sources. To achieve long-term ES, fuel diversification will be the most valuable.
- There is an overall degradation in the environmental sustainability dimension. This is alarming regarding climate change impact and Indonesia's environmental and energy policies, greenhouse gas (GHG) emission management and mitigation strategies. Government stakeholders would be responsible for shaping the future direction of this dimension with vision and policies towards climate change
- Access to energy in terms of electricity and fuel to households and the population in different sectors have shown the greatest improvement, and it is a positive takeaway from Indonesia's ES assessment of this study.

3.3 Singapore

In Table 5, it illustrates the dimensional values for Indonesia's ESI from the year 2010–2018.

Figure 4 shows the overall ESI level of Singapore for 8 years. The most notable point in the assessment of Singapore's ES is the consistency in its level. However, the ESI score was 6.44 in 2010 and increased to only 6.84 in 2018 after 8 years. In the years between, the lowest score was 5.77, followed by the highest in 2017 with 7.17. A consistent ESI level shows lower fluctuations and improved stability that prevents exogenous shocks to the energy demand-supply and price, but the ESI scores are still not highly satisfactory in this case. Ideally, Singapore's ESI score should be high due to its accessibility which has been 100% or a score of 10, indicating that all the population within the country has access to electricity and fuel at all times. This is possible with a small population allowing it to be maintained consistently and with overall perseverance by Singapore's effort to ensure good access.

The most noteworthy issue within the ESI framework for Singapore is the unavailability of reserves and energy production, indicating the heavy reliance on energy imports to meet the needs. This is coupled with a strong geopolitical relationship with its neighboring countries where the imported fuels are received by Singapore. Hence, the long-term

Table 5. Dimensional values for Singapore

Year	AV	AF	ENV	APE	AC
2010	2.76	6.10	3.63	9.68	10.00
2011	1.00	6.08	2.15	9.63	10.00
2012	3.81	5.19	4.91	6.61	10.00
2013	6.31	4.57	5.81	4.03	10.00
2014	7.06	3.99	7.17	1.92	10.00
2015	8.18	4.22	6.17	1.05	10.00
2016	7.45	6.98	7.44	2.42	10.00
2017	9.00	5.95	8.32	2.57	10.00
2018	7.53	6.46	8.56	1.66	10.00

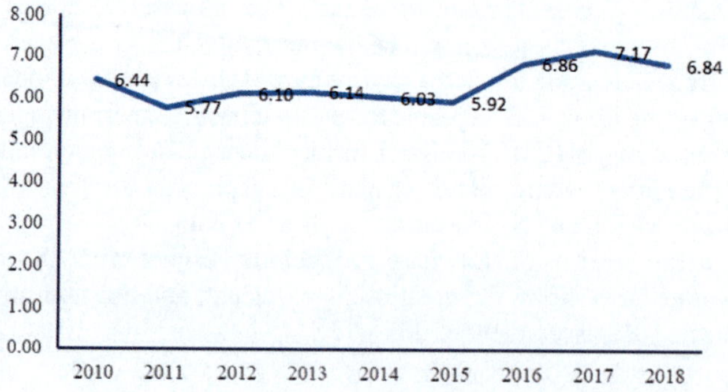

Fig. 4. ESI level for Singapore from the year 2010–2018

ES of the nation is heavily affected by this within this current assessment. Environmental sustainability has improved drastically, and this is in line with the nation's efforts to mitigate GHG emissions and exemplary implementation of existing measures within the energy policies of Singapore.

3.4 Thailand

The following Table 6 shows the dimensional values for Thailand's ESI from the year 2010–2018.

Figure 5 shows the ESI level for Thailand from 2010–2018. A similar trend has been observed in Malaysia and Indonesia, but with a higher overall score. The ESI level has increased and improved by a significant margin of 58.6% from 2010 to 2018, indicating a positive sign for Thailand's ES. Thailand also achieved the highest overall score in comparison to the three countries in this assessment. Environmental sustainability, the applicability of technology, and accessibility have shown increasing trends hinting at successful implementation of environmental policies and better frameworks for applying

Table 6. Dimensional values for Thailand

Year	AV	AF	ENV	APE	AC
2010	4.28	7.08	2.82	4.24	4.36
2011	2.77	5.40	3.99	5.76	3.53
2012	4.93	3.51	3.42	4.61	2.93
2013	6.51	4.35	3.05	4.98	5.75
2014	6.07	4.05	3.73	4.65	5.99
2015	7.47	4.93	3.79	4.15	8.08
2016	6.52	7.42	6.01	5.17	9.72
2017	6.64	6.79	7.31	5.69	10.00
2018	6.66	6.51	8.25	5.61	9.09

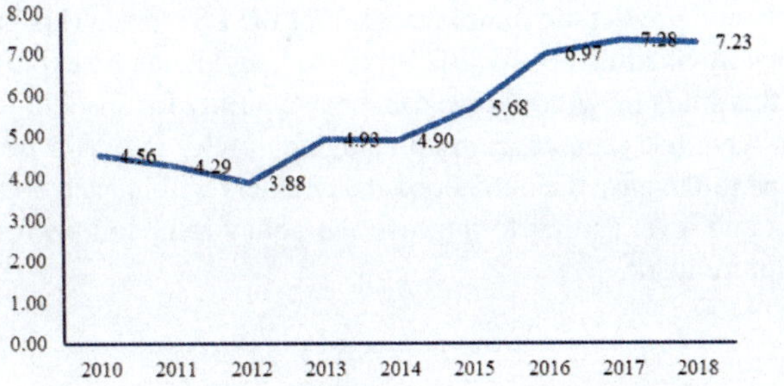

Fig. 5. ESI level for Thailand from the year 2010–2018

new and efficient existing technologies. Although affordability has deteriorated, and the availability of energy does not indicate an outstanding share of renewable energy (RE) within the energy mix of Thailand. The affordability of energy can be increased with a higher subsidy to the electricity tariff to compensate for high tariffs or lower ceiling prices for fossil fuels. Imported fossil fuels need to be of the same price or cheaper than the locally produced fuels in the reserves. The reserve to production ratio should improve availability while integrating RE as alternative sources for a clean start with lower GHG emissions.

4 Conclusion

The assessment in this study has shown an increase in ES levels for all four countries. Thailand is the country with the most significant improvement, followed by Malaysia, Indonesia and Singapore. Singapore's consistency in ES level shows a stable overall outlook with lower exogenous energy shocks. However, to secure long-term ES, there needs to be alternative plans and policies that can generate alternative fuel sources efficiently within Singapore and not heavily rely on imports. For Malaysia, Thailand, and

Indonesia, the stakeholders within the study have stated the confidence in a developing economy, as the use of energy for boosting economic growth takes up the priority in the energy trilemma followed by equity and lastly, environment. The perfect balance within these three is challenging because the primary aim is to ensure energy availability at all times at an affordable price. This prioritizes the two dimensions over the other three and hence leading to higher negligence and poor performance.

ASEAN as a region has abundant energy availability. Thus, the reserves are high, excluding Singapore, while other nations have been able to produce within the country. This secures the dimension. However, it comes at the cost of lower RE share and lower non-carbon-emitting sources share, hence higher risks of climate change and degradation. New and existing technologies can be made efficient with better research and development funds while ensuring access to energy can be reached 100% within the population throughout the countries. These can overall change the ES scenario of the ASEAN region, and with stronger geopolitical relationships and regional bonds, it will eventually be beneficial to all the countries in long-term ES.

The same set of dimensions and indicators can be developed further to create systems models to study and predict the future direction of the ESI level. This has been done in previous studies by Shadman et al. (2021) [40, 41], and it can be explored further. The limitations of this study lie within some data unavailability for specific years and certain indicators that have not been covered up using any proxy indicator or predicted data. There is a scope to increase the number of stakeholders within each country to validate the ESI scores and level further to improve the policy implications for the respective countries for future work.

References

1. García-Gusano, D., Iribarren, D., Garraín, D.: Prospective analysis of energy security: a practical life-cycle approach focused on renewable power generation and oriented towards policy-makers. Appl. Energy **190**, 891–901 (2017)
2. Balat, M.: Security of energy supply in Turkey: challenges and solutions. Energy Convers. Manag. **51**(10), 1998–2011 (2010)
3. Su, W., Zhang, D., Zhang, C., Streimikiene, D.: Sustainability assessment of energy sector development in China and European Union. Sustain. Dev. **28**(5), 1063–1076 (2020)
4. Sutrisno, A., Nomaler, Ö, Alkemade, F.: Has the global expansion of energy markets truly improved energy security? Energy Policy **148**, 111931 (2021)
5. Chester, L.: Conceptualising energy security and making explicit its polysemic nature. Energy Policy **38**(2), 887–895 (2010)
6. Cherp, A., Jewell, J.: The concept of energy security: Beyond the four as. Energy Policy **75**, 415–421 (2014)
7. Jewell, J., Cherp, A., Riahi, K.: Energy security under de-carbonization scenarios: an assessment framework and evaluation under different technology and policy choices. Energy Policy **65**, 743–760 (2014)
8. Liu, Y., Sheng, Z., Azhgaliyeva, D.: ASEAN: impacts of regional, no. 1041, pp. 1–23. Asian Development Bank Institute (2019)
9. Azzuni, A., Breyer, C.: Definitions and dimensions of energy security: a literature review. Wiley Interdiscip. Rev. Energy Environ. **7**(1), 1–34 (2018)

10. Shepard, J.U., Pratson, L.F.: Hybrid input-output analysis of embodied energy security. Appl. Energy **279**, 115806 (2020)
11. Dincer, I., Acar, C.: A review on clean energy solutions for better sustainability. Int. J. Energy Res. **39**(5), 585–606 (2015)
12. Safari, A., Das, N., Langhelle, O., Roy, J., Assadi, M.: Natural gas: a transition fuel for sustainable energy system transformation? Energy Sci. Eng. **7**(4), 1075–1094 (2019)
13. Ritchie, H.: Energy mix. Our World in Data (2019)
14. Bandala, E.R., Berli, M.: Engineered nanomaterials (ENMs) and their role at the nexus of Food, Energy, and Water. Mater. Sci. Energy Technol. **2**(1), 29–40 (2019)
15. Sahid, E.J.M., Sin, T.C.: Energy security in ASEAN region: a case study of Malaysia energy security performance with renewable energy implementation. J. Adv. Res. Fluid Mech. Therm. Sci. **61**(2), 190–201 (2019)
16. Kanchana, K., Unesaki, H.: ASEAN energy security: an indicator-based assessment. Energy Procedia **56**(C), 163–171 (2014)
17. Tongsopit, S., Kittner, N., Chang, Y., Aksornkij, A., Wangjiraniran, W.: Energy security in ASEAN: a quantitative approach for sustainable energy policy. Energy Policy **90**, 60–72 (2016)
18. Sovacool, B.K., Mukherjee, I., Drupady, I.M., D'Agostino, A.L.: Evaluating energy security performance from 1990 to 2010 for eighteen countries. Energy **36**(10), 5846–5853 (2011)
19. Shadman, S., Chin, C.M.M.: The role of current and future renewable energy policies in fortifying Malaysia's energy security: PESTLE and SWOT analysis through stakeholder engagement. Prog. Energy Environ. **16**, 1–17 (2021)
20. Yao, L., Chang, Y.: Energy security in China: a quantitative analysis and policy implications. Energy Policy **67**, 595–604 (2014)
21. Malik, S., Qasim, M., Saeed, H., Chang, Y., Taghizadeh-Hesary, F.: Energy security in Pakistan: perspectives and policy implications from a quantitative analysis. Energy Policy **144**, 111552 (2020)
22. Sovacool, B.K.: Assessing energy security performance in the Asia Pacific, 1990–2010. Renew. Sustain. Energy Rev. **17**, 228–247 (2013)
23. Shadman, S., Chin, C.M.M., Sakundarini, N., Yap, E.H., Velautham, S.: Methodological review of Malaysia's energy security measurement: a systems approach using stakeholder engagement. IOP Conf. Ser. Mater. Sci. Eng. **1092**(1), 012032 (2021)
24. Mohamad-Ali, N., Ghazilla, R.A.R., Abdul-Rashid, S.H., Sakundarini, N., Ahmad-Yazid, A., Stephenie, L.: End-of-life vehicle recovery factors: Malaysian stakeholders' views and future research needs. Sustain. Dev. **26**(6), 713–725 (2018)
25. Hesse-Biber, S.N., Leavy, P.: Approaches to Qualitative Research: A Reader on Theory and Practice. Oxford University Press, Oxford (2004)
26. Energy Statistics: Handbook Malaysia Energy Statistics, p. 84 (2019)
27. Ministry of Natural Resources and Environment Malaysia, Malaysia Biennial Update Report to the United Nations Framework Convention on Climate Change (UNFCCC) (2015)
28. Robinson, F.A.: National energy. Phys. Bull. **27**(9), 374–375 (1976)
29. Malaysia - Countries & Regions - IEA. https://www.iea.org/countries/malaysia. Accessed 25 Feb 2021
30. Malaysia Energy Production and Consumption. https://www.ceicdata.com/en/malaysia/energy-production-and-consumption. Accessed 25 Feb 2021
31. U.S. Energy Information Administration (EIA). https://www.eia.gov/. Accessed 21 Oct 2021
32. Statista - The Statistics Portal for Market Data, Market Research and Market Studies. https://www.statista.com/. Accessed 21 Oct 2021
33. ERIA: Economic Research Institute for ASEAN and East Asia. https://www.eria.org/. Accessed 21 Oct 2021

34. Our World in Data. https://ourworldindata.org/. Accessed 21 Oct 2021
35. Megel, M.E., Heermann, J.A.: Methods of data collection. Plast. Surg. Nurs. **14**(2), 109–110 (1994)
36. Qadir, Z., et al.: Predicting the energy output of hybrid PV–wind renewable energy system using feature selection technique for smart grids. Energy Rep. **7**, 8465–8475 (2021)
37. Murshed, M.: Can regional trade integration facilitate renewable energy transition to ensure energy sustainability in South Asia? Energy Rep. **7**, 808–821 (2021)
38. Raza, M.Y., Wang, X., Lin, B.: Economic progress with better technology, energy security, and ecological sustainability in Pakistan. Sustain. Energy Technol. Assess. **44**, 100966 (2021)
39. Danish, Ulucak, R., Khan, S.: Relationship between energy intensity and CO_2 emissions: does economic policy matter? Sustain. Dev. **28**(5), 1457–1464 (2020)
40. Nair, K., Shadman, S., Chin, C.M.M., Sakundarini, N., Yap, E.H., Koyande, A.: Developing a system dynamics model to study the impact of renewable energy in the short- and long-term energy security. Mater. Sci. Energy Technol. **4**, 391–397 (2021)
41. Shadman, S., Chin, C.M.M., Sakundarini, N., Yap, E.H.: Quantifying the impact of energy shortage on Malaysia's energy security using a system dynamics approach. In: Zahid, M.N.O., Sani, A.S.A., Yasin, M.R.M., Ismail, Z., Lah, N.A.C., Turan, F.M. (eds.) Recent Trends in Manufacturing and Materials Towards Industry 4.0. LNME, pp. 143–154. Springer, Singapore (2021). https://doi.org/10.1007/978-981-15-9505-9_14

Role of Geomechanics in Petroleum Production in Xinjiang Oil Field, China

Biao Li[1(✉)], Bin Xu[2,3], and Baohong Yang[4]

[1] Department of Building, Civil and Environmental Engineering, Concordia University, Montreal, QC H3G 1M8, Canada
biao.li@concordia.ca
[2] Origin Geomechanics Inc., Calgary, AB T3H 0X6, Canada
[3] Department of Civil Engineering, Schulich School of Engineering, University of Calgary, Calgary, AB T2N 4V8, Canada
[4] TRS Energy Consultants Ltd., Calgary, AB, Canada

Abstract. Xinjiang oil field is at the heart of the Belt and Road Initiative (BRI), and special technologies like hydraulic fracturing treatments and thermally enhanced oil recovery technology were applied for unconventional petroleum productions in the Xinjiang oil field. Geomechanics is particularly important in the petroleum productions in these unconventional reservoirs. Key points include the determination of 3D in-situ stress, laboratory characterization of poromechanical properties of reservoir rocks, and thermal-hydro-mechanical coupled modeling on the dilation yield process and induced permeability enhancement. We use cases from two sites to demonstrate the effectiveness of using geomechanical dilation to promote petroleum productions in tight oil and oil sands reservoirs, respectively. For the case of a tight oil reservoir, where hydraulic fracturing was applied, the result indicates that rock dilation, if proactively utilized, can enhance the fluid deliverability from the reservoir into the main hydraulic fracture. For the case of heavy oil reservoir where the SAGD technology was applied, we show that the case with dilation start-up uses a shorter period to complete the steam circulation when compared to that from a conventional SAGD.

Keywords: Unconventional oil · Geomechanical modeling · Hydraulic fracturing · Dilation · Thermal enhanced recovery

1 Introduction

The Xinjiang oil field is in the Junggar Basin and is located in the northwestern part of China (Fig. 1) and it is at the heart of the Belt and Road Initiative (BRI). The increasing demand for oil combined with the reduction in conventional oil production drive more attention from petroleum industries onto deeply buried tight oil reservoirs and heavy-oil resources. Geomechanics is particularly important in the petroleum productions in these unconventional reservoirs.

Special technologies like hydraulic fracturing treatments and thermally enhanced oil recovery (EOR) technology are needed to produce petroleum resources from unconventional reservoirs in the Xinjiang oil field. During the production operations in geological

H.-H. Zhu et al. (Eds.): BRWSG 2021, LNCE 230, pp. 95–104, 2022.
https://doi.org/10.1007/978-981-16-9963-4_8

Fig. 1. Geographical location of the nine Chinese giant oil fields, modified after Höök et al. (2010).

formations, coupled thermal-hydro-mechanical processes are heavily involved. Thus, geomechanics plays a major role in petroleum production. Massive hydraulic fracturing (MHF) treatments have been used to produce hydrocarbons from tight glutenite reservoirs at great depths deeper than 4,000 m in the Xinjiang oil field. Previous results showed the target formation has adequate oil reserves, and some hydraulically fractured wells indeed yielded good commercial oil rates. However, the performances of some hydraulically fractured wells on specific geological intervals rich in laumontite are associated with low production oil rates. Chen et al. (2020) proposed a rock matrix dilation-assisted hydraulic fracturing (D-Frac) technique to enhance oil production in tight glutenite reservoirs. Additionally, the Xinjiang oil field is also very rich in heavy oil resources. A class of enhanced oil recovery (EOR) technologies have achieved great advances in the development of heavy oil (Silva-Oliver et al. 2020). Among those thermal EOR technologies, the steam-assisted gravity drainage (SAGD) method technology has been widely applied to the exploitation of oil sands around the world (Li et al. 2018; Baghernezhad et al. 2019; Gao and Chen 2020). In the SAGD process, high-pressure steam is injected continuously into the injector well to heat the heavy oil and reduce its viscosity, causing the heated oil to drain into the production well (Butler 1985). Similar to the concept of using D-Frac, Sun et al. (2021) proposed to use geomechanical SAGD dilation startup in a Xinjiang oil field heavy-oil reservoir. In this paper, we will draw an introduction to the theoretical background and application of using geomechanical dilation to promote petroleum productions in several unconventional reservoirs in the Xinjiang oil field.

2 The Theoretical Background of Geomechanics

2.1 In-Situ Stress Determination

The in-situ stress in an oil field is critical for the designs of hydraulic fracturing design or SAGD. In general, the original in-situ stress condition for a studying area can be defined by three principal normal stress components: the vertical stress, S_v, and two horizontal stresses, which are commonly denoted as the maximum and minimum horizontal stresses, Sh_{max} and Sh_{min}, respectively. Away from the near-wellbore stress concentration zone, hydraulically driven fracture propagation is generally perpendicular to the direction of the in-situ minimum stress, Sh_{min}. This direction represents the least resistance, which requires the smallest pressure to extend the fracture. The vertical stress (S_v) can be estimated from well density logs. We can apply a minifrac test to measure the minimum horizontal stress if the minimum horizontal stress is the least of the three principal stresses. On the other hand, if the vertical stress is the least of the three principal stresses, the minifrac test should measure the vertical stress. Figure 2a presents an illustration of the openhole mini-frac testing system. It includes two major parts: a surface injection system and a downhole straddle packer system. The mini-frac in-situ stress measurement first creates a hydraulic fracture from the test well by injecting high-pressure fluid and then continues the injection to propagate the fracture to a sufficient distance into the test formation. Subsequently, the injection is stopped (also named as well shut-in) to allow the pressure to decay via fluid leaking off into the test formation and the fracture to close on itself. Figure 2b illustrates a typical pressure/rate history curve of the minifrac test, which includes two major parts. The first part shows that an injection creates and propagates a fracture, and the second part presents the closure of fractures due to fluid leak-offs during well shut-in or flowback procedures. The fracture closure pressure is equal to the minimum in-situ stress acting perpendicular to the fracture. More background information on the mini-frac tests can be found at Zang and Stephansson (2009). Figure 3 shows an example of a mini-frac test conducted on Xinjiang's oilsands formation. Measured curves show the bottom hole pressure (BHP) and related fluid flow rate.

Based on the information of wellbore geometric configuration, Sh_{min}, fracture reopening pressure, poroelastic properties of formation rocks, and the initial formation pore pressure, Sh_{max} can be derived by fulfilling that the hoop stresses circumventing the borehole has a minimum value of zero. Treated as a plane strain problem, the poroelastic stress solution around a borehole can be used to estimate the maximum in-plane stress in the plane perpendicular to the borehole axis. A substantial expression of the poroelastic solutions in the Laplace transforms domain for the pore pressure and the hoop stress was given by Cui et al. (1997).

2.2 Thermal-Hydro-Mechanical Coupled Geomechanical Modeling

2.2.1 Governing Equations

Thermal-hydro-mechanical (THM) coupled numerical modeling is needed to carry out the geomechanical analysis of hydraulic fracturing or SAGD processes. We will highlight the poromechanical governing equations herein. We neglect the complicated multiphase flow equations of oil, gas, and water in porous media herein. Related detailed equations can be found in Peaceman (2000). In addition, thermal convection is ignored in the

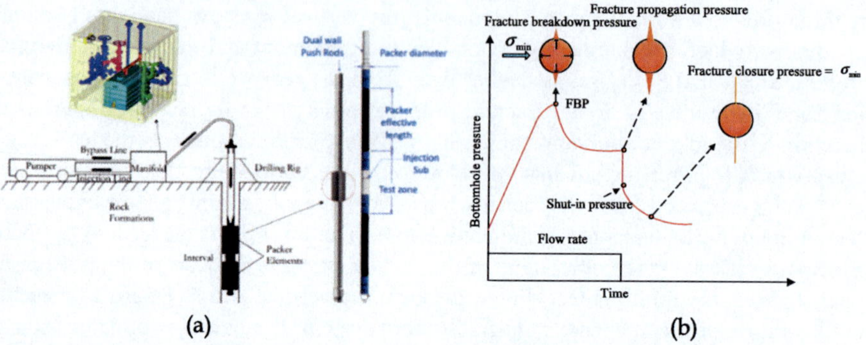

Fig. 2. (a) Illustration of the openhole mini-frac testing system and (b) crack mechanics in hydraulic fracturing demonstrated for the first pressurization cycle, modified after Zang and Stephansson (2009).

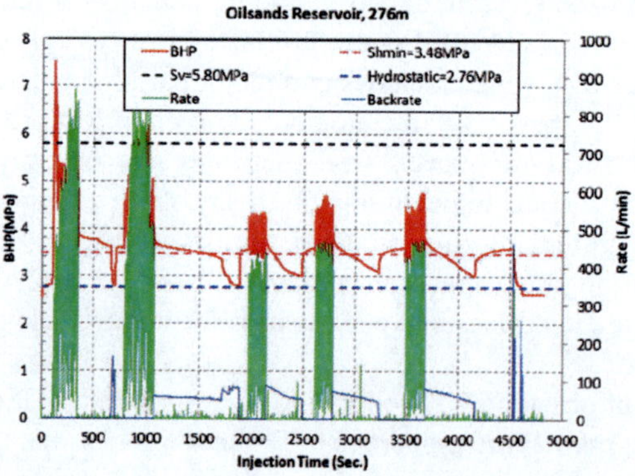

Fig. 3. Curves showing the bottom hole pressure (BHP) and fluid flow rate of a mini-frac test conducted at an oilsands reservoir in Xinjiang oil field (Buried depth = 276 m).

current formulation. According to Lewis et al. (1986), the governing equation can be formulated as:

$$\begin{cases} \mathbf{L}^T\boldsymbol{\sigma} + \rho\mathbf{g} = 0 \\ -\nabla^T\left\{\frac{\mathbf{k}}{\mu}\nabla(p + \rho_w gz)\right\} + \mathbf{m}\frac{\partial\boldsymbol{\varepsilon}}{\partial t} - \phi\beta_w\frac{\partial T}{\partial t} - (1-\phi)\beta_s\frac{\partial T}{\partial t} + \frac{\partial Q_e}{\partial t} = 0 \\ [(1-\phi)\rho_s C_s + \phi\rho_w C_w]\frac{\partial T}{\partial t} - \nabla^T\lambda\nabla^T - \frac{\partial Q_h}{\partial t} = 0 \end{cases} \quad (1)$$

where $\boldsymbol{\sigma}$ is the total stress tensor; $\mathbf{L}$ is the differential operator; $\mathbf{k}$ and μ are permeability matrix and viscosity of the total fluid phase, respectively. ϕ is the porosity. β_w and β_s are pressure and thermal expansion coefficient of water and solid, respectively. $\boldsymbol{\varepsilon}$ is the rock's strain, and $\mathbf{m} = [1, 1, 1, 0, 0, 0]^T$. λ is the thermal conductivity tensor. $\rho_s C_s$ and $\rho_w C_w$ are the heat capacities of the solid phase and the fluid phase, respectively. Q_e is the volumetric outflow of the fluid per unit volume of the solid. Q_h is the outflow of heat per unit volume of solid. Equation 1 describes a coupled THM process, and convection is not considered in the thermal transport equation. The thermal transport is decoupled from Eq. 1 by considering the assumptions: (1) The thermal convection effects can be ignored during the mechanical failure of geomaterials; (2) The fluid flow and mechanical deformation do not change the thermal properties. Thus, Eq. 1 can be simplified to two equation systems: the fully coupled Biot consolidation problem and the thermal conduction problem. ABAQUS can be applied to conduct finite element analysis considering above mentioned THM coupled process.

2.2.2 Plastic Models

For tight rock like the studied laumontite-rich tight rock, we can apply a hyperbolic Drucker-Prager (D-P) model with strain hardening (Abaqus 2016). The hyperbolic Drucker-Prager yield function is given in Eq. (2):

$$F = \sqrt{(d' - \sigma_t \tan\beta)^2 + q^2} - p \tan\beta - d' = 0 \tag{2}$$

where p is the mean effective stress, and q is the Mises shear stress; σ_t is the tensile strength; β and d are the friction angle and cohesion in $p - q$ space, respectively. It should be noted that d' is related to d by:

$$d' = \sqrt{(d' - \sigma_t \tan\beta)^2 + d^2} \tag{3}$$

For oil sands, the yield surface and the non-associative plastic flow rule of a soft rock plasticity model can be used and defined as (Crook et al. 2006):

$$\begin{aligned} F(\boldsymbol{\sigma}) &= \sqrt{\left(e_0 p_c^0 \tan\beta\right)^2 + (g(\boldsymbol{\sigma})q)^2} \\ &\quad -(p - p_t)\tan\beta\left(\frac{p-p_c}{p_t-p_c}\right)^{\frac{1}{n_y}} - e_0 p_c^0 \tan\beta \\ \Psi(\boldsymbol{\sigma}) &= \sqrt{\left(e_0 p_c^0 \tan\psi\right)^2 + (g(\boldsymbol{\sigma})q)^2} \\ &\quad -(p - p_t)\tan\psi\left(\frac{p-p_c}{p_t-p_c}\right)^{\frac{1}{n_y}} - e_0 p_c^0 \tan\psi \end{aligned} \tag{4}$$

where $F(\boldsymbol{\sigma})$ is the yield surface; $\Psi(\boldsymbol{\sigma})$ is the non-associative plastic flow rule function; $\boldsymbol{\sigma}$ is the effective stress tensor; p is the mean effective stress; q is the Mises deviatoric stress; p_c is the yield stress at the hydraulic static compression condition; p_t is the tensile strength; β is the angle of friction; n_y is a material parameter controlling the shape of the yield surface in the $p-q$ plane, and e_0 is a material parameter controlling the shape of the yield at the low effective stress region. $g(\boldsymbol{\sigma})$ captures the transition of the yield surface in the Π-plane from a rounded-triangular shape to a circular shape due to pressure increase.

2.3 Rock Dilation and Permeability Enhancement

A single-phase fluid flow model with permeability as a function of the porosity and current plastic strain state can be used to model the fluid flow during rock yielding or failure. The permeability (k) can be divided into two parts: the permeability before and after the dilation, namely k_a and k_f. The smooth transition from porosity dominated permeability (k_a) to dilation, or microcracking dominated permeability (k_f) is controlled by a state variable function H:

$$k = (1 - H) \cdot k_a + H \cdot k_f \tag{5}$$

where H was a function of the dilation plastic strain ($\tilde{\varepsilon}_v^p$). $\tilde{\varepsilon}_{v0}^p$ is defined as the critical dilative plastic strain energy above which the dilation makes a significant contribution to the permeability and m is a material parameter controlling the speed of the transition from k_a to k_f. H can be defined as:

$$H(\tilde{\varepsilon}_v^p) = \frac{\exp(m(\tilde{\varepsilon}_v^p - \tilde{\varepsilon}_{v0}^p)) - \exp(-m(\tilde{\varepsilon}_v^p - \tilde{\varepsilon}_{v0}^p))}{\exp(m(\tilde{\varepsilon}_v^p - \tilde{\varepsilon}_{v0}^p)) + \exp(-m(\tilde{\varepsilon}_v^p - \tilde{\varepsilon}_{v0}^p))} \tag{6}$$

k_a is modeled via the traditional Carmen-Kozeny type formula as:

$$k_a(\phi) = k_a^0 \left(\frac{\phi}{\phi_0}\right)^c \left(\frac{1 - \phi_0}{1 - \phi}\right)^2 \tag{7}$$

where k_a^0 is the initial formation absolute permeability, c is a material parameter; and ϕ and ϕ_0 are the porosity at the current time and initial time, respectively. The compaction-induced permeability change can be quantified through this porosity-dependent permeability (k_a). When the rock's stress state touches the rock's failure surface, the dilation creates additional porosities and fluid flow channels. It corresponds to a permeability increase k_f that can be modelled based on the cubic law:

$$k_f = \frac{f_a^2}{12\mu} \tag{8}$$

where f_a is the aperture of the microcrack and μ is the fluid viscosity.

3 Application in Petroleum Productions in Xinjiang Oil Fields

3.1 Dilation Assisted Hydraulic Fracturing in a Tight Oil Reservoir

A conventional hydraulic fracturing (C-Frac) tends to result in a major tensile fracture in the target reservoir. To reduce the reservoir rock compaction around the main fracture, we propose to create a dilation zone near the main fracture during the hydraulic fracturing process (Fig. 4). The method of Rock Matrix Dilation-Assisted Hydraulic Fracturing (D-Frac) was successfully applied to the hydraulic stimulation in a laumontite-rich tight oil formation. More details describing the reservoir geology, and D-Frac operations can be found in Chen et al. (2020). We present a numerical modeling result to demonstrate the significance of using D-Frac to promote petroleum production. The geomechanical dilation simulations were conducted by using ABAQUS software (Abaqus 2016) with FORTRAN user subroutines to include the permeability enhancement due to rock yielding or failure.

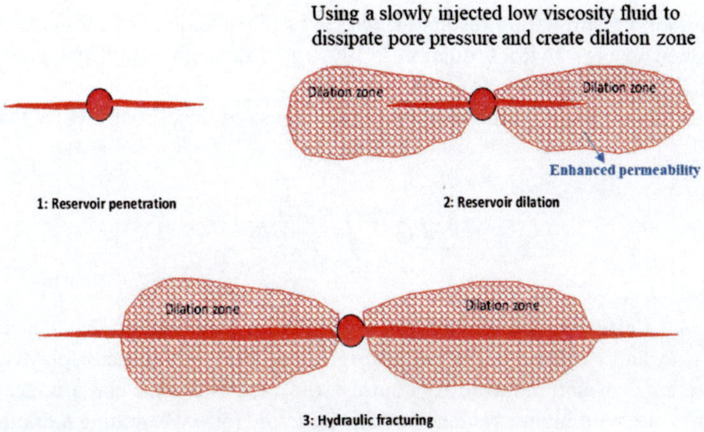

Fig. 4. Sketch illustrating the major procedures of rock matrix dilation assisted hydraulic fracturing.

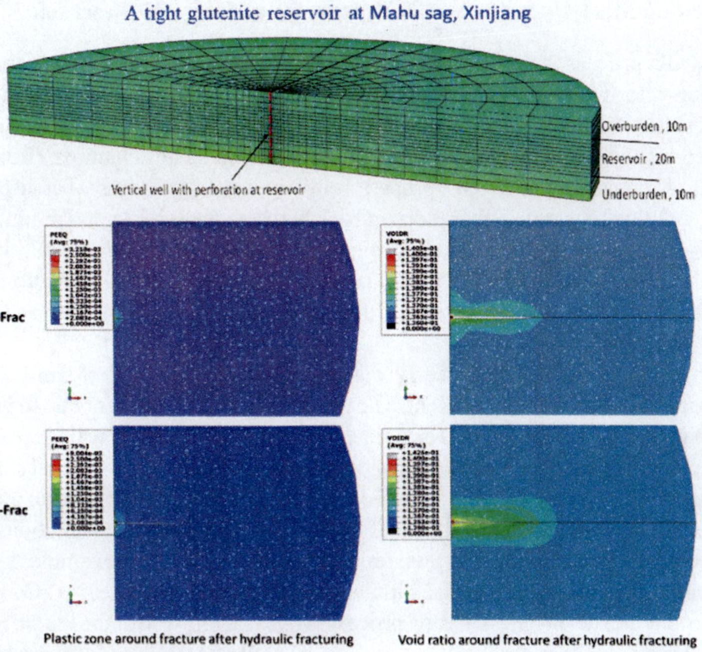

Fig. 5. Comparisons of simulated results from using C-Frac and D-Frac for a vertical well in a Xingjiang tight oil reservoir, after Chen et al. (2020).

The numerical simulation results of two cases (C-Frac and D-Frac) are conducted and presented in Fig. 5. Rock dilation is illustrated through equivalent plastic strain energy (PEEQ) and void ratio (e).

An increase in the void ratio indicates the increase of the porosity. PEEQ is defined as:

$$PEEQ = \int_0^t \frac{\sigma_{ij}^{'}\varepsilon_{ij}^{p}}{\overline{\sigma}_0} dt \tag{9}$$

where $\sigma_{ij}^{'}$ is the effective stress tensor, ε_{ij}^{p} is the plastic strain tensor, $\overline{\sigma}_0$ is the cohesion value of rock, and t is time. As shown in Fig. 5, the case of C-Frac displays a limited influencing zone around the well. By contrast, the case of D-Frac can produce a larger influencing zone with higher values of PEEQ and void ratio. By adding a dilation stage before the main hydraulic fracturing stage, D-Frac can dilate more rock near the main fracture and contribute to a larger oil production rate.

3.2 Dilation Start-Up in the SAGD Process for a Heavy Oil Reservoir

In the SAGD process, a pair of horizontal wells are drilled into the oil-sands reservoir, with the injector well a few meters above the producer. High-pressure steam is injected continuously into the injector well to heat the oil, causing the heated oil to drain into the lower wellbore steam-saturated zone known as the steam chamber (Butler 1985) (Fig. 6a). A conventional SAGD process relies on several months of non-productive steam circulation to warm up the reservoir rock between the injector and producer before the wells can be switched into normal production. In contrast, the SAGD dilation startup process relies on a short period (5 to 7 days) of high-pressure fluid injection to create a dilated zone in the oil-sands reservoir that connects the SAGD injector and producer vertically (Fig. 6b).

We also apply ABAQUS to carry out a geomechanical dilation simulation for a SAGD project in the Xinjiang oil field. The permeability enhancement due to yielding or failure of oil sands is considered as described in Sect. 2.3. In parallel with geomechanical modeling, we also conduct the reservoir simulation using a reservoir thermal simulator to investigate multiphase flow behaviors of oil, water, gas, vapor during steam circulation. We ignored the detailed descriptions of these reservoir thermal simulations, but the impact of dilation startup on heat transfer will be illustrated by an example.

The numerical reservoir thermal simulations considered two cases: (1) Conventional SAGD production (2) SAGD-Dilation process is used on this well at the toe side. Figure 7 gives the steam chamber shapes along the SAGD well after 4 years for these two cases. It is shown that the case with SAGD-Dilation is fully activated, and the steam chamber is growing at the toe of the SAGD well, which will contribute to the oil production rate.

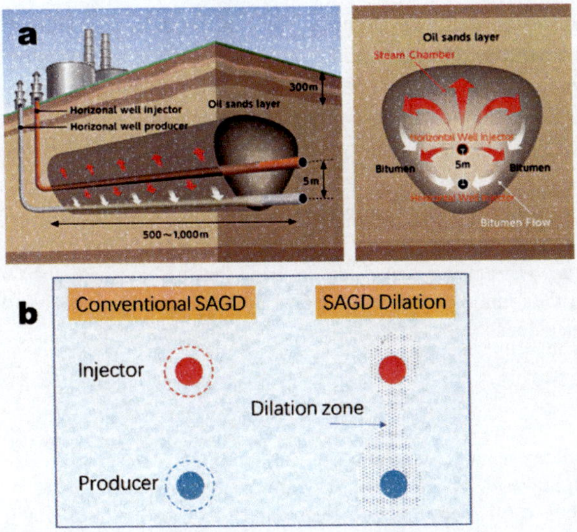

Fig. 6. Sketches showing (a) the schematic of a field-scale application of the SAGD process (courtesy of JAPEX), and (b) an illustration of a conventional SAGD and a SAGD dilation startup, after Sun et al. (2021).

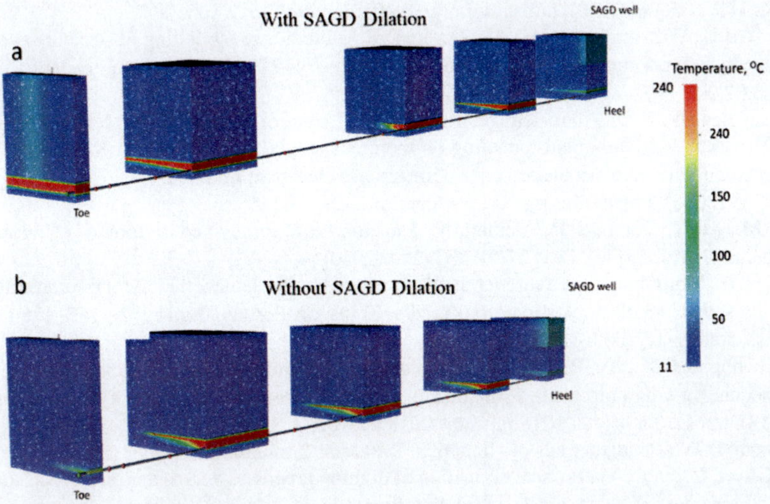

Fig. 7. Simulated steam chamber growth along with SAGD wells after 4 years of SAGD operation for cases of (a) with SAGD-Dilation, and (b) without SAGD-Dilation, after Yang et al. (2021).

4 Summary

We use cases from two sites in the Xinjiang oil field to demonstrate the effectiveness of using geomechanical dilation to promote petroleum productions. The determination of 3D in-situ stress, laboratory characterization of geomechanical properties of reservoir rocks is essential for the field scale thermal-hydro-mechanical coupled modeling. Rock dilation and permeability enhancement can be considered by introducing a state variable function, where the Carmen-Kozeny equation for porosity-dependent permeability and the cubic law for characterizing fluid flow in fractures are included. A Dilation-Assisted Hydraulic Fracturing Technique (D-Frac) is shown to be effective in promoting petroleum production in the tight oil reservoir. Similarly, SAGD-Dilation is also shown to have a great potential for increasing the oil production rate in heavy oil reservoirs.

Acknowledgments. The authors would like to thank Petro-China Xinjiang Oilfield Branch for sharing their field data and experiences. Support by Concordia University's Seed Grant (NO. VS1233) is acknowledged.

References

Abaqus: Abaqus theory manual, version 6.14. Simulia (2016)

Baghernezhad, D., Siavashi, M., Nakhaee, A.: Optimal scenario design of steam-assisted gravity drainage to enhance oil recovery with temperature and rate control. Energy **166**, 610–623 (2019). https://doi.org/10.1016/j.energy.2018.10.104

Butler, R.M.: A new approach to the modelling of steam-assisted gravity drainage. J. Can. Petrol. Technol. **24**(03), 42–51 (1985)

Chen, B., Xu, B., Li, B., Kong, M., Wang, W., Chen, H.: Understanding the performance of hydraulically fractured wells in the laumontite-rich tight glutenite formation. J. Petrol. Sci. Eng. **185**, 106600 (2020). https://doi.org/10.1016/j.petrol.2019.106600

Crook, A.J.L., Willson, S.M., Yu, J.G., Owen, D.R.J.: Predictive modelling of structure evolution in sandbox experiments. J. Struct. Geol. **28**(5), 729–744 (2006). https://doi.org/10.1016/j.jsg.2006.02.002

Cui, L., Cheng, A.H.D., Abousleiman, Y.: Poroelastic solution for an inclined borehole (1997)

Gao, Y., Chen, M.: Numerical modeling on thermoelastoplastic responses of Karamay oil sand reservoir upon steam circulation considering phase change of bitumen. J. Petrol. Sci. Eng. **187**, 106745 (2020). https://doi.org/10.1016/j.petrol.2019.106745

Höök, M., Xu, T., Xiongqi, P., Aleklett, K.: Development journey and outlook of Chinese giant oilfields. Petrol. Explor. Dev. **37**(2), 237–249 (2010)

Lewis, R.W., Majorana, C.E., Schrefler, B.A.: A coupled finite element model for the consolidation of nonisothermal elastoplastic porous media. Transport Porous Media **1**(2), 155–178 (1986). https://doi.org/10.1007/BF00714690

Li, B., Wong, R.C.K., Xu, B., Yang, B.: Comprehensive stability analysis of an inclined wellbore embedded in Colorado shale formation for thermal recovery. Int. J. Rock Mech. Min. Sci. (2018). https://doi.org/10.1016/j.ijrmms.2018.07.019

Peaceman, D.W.: Fundamentals of Numerical Reservoir Simulation. Elsevier (2000)

Silva-Oliver, G., et al.: Theoretical evaluation of dilution processes versus thermal effects induced on the transport of heavy oil. J. Petrol. Sci. Eng. **192**, 107246 (2020). https://doi.org/10.1016/j.petrol.2020.107246

Sun, X., Xu, B., Qian, G., Li, B.: The application of geomechanical SAGD dilation startup in a Xinjiang oil field heavy-oil reservoir. J. Petrol. Sci. Eng. **196**, 107670 (2021). https://doi.org/10.1016/j.petrol.2020.107670

Yang, Z., Sun, X., Luo, C., Xu, B., Yang, B., Li, B.: Vertical-well-assisted SAGD dilation process in heterogeneous super-heavy oil reservoirs: Numerical simulations. Undergr. Space **6**(6), 603–618 (2021). https://doi.org/10.1016/j.undsp.2020.12.003

Zang, A., Stephansson, O.: Stress Field of the Earth's Crust. Springer, Dordrecht (2010). https://doi.org/10.1007/978-1-4020-8444-7

Recent Innovations on Some Coastal Structures

Subramaniam Neelamani[1(✉)], K. Al-Salem[1], Altaf Taqi[1], K. Al-Banna[2], and Noor Al-Anjari[3]

[1] Kuwait Institute for Scientific Research, PO Box: 24885, 13109 Safat, Kuwait
nsubram@kisr.edu.kw
[2] General Electric Kuwait Technology Centre, AlHamra Tower, Kuwait City, Kuwait
[3] Department of Civil Engineering, College of Engineering and Petroleum, Kuwait University, Kuwait City, Kuwait

Abstract. This article discusses the technical, economical and environmental aspects of three different novel coastal structures used as wave dampers for different applications. One is floating breakwater with skirt walls; other one is vertical porous walls with an impervious back wall used for the construction of Ports, Harbors and marinas and the third one is on vertical porous slotted wave barrier as a substitute or for replacing offshore rubble mound breakwaters. The knowhow of these three US patents are revealed. The merits and demerits are also described. It is proved that these inventions are economical when compared to the conventional solution used for wave damping applications. It is encouraging to see that some of these products are better in terms of wave damping performance compared to rubble mound wave barriers. It is strongly believed that the clients from around the world would use these innovations for their future coastal infrastructure developments.

Keywords: Wave barrier · Coastal structures · Sustainable development

1 Introduction

Out of 257 countries in the world, 213 countries have coastline and sea fronts [1]. The total coastline of the world is about 620,000 km. A country with a coast is considered economically blessed because of its beneficial ecological services for the socio-economic development of the country. Coastal and offshore area attracts more development because of the following reasons:

- The sea is rich in protein and provides many varieties of Fish/Crabs/Shrimps/Oysters etc.
- Sea is the source of water for desalination and cooling of power plants
- Coastal space is widely used for the construction of ports and harbors for international trade and marinas for local economic activities
- Coast and quality beaches are the main attractions for tourism and entertainment
- Coastal area provides many different types of nonliving resources

H.-H. Zhu et al. (Eds.): BRWSG 2021, LNCE 230, pp. 105–120, 2022.
https://doi.org/10.1007/978-981-16-9963-4_9

- The climate near the coast is mostly moderate due to the presence of a cool breeze from the sea
- Ocean space acts as a sink for treated waste water
- Ocean is the source for future energy and for a wide range of life-saving medicines

The Ocean Conference, United Nations, New York, 5–9 June 2017 reveals that more than 600 million people (around 10% of the world's population) live in coastal areas that are less than 10 m above mean sea level. Nearly 2.4 billion people (almost 40% of the population of the world) live within 100 km of the coast. The ocean economy, which includes employment, ecosystem and cultural services provided by the ocean, is estimated at between US$ 3–6 trillion/year.

Most of the ocean and coastal countries are working at a fast phase to improve the blue economy [2], which demands high investment for different types of marine structures for sustainable exploitation of different kinds of living and nonliving resources. It is essential to innovate marine structures with the following strategies: a. It must be economically viable; b. Environmentally acceptable; c. Easy to build; d. Appear elegant; e. Do better performance than the conventional marine structures. This article reveals three different types of marine structures, which almost fulfill these requirements for modern infrastructure development in the coastal area. Three US patents are obtained for the innovations [3, 4] and [5].

This article provides the brief technical, economic and environmental aspects of three innovative coastal structures from these three US patents, and the knowledge is believed to be useful for the coastal space development around the world. Further details for commercialization of these inventions can be obtained from the Scientific and Technical Innovations Department (STID), Commercialization Division (CD), Kuwait Institute for Scientific Research (KISR), Tel: +965-24989955, Mob: +965-99014472, E-mail: ahaji@kisr.edu.kw. This is an attempt to market these products to the world.

2 Technical, Economic and Environmental Aspects of Three Different Coastal Structures

First, the technical, economic and environmental aspects of Floating breakwater [3] will be discussed. Next similar aspect for the Method of Dissipating Water Wave Energy [4] will be revealed. Finally, these aspects from Method for Damping Ocean Waves in Coastal Area [5] will be elaborated.

2.1 Technical, Economic and Environmental Aspects of Floating Breakwater

Technical Aspects. Floating breakwaters (FBW) are typical marine structures used for wave damping applications. A typical floating breakwater is shown in Fig. 1.

FBW is technically suitable for marine locations with moderate wave climates, especially when the wave periods are less than 6 s and wave heights are less than 1.0 to 2.0 m [6]. It is more suitable in locations with no severe cyclone and Tsunami activities. They are suitable for wave damping in deep waters such as open sea oil and gas loading stations, offshore fish farms etc. The floating breakwaters are moored with the sea bed.

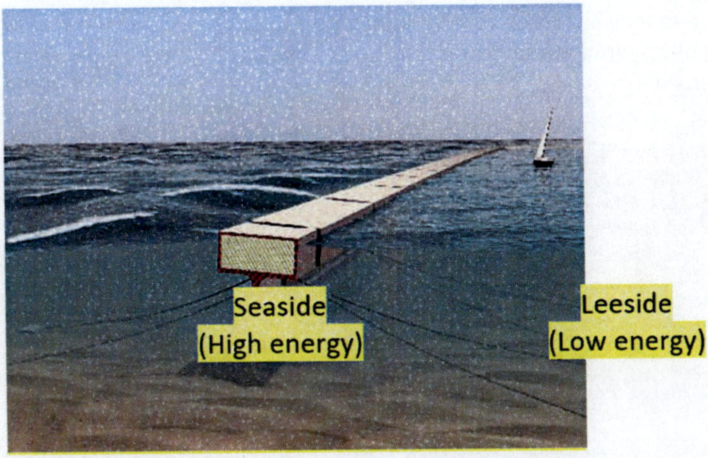

Fig. 1. A typical floating breakwater with pontoon and a single slotted barrier.

Studies on many different types of floating breakwaters are available [7]. The wave transmission performance of FBW is mainly a function of its dimensions, such as its width along the direction of wave propagation (B), water depth (d) and a draft of the breakwater (Δ). Pontoon type floating breakwaters (Fig. 2) are used at the beginning. The width of this breakwater should be about 50% of the design wave length to reduce 50% wave energy transmission, and hence it is expensive.

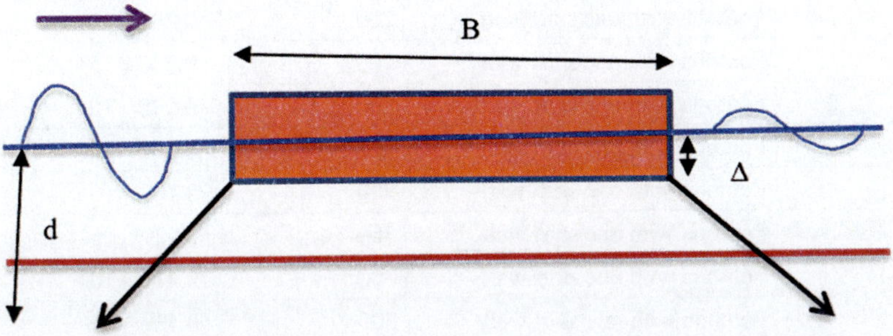

Fig. 2. A pontoon type floating breakwater with width 'W' and draft 'Δ'.

Hence, it is felt that modification of its cross-sectional shape is required to reduce wave transmission. Hence, few cross-sections, as shown in Fig. 3a to 3d are selected, and detailed physical model studies were carried out at Kuwait Institute for Scientific Research, Kuwait. 29 different FBW configurations are tested. Tests are also carried out on a fixed pontoon without any skirt wall. The details of these configurations are provided in Table 1. The wave transmission, reflection, mooring forces, dynamic responses of these models were measured and assessed for a wide range of wave conditions [8]. For 3a, different depths of skirt walls, 'h' were used, as shown in Table 1. Porosities are introduced in the skirt wall and vary from 0% to 20%. Porosity in the skirt wall helps to change the hydrodynamic performance due to a change in energy dissipation during wave interaction.

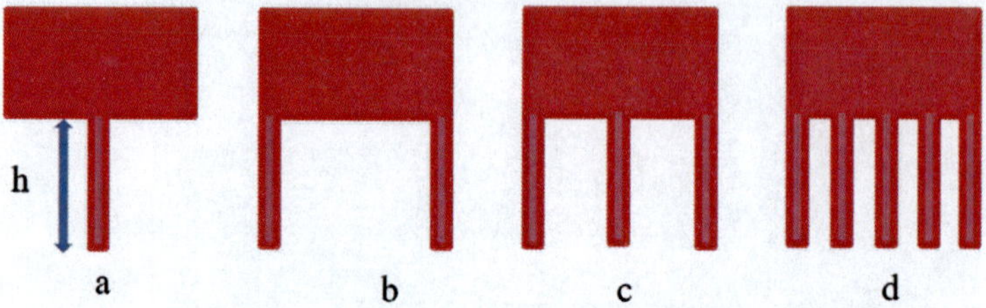

Fig. 3. Pontoon floating breakwater with different number of skirt walls.

Table 1. Details of different floating breakwater configurations

Configuration	Type	Skirt wall depth, h (mm)	h/d	Porosity of the skirt wall (%)
1	Pontoon (Reference case)	No skirt wall	0	-
2	Pontoon with one skirt wall	100	0.143	0
3	Pontoon with one skirt wall	100	0.143	5
4	Pontoon with one skirt wall	100	0.143	10
5	Pontoon with one skirt wall	100	0.143	20
6	Pontoon with one skirt wall	200	0.286	0
7	Pontoon with one skirt wall	200	0.286	5
8	Pontoon with one skirt wall	200	0.286	10
9	Pontoon with one skirt wall	200	0.286	20
10	Pontoon with one skirt wall	300	0.429	0
11	Pontoon with one skirt wall	300	0.429	5
12	Pontoon with one skirt wall	300	0.429	10
13	Pontoon with one skirt wall	300	0.429	20
14	Pontoon with one skirt wall	400	0.572	0
15	Pontoon with one skirt wall	400	0.572	5
16	Pontoon with one skirt wall	400	0.572	10
17	Pontoon with one skirt wall	400	0.572	20
18	Pontoon with two skirt walls	200	0.286	0
19	Pontoon with two skirt walls	200	0.286	5
20	Pontoon with two skirt walls	200	0.286	10
21	Pontoon with two skirt walls	200	0.286	20
22	Pontoon with three skirt walls	200	0.286	0

(*continued*)

Table 1. (*continued*)

Configuration	Type	Skirt wall depth, h (mm)	h/d	Porosity of the skirt wall (%)
23	Pontoon with thee skirt walls	200	0.286	5
24	Pontoon with three skirt walls	200	0.286	10
25	Pontoon with three skirt walls	200	0.286	20
26	Pontoon with five skirt walls	200	0.286	0
27	Pontoon with five skirt walls	200	0.286	5
28	Pontoon with five skirt walls	200	0.286	10
29	Pontoon with five skirt walls	200	0.286	20
30	Fixed pontoon, without skirt wall	0	0	0

A typical comparison of the hydrodynamic performance of these different FBW configurations is shown in Fig. 4. Here, K_t, K_r and K_l are called transmission coefficient (H_{ts}/H_s), reflection coefficient (H_{rs}/H_s) and dissipation coefficient ($K_l = (1 - K_t^2 - K_r^2)^{1/2}$) respectively for a typical wave condition ($B/L_p = 0.646$ and $H_s/d = 0.214$), where H_s is the significant incident wave height, H_{ts} is the significant transmitted wave height, H_{rs} is the significant reflected wave height, and L_p is the wave length corresponds to peak wave period. It can be seen that wave transmission can be changed by introducing skirt walls of varying numbers and porosity. The user can select the configuration, which provides minimum wave transmission for this wave condition, which is 0.23 for configuration 29 (pontoon with five skirt walls and 20% porosity, as shown in Table 1).

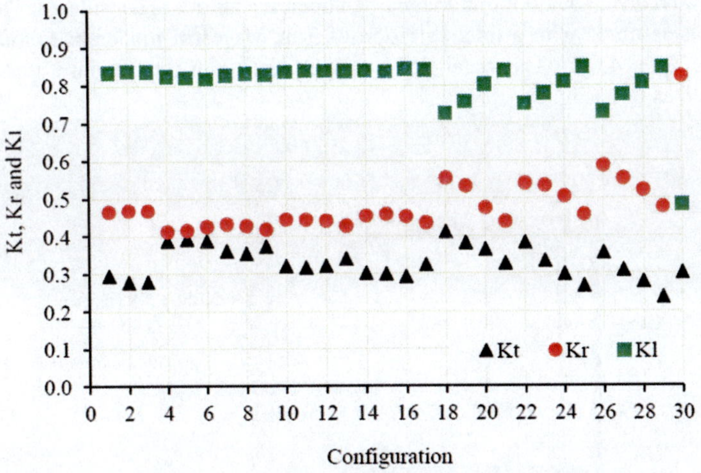

Fig. 4. Effect of the Floating Breakwater Configuration on K_t, K_r, and K_l for $B/L_p = 0.646$ with $H_s/d = 0.214$.

For the same input condition, the normalized seaside and leeside mooring forces are provided in Fig. 5 for all configurations. It is to be noted that the seaside mooring force is always higher compared to the leeside force, and the mooring force value increases with more number of skirt walls.

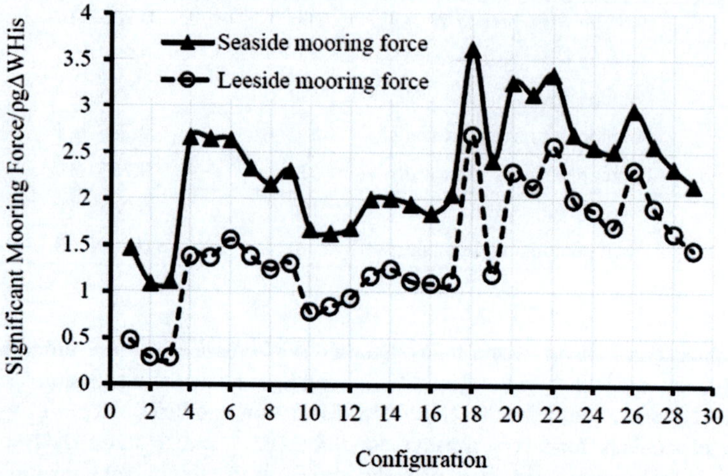

Fig. 5. Effect of the floating breakwater configuration on normalized mooring forces for $B/L_p = 0.646$ with $H_s/d = 0.214$.

More information is available in [6] and [8].

Economics and Environmental Analysis. A detailed economic and environmental analysis was carried out. For this analysis, an open sea marina is considered. The marine location is assumed with a tidal variation of 3 m, waves of moderate conditions ($H_s = 1.0$ m, $T_p = 4.0$ s). The water depth at the location of the offshore breakwater is considered as 6 m (Fig. 6).

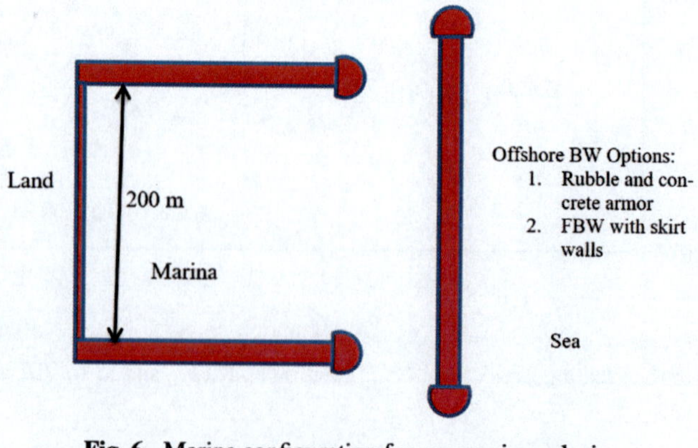

Fig. 6. Marina configuration for economic analysis.

For economic analysis, two different options of the offshore breakwater are considered. Option 1 is a rubble mound offshore breakwater with concrete armor. The seaside slope is 1 V: 2 H, and the leeside slope is 1 V: 1.5 H. The top level of the breakwater is at 5 m, and the sea bottom is at-6 m. For option 2, a floating BW with 5 skirt walls is considered. Each FBW is 8 m wide × 12 m long and 5 m deep. Skirt walls are 2 m deep and 20 cm thick. The FBW bottom thickness is 0.4 m, and all vertical wall thickness is 0.3 m. The details of material quantity, unit costs, economics and environmental issues are described in Table 2. For the economic analysis, the currency of Kuwait, Kuwaiti Dinar (KD) is used (1 KD = 3.32 US $).

It is found that floating breakwater of suitable configuration can be elected and designed economically compared to a conventional rubble mound breakwater. It also offers better environmental benefits.

Table 2. Material estimate, cost and environmental analysis for rubble mound and floating breakwater

Items	Rubble mound with concrete armor	FBW with five skirt walls
Material estimate	• Stone volume for 200 m run = 40916.6 cubic m • Plain concrete volume for the armor for 200 m run = 10230 cubic m	• Volume of reinforced concrete for each caissons, 12 m long = 174.28 cubic m • Total volume for 200 m = 2905 cubic m
Unit costs	• Cost of stone/cubic m = KD 20 • Cost of plain concrete/cubic m = KD 40 • Cost of reinforced concrete/cubic m = KD 80	
Cost estimate	• Cost of rubble for 200 m = KD 818,332 • Cost of plain concrete for 200 m = KD 409,200 • Sea bed preparation (10%) + Engineering study (3%) = KD 159,579 • Total cost for 200 m offshore breakwater = KD 1,387,111	• Cost of reinforced concrete for caissons to cover 200 m = KD 232,368 • Mobilization (50%) + Mooring cost (20%) + Scaffolding cost (20%) + Transport cost (10%) + Cranes for erection (20%) + Engineering study (10%) + Miscellaneous cost (20%) = KD 348,552 • Total cost of FBW = KD 580,920
Cost of FBW/Cost of conventional BW	0.42 (This value may vary depend upon the country, where the unit cost of materials and manpower are different)	

(*continued*)

Table 2. (*continued*)

Items	Rubble mound with concrete armor	FBW with five skirt walls
Environmental merits and demerits	Merits • Maintenance will be less expensive (less than 1% of construction cost) • Wave climate at the harbor side will be very less Demerits • Water quality at the marina will be poor • If the sea bed is coral dominated, this solution is not good	Merits • Better water quality at the rear side • Seabed strength is not an issue • Coral on the seabed will not get damaged • Easily rearranged based on field monitoring Demerits • Maintenance will be expensive (About 5% of initial cost) • There will be some waves on the rear side

2.2 Technical, Economical and Environmental Aspects of Designing Port, Harbor and Marina Using Vertical Porous Walls Instead of Rubble Mound Breakwaters

Motivation and Technical Details. The volume of rubble mound required to build breakwaters for Port, Harbor or marina are usually very high. The motivation for replacing rubble mound breakwater for Port, Harbor or marina is because of a few important reasons: a. to reduce the volume of stones needed for construction; b. Increase the area of boat mooring space; c. Elegance; d. Environmental aspects and e. Economics.

Normally, when a sloped wave barrier is used inside the marina, it occupies significant space as shown in Fig. 7. The breakwater here is for damping the waves.

Fig. 7. Inner part of a typical marina with sloped rubble mound breakwaters for dissipating the wave energy.

A detailed study is carried out by Al-Salem et al. [9] and the promising results of this study is available in Neelamani et al. [10]. It is found that the conventional sloped rubble mound breakwater can be replaced by an array of slotted vertical barriers (Fig. 8) with the same or better wave energy dissipation characters (Neelamani et al. [10]).

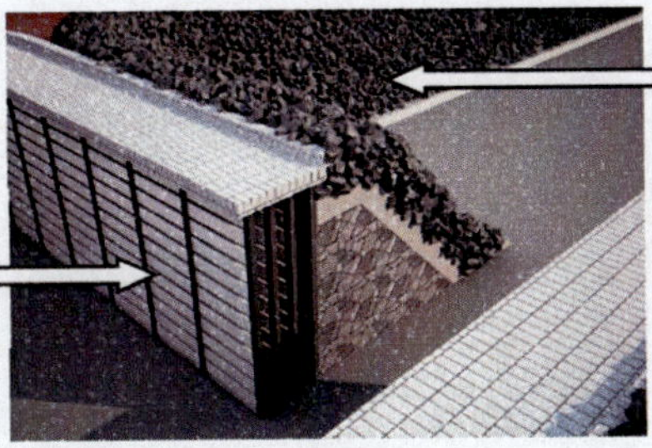

Fig. 8. A comparison of the inner part of a marina built using the slotted vertical barrier and conventional rubble mound breakwater.

A 3D view of a marina built using this innovative solution is shown in Fig. 9.

Fig. 9. Marina to be built using the slotted vertical barrier.

The main advantage of this innovation is the significant reduction in the volume of materials compared to rubble mound breakwater. The details about the volume of material required for the slotted barrier type wave barrier compared to conventional rubble mound breakwater are provided in Table 3.

The performance of the present wave barrier is explained by its wave reflection characteristics for different d/L_p values and for $H_s/d = 0.214$, which is provided in Fig. 10. X-axis represents the type of wave barrier. (1,0%) means a single vertical wall with 0% porosity. RM-1 is a rubble mound breakwater with the slope of 0.7 V: 1.2 H. Seabee-1 and Dolos-1 are breakwater with Seabee and Dolos as armour, respectively, with 0.7 V:1.2H slopes. Similarly, RM-2, Seabee-2, Dolos-2 are similar structures with 0.9 V: 1.2H slope. (1,10%) means single vertical porous barrier with 10% porosity and

Table 3. The percentage volume of material required for different number of slotted wall and porosity combination, when compared to the volume required for 0.9 m V: 1.2 m H sloped rubble mound breakwater

Number of perforated walls	10% porosity	20% porosity	30% porosity	40% porosity	50% porosity
1 Wall	3.50%	3.11%	2.72%	2.33%	1.95%
2 Walls	7.00%	6.22%	5.45%	4.67%	3.89%
3 Walls	10.50%	9.34%	8.17%	7.00%	5.84%
4 Walls	14.00%	12.45%	10.89%	9.34%	7.78%
5 Walls	17.51%	15.56%	13.62%	11.67%	9.73%
6 Walls	21.00%	18.66%	16.33%	14.00%	11.67%

similar meaning is applicable for others. Smaller the reflection coefficient, the better it is for application. From this plot, one can select the needed number of slotted barriers and the porosity for a selected wave reflection. More information is available in Al-Salem et al. [9].

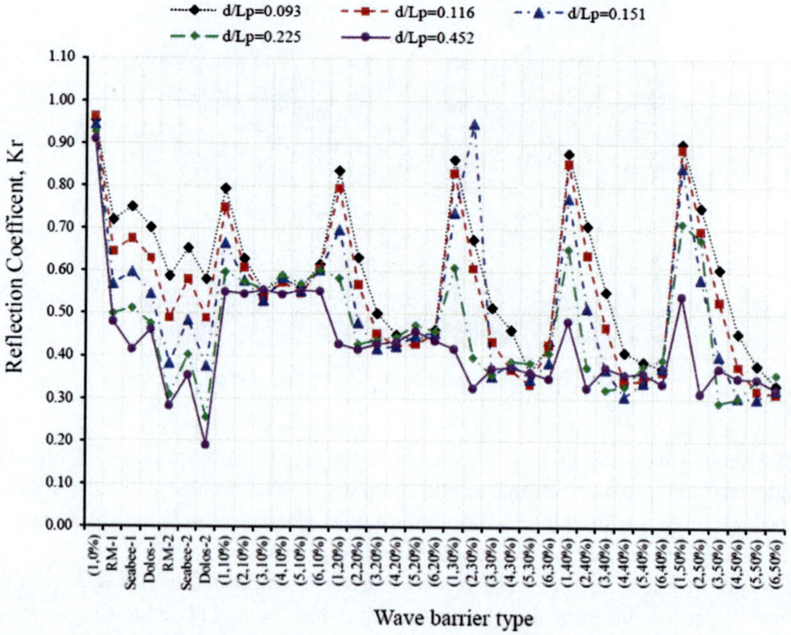

Fig. 10. The effect of different wave barriers on wave reflection for different d/L_p values and $H_s/d = 0.214$.

The main merits of slotted vertical barrier type breakwaters are a. the volume of material needed is only 3.5% to 21% (depends upon the number of slotted walls and porosity) when compared to a sloped rubble mound sea wall; b. If used, it will create more space for berthing; c. With proper design, the top level of the barrier can be used as a berthing facility, as shown in Fig. 8 or 9; d. It is easy to prefabricate and quickly install in a short span of time; e. High-strength polymer materials can be used for its construction, and hence, it offers relief from corrosion of reinforced concrete panels and f. The load on the seabed is less, and the effect on benthic life is also less.

This experimental study is carried out for a wide range of wave conditions. Kuwait Institute for Scientific Research (KISR) can provide solutions for two types of clients: Client 1, mainly from economically strong countries, who wish to invest more at the beginning with less risk and Client 2, the one from developing countries, who wish to spend less for initial construction and wish to take a bit more risk than client 1, but willing to spend more for annual maintenance and any future partial failures. KISR can provide optimized configuration (minimum number of porous walls, optimized porosity), which performs better than sloped sea walls in terms of wave energy dissipation. It is found from this study that a slotted barrier is a better wave energy dissipator for long waves [9, 10], which is very attractive for coastal engineers.

Economics and Environmental Analysis. The marine conditions assumed for the floating breakwater are considered for the economic analysis for this structure also. Option 1 is the rubble mound breakwater, and option 2 is the wave barrier with porous vertical walls. The cross-section and plan view of each unit can be as shown in Fig. 11a and 11b. Each unit can be 14.7 m wide × 8 m long; 25 units is required to cover the 200 m width of the coastal length. This type of structure can be prefabricated and assembled quickly. For its construction, a floating barge with fabrication and crane facilities are needed.

Table 4 provides the details of material estimate, economic and environmental analysis.

The knowhow, scientific results, economic analysis and the merits and demerits can be considered for appropriate design of this new type of breakwater with slotted walls, which can be used for construction of new Ports, Harbors and marina. Significant savings can be achieved with attractive environmental merits and elegance.

2.3 Technical, Economical and Environmental Aspects of Method for Damping Ocean Waves in Coastal Area

Motivation and Technical Details. The main difference between the previous wave barrier and this one is that in the previous case, the waves were not allowed to transmit since the rear side was provided with an impervious vertical wall. In the present case, waves are allowed to transmit. This study is aimed as an alternative for rubble mound offshore breakwater by using a vertical slotted barrier. The motivation is to reduce or save materials during construction, improve the hydrodynamic performance, add more environmental benefits compared to rubble mound offshore breakwater. The complete scientific details of the experimental study is available in [11]. The results of wave

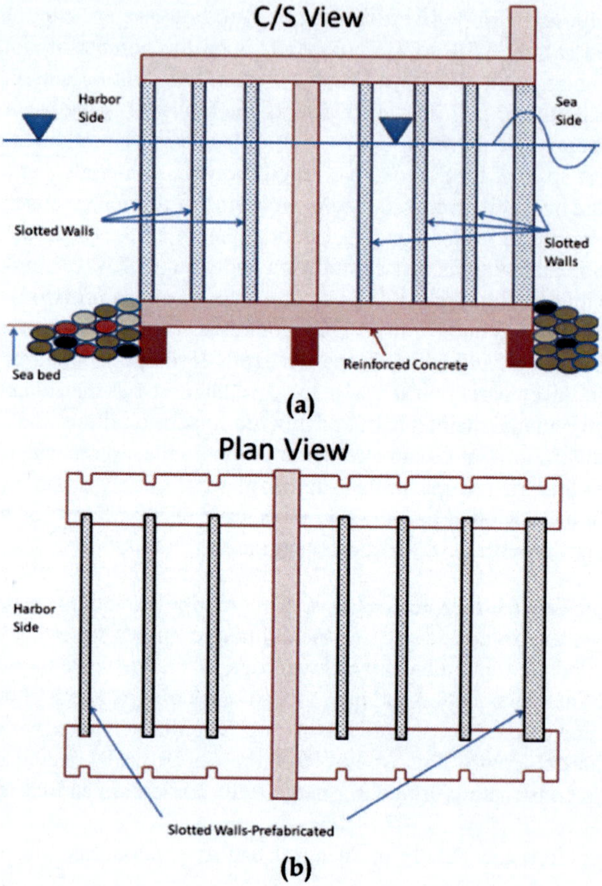

Fig. 11. a. Typical cross section of the breakwater with slotted wall. b. Typical plan view of the breakwater with slotted wall.

pressures on the panels are available in [12]. From this study, it is possible to select the number of slotted wave barriers and their porosity for any required wave transmission coefficient. This is an important merit of the slotted vertical barrier, which is not easily possible if a rubble mound offshore breakwater is selected. The typical conventional rubble mound offshore breakwater used is as shown in Fig. 12.

The motivation is to replace the rubble mound offshore breakwater with slotted wave barrier as shown in Fig. 13.

Figure 14 provides the characteristics of wave transmission coefficient for different d/L_p values and for $H_s/d = 0.214$. From this plot, the user can select a suitable slotted wave barrier configuration for the required wave transmission conditions.

Similar to the other patents, KISR can provide the needed design information for both the type of clients as discussed in the previous section. The information needed for optimized design are the optimized configuration (No. of porous walls, optimized

Table 4. Material estimate, cost analysis and environmental issues

Items	Rubble mound with concrete armor	New wave energy dissipater (25 units; each unit of 8 m long)
Material estimate	• Stone volume for 200 m run = 40916.6 m^3 • Plain concrete volume for the armor for 200 m run = 10230 m^3	• Top slab (8 m × 14.7 m × 0.3 m) = 35.28 m^3 • Wave screen (8 m × 1.0 m × 0.3 m) = 2.40 m^3 • Middle wall (8 m × 9.4 m × 0.3 m) = 22.56 m^3 • Bottom slab (8 m × 14.7 m × 0.3 m) = 35.28 m^3 • Toe beams 3 Nos. for stability against sliding (8 m × 0.5 m × 0.3 m) × 3 = 3.60 m^3 • Slotted barrier, 40% porosity {[(7.8 m × 9.4 m × 0.2 m) × 6] + (7.8 m × 9.4 m × 0.3 m)} × 0.6 = 65.988 m^3 • Walls in the direction of wave (14.7 m × 9.4 m × 0.2 m) × 2 = 55.272 m^3 • Total volume of reinforced concrete for unit of 8 m long = 220.38 m^3 • Total concrete volume for 25 units (220.38 × 25) = 5509.5 m^3
Unit costs	• Cost of stone/cubic m = KD 20 • Cost of plain concrete/cubic m = KD 40 • Cost of reinforced concrete/cubic m = KD 80	
Cost estimate	• Cost of rubble for 200 m = KD 818,332 • Cost of plain concrete for 200 m = KD 409,200 • Sea bed preparation (10%) + Engineering study (3%) = KD 159,579 • Total cost for 200 m offshore breakwater = KD 1,387,111	• Cost of reinforced concrete for 25 units to cover 200 m (5509.5 × 80) = 440,760 KD • Mobilization (10%) + Hiring cost for floating barge (15%) + Scaffolding for prefabrication (10%) + Crane facility (10%) + Seabed preparation (20%) + Engineering investigation (10%) + Miscellaneous cost (15%) = KD 396,684 • Total cost of the new wave energy dissipater = KD 837,444
Cost of new barrier/Cost of Rubble BW	0.60 (This can vary for different countries based the unit cost of material and manpower cost)	

(*continued*)

Table 4. (*continued*)

Items	Rubble mound with concrete armor	New wave energy dissipater (25 units; each unit of 8 m long)
Environmental merits and demerits	Merits • Maintenance will be less expensive (less than 1% of construction cost) • Wave climate at the harbor side will be very less Demerits • Water quality at the marina will be poor • If the sea bed is coral dominated, this solution is not good	Merits • Same as rubble mound • Less volume of material • Quick construction • More environment friendly (Act like artificial reef) • More water plan space • Can be used as berthing structure Demerits • Similar to Rubble mound structure but it is easy to make holes on the main wall for better water circulation

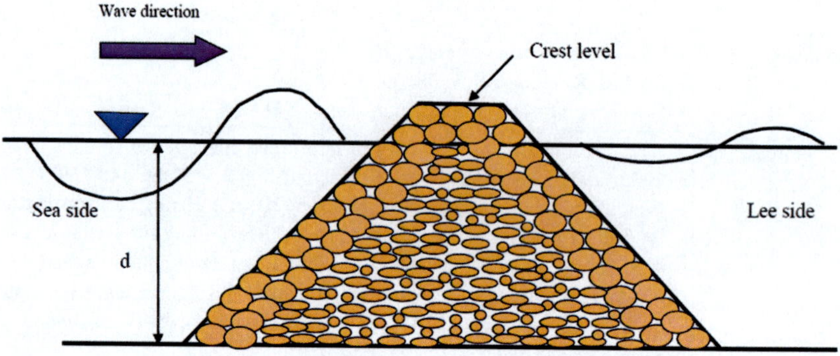

Fig. 12. Typical rubble mound offshore breakwater.

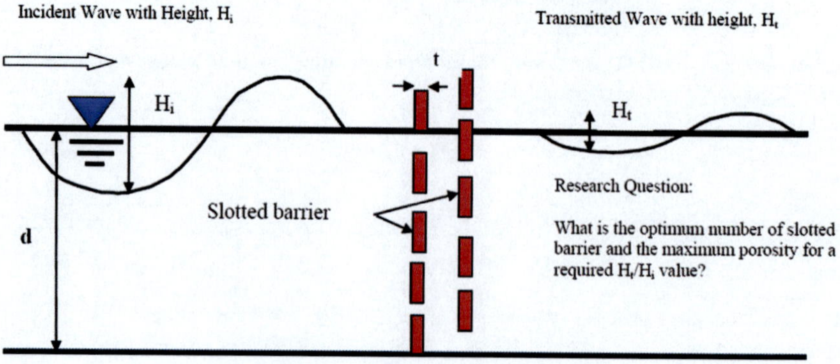

Fig. 13. Typical slotted wave barrier.

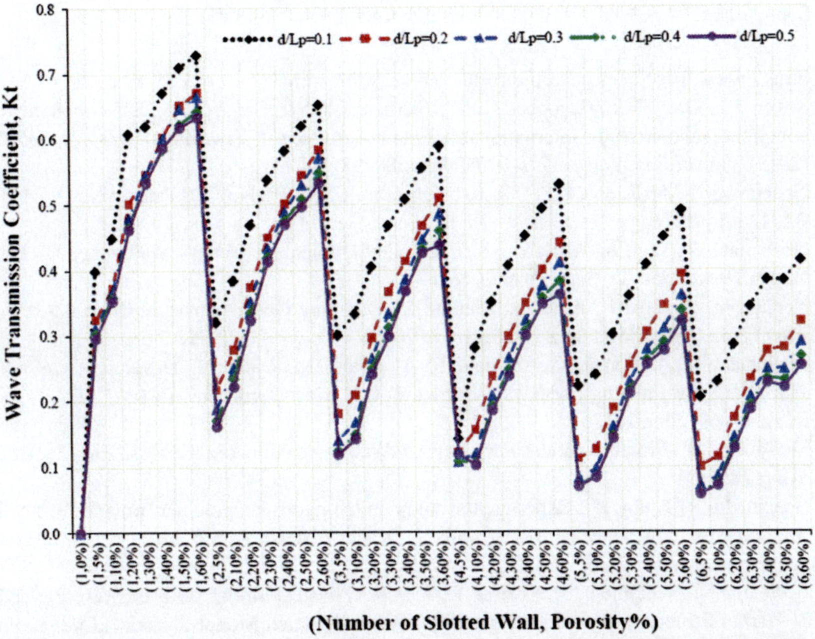

Fig. 14. The wave transmission coefficient for different d/L_p values and $H_s/d = 0.214$.

porosity), which performs better than rubble mound offshore breakwater in terms of wave transmission; design wave forces, moments and wave pressures on the barrier plates.

3 Conclusions

Innovation on marine structures is a key and critical issue today and in the future. The focus is required on reducing the volume of materials used for construction (overall cost saving); Improved performance; more positive environmental impact; improved aesthetic value; reduction in the construction time, and improvement in the intended functions. Three different US patents are secured on three different coastal structures, which fulfills these requirements. The detailed knowhow can be obtained from the final reports and publications as provided in the reference section.

Acknowledgments. The authors acknowledge the Kuwait Foundation for Advancement of Sciences for their financial support for the research work through Project No. (2009-1401-03) and Project No. 07-0818-007). Our sincere acknowledgment for Kuwait Institute for Scientific Research for financial support through Project No. EC089K. The support by the technicians of the Coastal Management Program for the model fabrication, instrumentation, execution of physical modeling and data analysis is gratefully acknowledged.

References

1. https://www.listofcountriesoftheworld.com/coastline.html
2. Silver, J.J., Gray, N.J., Campbell, L.M., Fairbanks, L.W., Gruby, R.L.: Blue economy and competing discourses in international oceans governance. J. Environ. Dev. **24**(2), 135–160 (2015). https://doi.org/10.1177/1070496515580797
3. Neelamani, S., Al-Banaa, K.: Floating breakwater. US Patent No.: US Patenet No.: 9,340,940 B2, 17 May 2016
4. Neelamani, S., Taqi, A., Al-Salem, K.: Method of Dissipating Water Wave Energy. US Patent No.: US 9,447,554, 20 September 2016
5. Neelamani, S. and Al-Anjari, N.: Method for Damping Ocean Waves in Coastal Area. US Patent No.: US 10,550,534 B1, 4 February 2020
6. Neelamani, S., Al-Banaa, K., Al-Shatti, F., Al-Khaldi, M., Ljubic, J.: Development of a Feasible Prototype Floating Breakwater for Kuwaiti Marine Conditions, EC037C, Final report, KISR 11669, June 2013
7. McCartney, B.: Floating breakwater design. J. Waterway Port Coastal Ocean Eng. **3**(2), 304–318 (1985)
8. Neelamani, S., Ljubic, J.: Experimental study on the hydrodynamic performance of floating pontoon type breakwater with skirt walls. J. Offshore Mech. Arctic Eng. 140(2), 021303-9. Paper No. OMAE-17-1029 (2017). https://doi.org/10.1115/1.4038343
9. Al-Salem, K., Neelamani, S., Al-Essa, F., Taqi, A.: Physical Model Study on Wave Reflection by Harbor Structures for Validating KISR's Wave Agitation Model-Phase I, EC089K. Final Report, KISR 12252, July 2014
10. Neelamani, S., Al-Salem, K., Taqi, A.: Experimental investigation on wave reflection characteristics of slotted vertical barriers with an impermeable back wall in random wave fields. J. Waterway Port Coastal Ocean Eng. ASCE **143**(4), 06017002 (2017). https://doi.org/10.1061/(ASCE)WW.1943-5460.0000395
11. Neelamani, S., Al-Anjari, N.: Hydrodynamic studies on slotted vertical wave barriers, EC097C, Final Report to Kuwait Foundation for the Advancement of Sciences, KISR 15248, December 2018
12. Neelamani, S., Al-Anjari, N.: Experimental investigations on wave induced dynamic pressures over slotted vertical barriers in random wave fields. Ocean Eng. **220**, 108482 (2021)

Performance of Expansive Soil Treated with Fiber Reinforced Alkaline-Activator Binder for Pavement Subbase Applications

Anasua GuhaRay([✉]) and Mazhar Syed

BITS-Pilani Hyderabad Campus, Hyderabad 500078, Telangana, India
guharay@hyderabad.bits-pilani.ac.in

Abstract. When exposed to persistent moisture fluctuations, expansive clayey soil is mostly impermeable and has a dual tendency (swelling/shrinkage). The heave motion causes such clayey soils to lose volumetric and mechanical strength and is difficult to regulate the differential settlement. The current study focuses on the stability of black expansive clayey soil (BES) using alkaline-activator binders (ABs) reinforced with natural and artificial fibers instead of standard cement-based binders. The effects of different concentrations of ABs at varying curing durations are examined by conducting a series of index and shear strength property tests on treated BES. The treatment improves the geoengineering behavior of BES and helps to increase the unconfined compressive strength (UCS) values by regulating the heave and void ratios effectively. Mineral alteration caused by BES treatment with AB is responsible for reducing volumetric swelling and plasticity index. This study could be very useful in developing a cost-effective solution to the problem associated with BES.

Keywords: Expansive soil, stabilisation · Alkaline binders · Fiber

1 Introduction

Because of the unique character of expansive clayey soil, it has limited volumetric constancy when there is a moisture imbalance, which causes considerable damage to roadways, airport paved surfaces, and lightweight structures built on top of it. The active occurrence of montmorillonite and other smectite-related clay minerals is primarily responsible for periodic swell-shrinkage activity due to a high retain moisture tendency, resulting in significant distress when the load is applied [1, 2]. The major issues of this type of soil (especially at the subgrade layer) are formed the desiccating cracking, which typically impacts the performance against the rutting and fatigue failures of the asphalt or concrete layers. If these clayey soils are not adequately treated before constructing a substructure, it may lead to natural hazards resulting in millions of dollars in restorations and rehabilitation costs [3, 4]. The annual losses caused by construction on expansive soil are estimated to reach $9000 million in the United States, $1000 million in China, and 500 million in the United Kingdom. The overall annual loss increased by

H.-H. Zhu et al. (Eds.): BRWSG 2021, LNCE 230, pp. 121–136, 2022.
https://doi.org/10.1007/978-981-16-9963-4_10

nearly 141% between 1970 and 2000, with 4000 million USD in lightweight structures alone in America [5, 6]. Before any construction is built on top of the soil, chemical stabilization can significantly increase the interfacial bonding and mechanical qualities. Lime and cement are calcium-based chemical binders that efficiently limit soil swelling potential and compressibility through pozzolanic and hydration processes. Nevertheless, the production of these cementitious binders has a substantial environmental impact by releasing carbon dioxide (CO_2) and other greenhouse gases. By 2050, demand for traditional binders will have increased by up to 200% globally [7, 8]. Due to the quick sulfate reaction, excessive levels of lime/cement in the soil can induce carbonation shrinkage fractures and make it impossible to achieve the desired geomechanical strength without the admixture inclusions.

Alkaline-activator Binders (ABs) have recently acquired popularity as an alternative cement binder for soil stabilization, particularly in expansive clayey-type soils. ABs is a low carbon emission alkaline agent with a long-chain polymeric sodium aluminosilicate compound upon the activation of pozzolanic precursors. The high performance of alkaline activation in organic clay (geopolymerization) generated from silica (SiO_2) and alumina (Al_2O_3) has proven to be an effective ground improvement approach. Using an environmentally safe alkaline binder has dual advantages: it eliminates the need for traditional cement binders on the one hand, and it avoids the cost of dumping fly ash/slag into landfills (Provis & Deventer, 2014). It surpasses PC-based binders in terms of mechanical strength, workability, and durability while emitting up to 75–80% less carbon dioxide and having a 70–75% lower global warming potential. Although alkaline mixed clayey soil effectively strengthened the peak compressive strength, it exhibited poor flexural and tensile forces resistance. In the subgrade layer, where this type of clayey soil is likely to shrink in the peak summer season, the occurrence of linear shrinkage, tensile and flexural cracking is a critical concerned [9]. The use of high tensile and stiffness-bearing fiber reinforcement in soft soil mixtures is a well-known geoengineering technique for regulating the width and number of cracks that propagate. Several studies have shown the effectiveness of discrete fiber (artificial/natural) reinforcement in chemically stabilized soil and considered good earth reinforcing material [10–16].

In the proposed stabilization procedure, fly ash and slag are used as binding ingredients on a poor subgrade soil surface. After China and the United States, India is the world's third-largest coal-fired power producer. According to a study by India's central electrical regulator, power stations produce around 200 million tonnes of fly ash each year. More than 75% of the ash is very fine in nature. Also, every year 20% of fly ash goes unused. Moreover, India has 60 thousand acres of fly ash and other pond ash, with production estimated to reach 230 million tonnes by the end of 2022 [14]. Furthermore, toxic elements mobilized in deposited fly ash-slag have the potential to damage surface and groundwater, soil, and vegetation. The cost of producing alkaline binder is less than Portland cement, although most plants donate fly ash and slag. India is the world's second-largest producer of fiber. In alkaline soils, these fibers (synthetic and chemically treated fibers) are unaffected by salts and can function at high temperatures with negligible biodegradation. The use of alkaline soil fly ash, slag, and fibers in combination will provide a green and long-term solution for addressing the subgrade's inferior characteristics.

This research focused on using green and sustainable industrial by-products as an alternative to traditional cement binders for soil stabilization. The primary goal of this research is to determine the optimal AB dose for different water to solid (w/s) ratios and aluminosilicate precursors in the alkaline mixture. According to the literature study, considerable research on the effectiveness of fiber reinforcing (type, size, and quantity) with conventional binders on soil subjected to geomechanical strength has been undertaken. However, the effect of a combination of pozzolanic precursors and discrete fibers (natural/manufactured) in an alkaline binder on expansive clayey soil performance against linear shrinkage and tensile cracking has not been adequately studied. The major purpose of this study is to determine the geotechnical reaction of an alkaline binder mixed clayey soil reinforced with a random inclusion of synthetic and chemically treated fibers at the varying proportion of fly ash and slag.

2 Materials and Experimental Methodology

2.1 Materials

Soil. The black expansive clayey soil (BES) utilized in this study was gathered at a depth of a minimum of 10 cm below the surface level in the Kazipet region of the Warangal district of Telangana, India, to prevent the collecting of stems and vegetation along with the soil. The soil is dark brownish in color and has an in-situ moisture content of around 20%. The collected soil is categorized to a compressible clay (CH) according to the Unified Soil Classification System (USCS), with about 20–25% of the fine content being smaller than 75 μ. The high silica content detected in the expansive clayey soil and can be a prime chemical modification requirement before mechanical load application. The basic geoengineering (both physical and chemical) properties of soil are presented in Table 1. The main crystalline phases of clayey soil detected from X-ray diffraction (XRD) are Quartz (Q), Montmorillonite (M), and Muscovite (Ms).

Fly Ash and Slag. The national thermal power plant (NTPC) Ramagundam, Telangana, India, provided Class F (low calcium) fly ash for this investigation, conforming to ASTM C618-17a. JSW Cement Ltd. in Andhra Pradesh supplied the blast furnace slag (conforming to ASTM C989). X-ray fluorescence analysis (XRF) is used to determine the chemical composition of fly ash and slag. In fly ash, the principal constituents are aluminum oxide (Al_2O_3), silicon dioxide (SiO_2), ferric oxide, and magnesium oxide (MgO). Slag attained a high amount of calcium oxide (CaO) and another element (Al_2O_3, MgO, and SiO_2) compounds. The amount of heavy metals leaching from the mixture of fly ash and slag is found to be negligible.

Fiber. As reinforcing agents of geopolymerized expansive soil, two different kinds of fibers, polypropylene (PF) and glass (GF), and two forms of natural fibers, hemp (HF) and coir (CF), are used. Kanaka Durga Industries in Hyderabad and Go-green Industries in Tamil Nadu, respectively, provide a consistent length of 12 mm for polypropylene (PF), glass (GF), and 24 mm for hemp (HF) and coir (CF) fiber. PF and GF with a length of 12 mm have greater soil-fiber interaction with good frictional bonding, improved interlocking, and a high tendency to regulate surface cracking when compared to other synthetic fibers. Due to their length and diameter, natural fibers (coir and hemp) significantly impact the soil reinforcing mechanism. It's also worth noting that fiber lengths below 10 mm and greater than 40 mm have a low contact area with the soil, making higher pullout intensity difficult during loading. As a result, the ideal fiber lengths in the AB-soil mixture are 12 mm (PF and GF) and 24 mm (CF and HF). Before being used as a sustainable soil reinforcing material, cellulose and hemicellulose-rich natural fibers are thoroughly treated by using sodium hydroxide (NaOH-solution), maintaining 10M [13, 17, 18].

Table 1. Basic engineering properties of soil, fly ash (FA), and slag materials

Properties	BES	FA	Slag	Test method
Specific gravity	2.58	1.92	2.65	ASTM D-854
Free swell index (%)	89.0	25	–	IS:2720 40
Dry density (kg/m^3)	1618	–	–	ASTM D-698
Moisture content (%)	24.0	–	–	ASTM D-698
Liquid limit (%)	63	19	–	ASTM D-4318
Linear shrinkage (%)	24	–	–	AS-1289 C4
Plasticity index (%)	41	NP	NP	ASTM D-4318
Unconfined compressive strength (kPa)	186	–	–	ASTM D-2166
Soaked CBR (%)	1.93	–	–	ASTM D-1883

Alkaline-Activator Binder (AB). The alkaline binder is prepared by mixing the activating solution (consisting of sodium silicate, sodium hydroxide, and water) and a dry pozzolanic precursor (Class F fly ash/slag). Both chemicals (NaOH pellets and Na_2SiO_3 solution) are obtained from Hychem Laboratories, Hyderabad. The food-grade of NaOH pellets with 99% purity and the industrial-grade Na_2SiO_3 contained 29.5% SiO_2 and 14.7% Na_2O by weight is maintained. The mass ratio of 1.057:12.9:40.00 to NaOH, Na_2SiO_3, and slag-fly ash precursors are utilized for binder preparation [13–15, 19]. The amount of water in the alkaline mixture is varied from 0.3 to 0.5 as water to solid ratios (w/s). The different slag and fly ash percentages are applied in the alkaline solution to obtain the optimum mix.

2.2 Experimental Methodology

Microstructural Characterization. A series of X-ray diffraction (XRD), Fourier transform infra-red (FTIR), scanning electron microscopy (SEM/EDS) are carried out on BES, alkaline mixed clayey soil, and fiber mixed AB soil at different slag-fly ash doses. These tests are carried out to quantify the phases and morphological changes due to cementitious formation in both unreinforced alkaline and fiber-reinforced soil. Moreover, the microstructure of a material (particle size, molecular bonding, element, and mineral compositions) can strongly influence soil mechanical properties, such as shear strength, volumetric stability, and bearing capacity. Microstructural studies allow full recognition of fiber-soil interaction, soil surface crack propagation, and their stiffness behavior on a macro-scale.

Powder XRD examination is carried out on a RIGAKU Ultima IV diffractometer with a 2θ value spanning from 0° to 80°. The CuK rays were generated at 40 mA with 40 kV at 0.02 step of 2θ values and integrated at a rate of 2 s for evaluating the samples. The transmittance spectroscopy was performed using a JASCO FTIR 4200 setup. K.Br pellets are used to coat samples, and wavelengths are kept between 4000 and 400. Thermo Scientific Apreo SEM machines are used at varying magnifications to analyze the surface morphological behavior of soil before and after fiber reinforcement.

Geoengineering Characterization. Detailed geotechnical tests such as compaction, consistency limit, linear shrinkage, consolidation, and swelling characteristics are performed on expansive untreated soil and alkaline mixed clayey soil at varying proportions of fly ash-slag content. The findings of these tests are used to evaluate the effectiveness of stabilizers in controlling volumetric strain and cracking behavior in expansive soil. All proportions of alkaline-soil mixes based on fly slag were prepared at their MDD/OMC values and tested according to ASTM standards. Soil linear shrinkage test is measured in a semi-cylindrical shrinkage mold of 25 cm in length with a radius of 1.25 cm as per the Australian standard (AS-1289.C4-1). Indirectly, this linear shrinkage can help estimate the shrinkage properties of clayey soil, especially in road construction materials. It is defined as the ratio of soil length specimen in a bar mold oven drying to its original length before drying. The soil samples are kept at room temperature followed by oven drying at 110 °C for 24 h, and the size of dried specimens is determined. The linear shrinkage L_S (%) is given by

$$L_s = \frac{L_i - L_f}{L_i} * 100 \tag{1}$$

where L_i and L_f are the initial and final lengths of soil samples, respectively (Fig. 1).

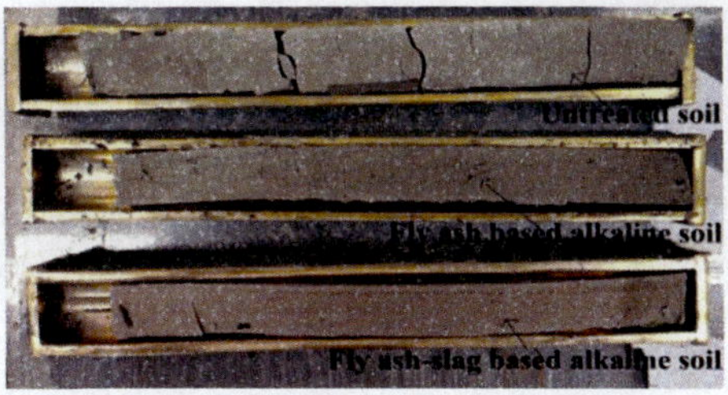

Fig. 1. Linear shrinkage of oven-dried expansive untreated and alkaline mixed soil

The 1D-Oedometer test is used to evaluate the compressibility and swell properties of expansive soil supplemented with various alkaline binder dosages at various w/s ratios and fly ash slag content, according to ASTM D-2435. Both untreated and alkaline treated clayey soil is compacted in a cylindrical consolidation metal ring with a diameter of 6 cm and a height of 2 cm. For all samples, swell percentages are calculated using a seating pressure of 6.25 kPa, and the e-log (p) curves are approximated for each load increment. All composite soil undergoes multiple UCS testing following ASTM D-2166. Before testing, different dosages of the fibers-fly ash-slag mixture in alkaline mixed soil specimens are stored for 30 min in an airtight vacuum desiccator to confirm maturation. The compression machine is a strain-controlled automatic compression machine with a maximum capacity of 20 KN and a fixed strain rate of 1.25 mm/min.

3 Results and Discussions

3.1 Microstructural Characterization

X-Ray Diffraction (XRD). XRD tests are carried out on both untreated clayey soil and fiber mixed alkaline clayey soil in order to: (1) examine mineralogical changes in the soil structure owing to the presence of the alkaline stabilizing agent, and (2) evaluate the formation of crystalline pozzolanic products during geopolymerization. Montmorillonite (M), the major clay mineral in the expansive soil, is detected around $2\theta = 19.7°$, 27.1°, 36.2°, and 60.8° values, which mainly affects the volumetric stability behavior. Around 26.1°, 27.6°, and 59.4°, a minor amount of Muscovite (Ms) is also be found. The sharp crystalline peaks indicate the presence of Quartz (Q) around the 2θ values of 27.3°, 36.1°, and 61°. The diffractogram patterns reveal that different percentages of alkaline binder and cure times result in noticeable changes in the crystalline structure of the soil. As seen in Fig. 2(a–b), the peak intensity of Montmorillonite (M) decreases while the crystallinity of Quartz (Q) remains unchanged.

The diffraction pattern of alkaline mixed soil samples shows slag as one of the pozzolanic components in the soil matrix, as shown in Fig. 2b. In the 10% slag-based AB-clay mixture, a new crystalline pozzolanic reaction material, Augite (A), was identified at 22.9° and 36.3°. Interestingly, after 20% fly ash replacement with slag in the alkaline binder, ion participation improves significantly. The addition of slag is beneficial from the fact that as the amount of slag concentration in the alkaline soil increases, another notable crystalline peak Gehlenite (G) is identified at around 2θ values of 28.8°, and 39.1°, which also correspond to pozzolanic activity [20]. The rapid emergence of pozzolanic reactions in the soil also indicates active cementitious compound formation. Furthermore, the new cementitious compound calcium silicate hydrate (CSH) has been identified as a 30% slag substitute. These minerals may form due to rapid flocculation and vigorous cation exchange occurrence in silica-rich soil. As shown in Fig. 2b, the simultaneous mixing of alkaline binder and fiber reinforcement in clayey soil samples did not affect the mineralogical crystalline pattern. Flatter portions of the diffractogram indicate the amorphous phases in the hardened paste of alkaline binders.

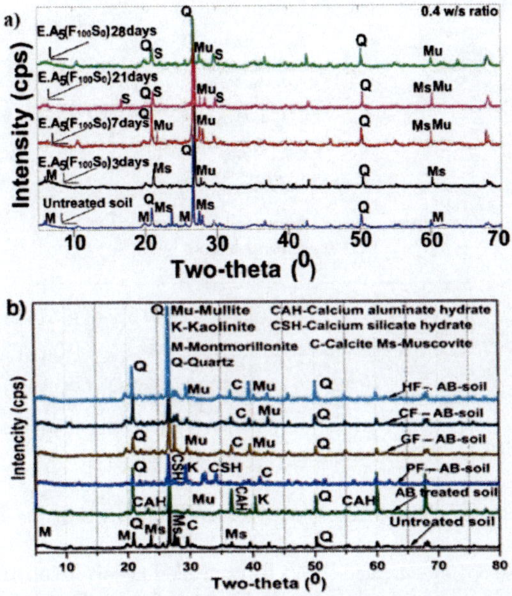

Fig. 2. XRD peaks of untreated clayey soil and alkaline mixed soil **a)** Dosage of 10% AB with 0.4 w/s ratio **b)** Fiber-reinforced alkaline soil.

Fourier Transfer Infra-Red (FTIR) Spectroscopy. The FTIR technique measures the transmittance spectroscopy of chemical bonds that vibrate independently at distinctive wavelengths. The molecular vibration spectrum curve in untreated soil is widely employed by montmorillonite with a strong band at 3640 cm^{-1} as O-H stretching in Portlandite [$Ca(OH)_2$]. After alkaline treatment, there is a slight drop in montmorillonite peak intensity (O-H stretch bond), as shown in Fig. 3a. The C = O carbonyl bond is seen about 1740 cm^{-1} in slag-based AB mixture and fiber-reinforced alkaline soil containing (fly ash replacement with slag) followed by the bending vibration of the = CH2 group (carbonation reaction) broadband in the region of 1470 cm^{-1}. An excess of Na in the alkaline solution may be associated with the cause of this carbonation reaction. Both alkaline mixed and untreated soil show a strong band at 1050 cm^{-1}, which corresponds to the asymmetric vibration of the Si(Al)-O group. As the w/s ratio in the AB mixture fraction grows, the intensity of the different band's increases due to a greater extent of polymerization reaction between the fly ash and alkali-activating solution [9, 19, 21]. The weaker peaks between 880 cm^{-1} and 600 cm^{-1}, ascribed to Si–Al in-plane bending vibration, are attributed to amorphous silica. Apart from that, Si-O-Al bending vibrations are identified with a wavenumber of nearly 540 cm^{-1}. Thus, the spectra peaks from untreated and alkaline mixed soil exhibit identical linkage with a chemical shift of about 20 cm^{-1}.

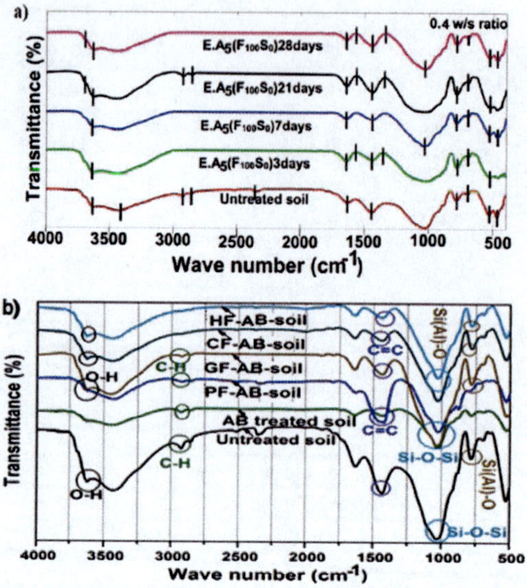

Fig. 3. FTIR spectroscopy of untreated clayey soil and alkaline mixed soil **a)** Dosage of 10% AB with 0.4 w/s ratios at different curing periods, **b)** Fiber-reinforced alkaline soil

Scanning Electron Microscope (SEM)

An SEM image of untreated clayey soil at 40 μm resolution displays a flaky microstructure that mostly corresponds to smectite groups [22, 23]. The micrograph of the AB-soil samples showed better homogeneity after 20% slag content in the alkaline binder. Slag is beneficial in the AB-soil because it effectively binds the clay particles and produces silicate hydrates surrounding the clay particles in a cement paste coating, leading to reduced fluidity (low moisture affinity) and swelling [24, 25]. Thus, the dissolution of precursors can easily attribute to forming the dispersed gel around the clay particles, resulting in a more compact matrix with increased strength and stiffness. The surface morphology of polypropylene fiber (PF) reinforced alkaline mixed soil at 100 μm, and 500 μm resolution is shown in Figs. 4(a–b). A series of discrete polypropylene fiber (PF) reinforcement bounded by pozzolanic material enhances the bond strength in an alkaline clay matrix [13]. Furthermore, by interlocking the clayey particles, the combined inclusions of the PF-AB mixture act as a spatial groove network, thereby regulating the shrinkage crack propagation. The addition of glass fibers (GF) in the alkaline soil mixture appears like a rod-like ettringite needle structure (Figs. 4c–d), contributing to achieving higher interfacial bonding and soil stiffness by enhancing frictional resistance [16, 26]. By retaining the soil around the fiber surfaces, the discrete GF serves as a thread network, overcoming the soil's brittle nature. As a result, the PG-GF-AB soils serve as a complex bracing network, generating efficient bonding and interfacial friction between soil-fiber matrixes and increasing mechanical stress resistance.

Similarly, Figs. 4(e–h) show the coir (CF) and hemp (HF) fiber-reinforced alkaline soil mixture at 100% and 60% fly ash dosages. The simultaneous presence of HF-AB in the soil results in forming a bridge fabric network that effectively limits the number and size of linear shrinkage fractures, as seen in Figs. 4(e–f). The spatial groove matrix is formed when the rough surface of hemp fiber is tightly encircled by the dense cementitious gel layers surrounding it. In the alkaline soil, the discrete HF is intertwined and operates as a spatial stress network that enhances the reinforced soil's stiffness and tensile strength. When a relative displacement trend arises between one fiber in the network and soil particles under external load, the other nearby fibers help to restrict that displacement. As a result, the external strain is shared by all nearby threads, thereby increasing the total strength of the reinforced soil [13–15]. Figures 4(g–h) show a 500 μm resolution micrograph of the coir alkaline sample reinforced chemically treated fiber in alkaline soil. A strongly compacted micrograph exists in a bar-like wire construction with several coir fibers confined with cementitious materials around the clay minerals [27]. Furthermore, isolated coir fiber-connected in alkaline soils enhances interlocking friction and penetration resistance by increasing interfacial bonding and preserving the clay pocket around their surfaces [17, 18]. As a result of the increased mobilization of interfacial friction formed around the fiber surfaces, the combined fly ash-slag-based alkaline binder with fiber reinforcement in the expanding soil effectively improves compressive shear, flexure, and stiffness resistance outcome.

3.2 Geotechnical Characterization

Linear Shrinkage. The linear shrinkage results for PF-AB-soil with different quantities of fly ash and slag in the fiber-alkaline-soil mixture are shown in Fig. 5. The linear shrinkage and shrinkage limit of expansive untreated soil was calculated to be 15% and 24%, respectively, indicating significant shrinkage. After adding a fly ash-based alkaline binder, the expansive clayey soil's linear shrinkage was reduced from 24 to 14%, attributed to the production of pozzolanic compounds. The shrinkage behavior improved as the slag substitution increased in the alkaline mixture; this improvement could be caused by poor liquidity and capillary forces hastening the pozzolanic process. Furthermore, the fiber-AB paste containing 100% fly ash reduced linear shrinkage percentages significantly, indirectly lowering the plasticity index. As a result, the pure fly ash-based alkaline soil attains less shrinkage compared to based mixed AB-soil.

Consolidation. The results of a series of e-log p and heave curves are shown in Fig. 6(a–b). The apparent correlation of soil volumetric stability behavior was demonstrated in the slope of void ratio curves. In comparison to alkaline mixed soils, untreated clayey soil showed the highest heave height. The presence of a montmorillonite-illite group and a large specific surface area of clay particles impaired the interfacial particle bonding, resulting in substantial swelling [28]. As the amount of slag in the soil grew, the compressibility of the alkaline soil fell significantly. The active multivalent cationic generation between the clay particles lowered the void ratio after replacing it with slag. In addition, the active dissolution of pozzolanic compounds in silica-rich soil activates the aluminum silicate hydrate gel, affecting soil heaving behavior [26]. When the soil specimen was unloaded, the cementitious agents enwrapped the volumetric particle activity [21].

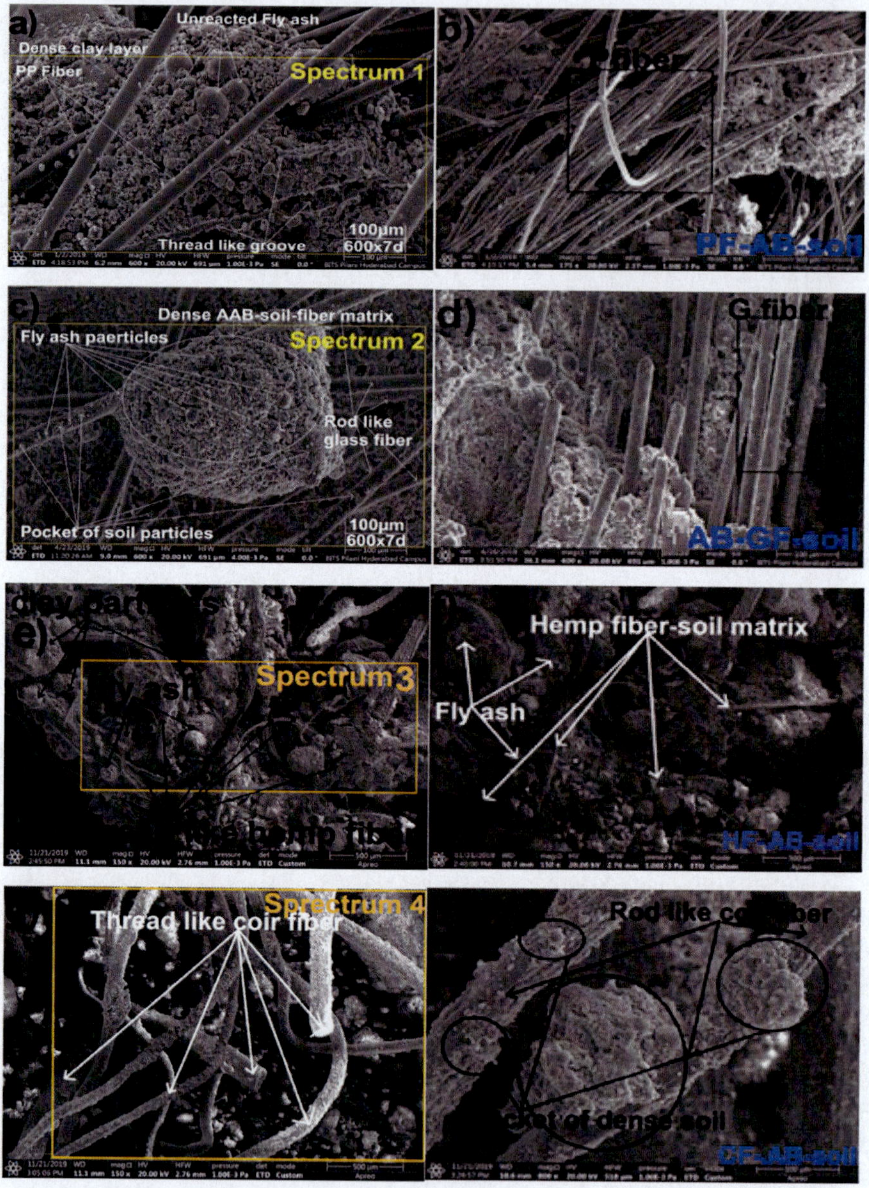

Fig. 4. SEM images of alkaline mixed soil strengthened with **a–b)** polypropylene fiber soil, **c–d)** glass fiber, **e–f)** chemically modified hemp, and **g–h)** coir fiber soil mixture.

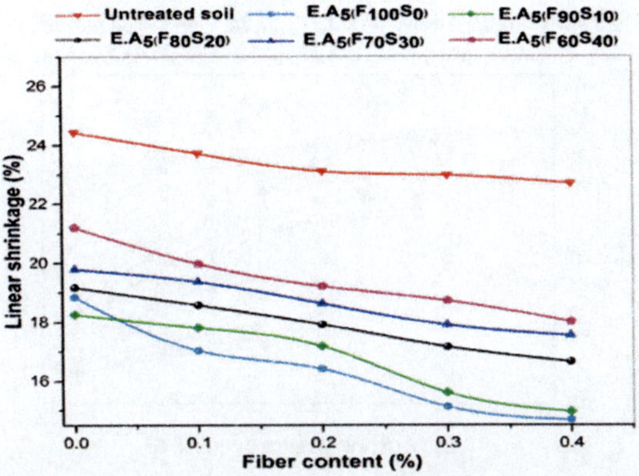

Fig. 5. Variation of linear shrinkage of fiber-reinforced alkaline soil

The change in heave rate for untreated and alkaline treated soil at various dosages of fly ash-slag was displayed in Fig. 6b. The highest heave for untreated soil under the seated load was 2.78 mm in 3 days. The findings revealed that when different quantities of aluminosilicate precursors were used, the heave in alkaline soil was dramatically reduced. The usage of a pure fly ash-based AB mixture in the soil helped to preserve heave effects (from 2.78 mm to 1.88 mm), which could be attributed to an improvement in clay composition and morphology (due to rapid cementitious reaction activation) [29]. The volumetric expansion of the alkaline binder decreased considerably as the slag content increased. It's also worth noting that replacing up to 40% of the fly ash in AB soil mixtures with an alkaline solution reduces the void ratio and swelling. The addition of slag to the alkaline solution aided the rate of geopolymerization reaction created around the clay matrix. As a consequence, the soil compressibility results showed that using alkaline binder as a cementitious additive can successfully improve interlocking particle density by preventing substantial soil settlement.

Unconfined Compressive Strength. The UCS of four various types of fiber reinforcement with different slag-fly ash content in the alkaline binder mixed soils is shown in Figs. 7(a–b). It was discovered that in silica-rich expansive soil, combining fiber reinforcing with an aluminosilicate-based alkaline binder had a considerable impact on shear strength behavior. UCS growth percentages increased with fiber doses in both figures regardless of fly ash-slag proportions and fiber type. Fiber-AB soil achieved a higher UCS value than the fly ash alone ($F_{100}S_0$) alkaline-fiber matrix and slag ($F_{60}S_{40}$). The low silica and alumina reactivity with fly ash in the alkaline solution compared to rapid slag hydration could explain these large variations in shear resistance. Similar findings on the improvement of expansive soil shear strength behavior reinforced with polypropylene fibers in cement and lime were reported [16, 30, 31].

The alkaline soil combination's interlocking bonding and frictional resistance rose as the slag and fiber content increased. As a result, the fibers restricted the relative shifting of their position from one step to another, enhancing shear resistance during loading [9]. The compressive shear strength of coupled polypropylene fiber and AB inclusions

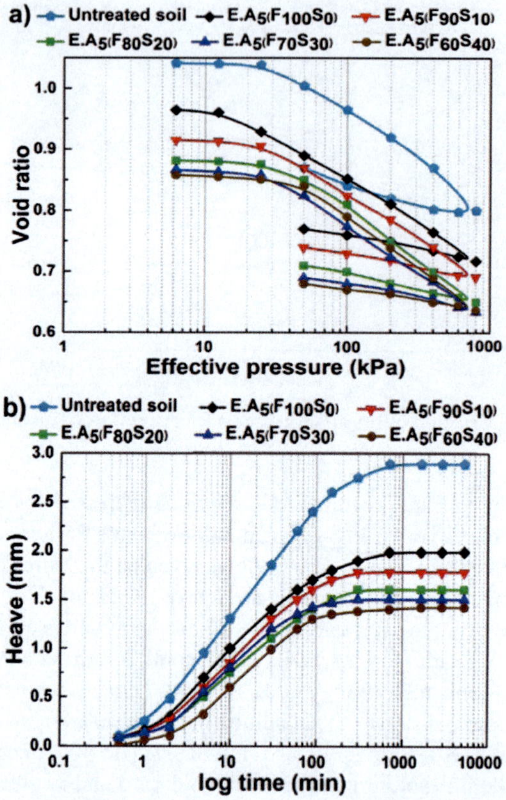

Fig. 6. Variations of **(a)** Void ratio-effective stress **(b)** Rate of heave of alkaline soil.

increased significantly more than glass fiber-AB inclusions. The PF-AB mixed UCS values also rose by roughly 26% as compared to the GF-AB matrix. This increase may be related to the increased frictional bonding resistance [18]. Fiber added to alkaline soil mixes (both natural and synthetic) improved confinement density and interfacial bonding during loading. Excessive fiber inclusion (more than 0.6% for synthetic and 0.8% for natural) mixed with alkaline biner-soil attained difficulty to compact, influencing stiffness behavior by forming smooth textures, weakening the particle holding efficacy. The flexible nature of fibers may indirectly dominate the resisting power of shear strength when a large amount of fiber is introduced to the soil. As a result, the fiber-reinforced alkaline soil reached its highest density between 0.6 and 0.8% fiber content.

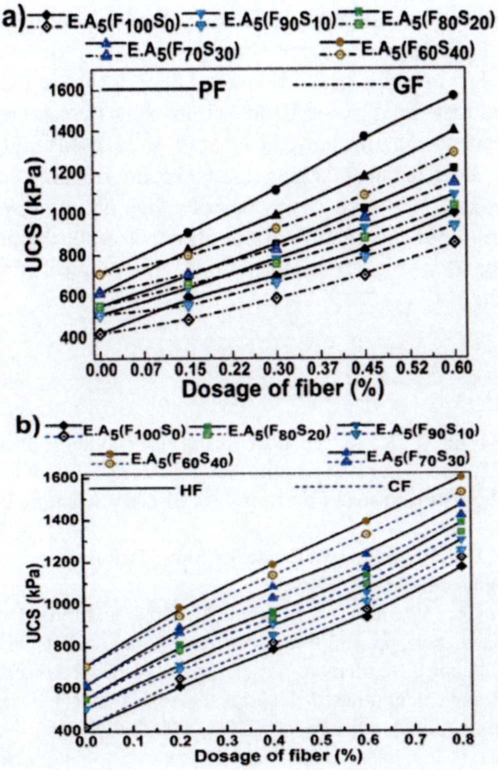

Fig. 7. UCS for **(a)** PF and GF alkaline soil, **(b)** HF and CF alkaline soil at a different slag-fly ash content

4 Conclusions

The present study examines the geotechnical and microstructural characteristics of untreated expansive clayey soil, alkaline mixed soil, and fiber-reinforced-AB soil with varying fiber and AB doses in the soil mixture. After 28 days of curing, XRD and FTIR measurements demonstrate the creation of new mineralogical phases and chemical vibrational bonding in the alkaline soil mixed with various slag-fly ash concentrations. At a minimum of 0.4 w/s ratios, a fiber-reinforced alkaline soil micrograph demonstrates a high interfacial bonding interaction around the soil-fiber matrix. With enough friction and stiffness, the fiber bridge effect can effectively transform brittle soil behavior to ductile. Furthermore, the physical surface morphology of fiber-AB-treated soil tightly encircling the clay particles surrounding the fiber.

The combined effect of an alkali activator with four kinds of fiber reinforcements on expansive soil treated at different fractions of aluminosilicate precursors in the alkaline solution is also investigated. The results of the experiments revealed that soil fiber reinforcement may effectively control shrinkage fractures. In pavement geotechnics, the combination of fiber-slag-fly ash as an envirosafe alkaline additive in geopolymer stabilized subgrade soil was innovative and noteworthy. Laboratory studies were used to assess the viability of fly ash-slag proportions, and fiber dosage in the alkaline stabilized material as a bonded pavement material. At various slag-fly ash proportions and fiber doses, geomechanical investigations revealed that PF and HF-AB-soil had stronger penetration, compressive shear, tensile cracking resistance, interlocking density, and stiffness than GF and CF-AB-soil. However, it may be noted that mixing beyong 0.8% of fibers with clay resulted in minor segregation of the fibers. Overall, the proposed envirosafe alkaline binder can be used as a subgrade stabilizer with reduced carbon emissions from industrial by-products.

References

1. Ackroyd, L.W., Husain, R.: Residual and lacustrine black cotton soils of north-east Nigeria. Géotechnique **36**(1), 113–118 (1986). https://doi.org/10.1680/geot.1986.36.1.113
2. Chen, F.: Foundations on Expansive Soils, p. 280. Elsevier Scientific Publishing Company, New York (1975)
3. Nelson, D., Miller, D.J.: Book Review: Expansive Soils - Problems and Practice in Foundation and Pavement Engineering, p. 17 (1993)
4. Puppala, A., Hoyos, L., Viyanant, C., Musenda, C.: Fiber and fly ash stabilization methods to treat soft expansive soils, pp. 136–145 (2008). https://doi.org/10.1061/40552(301)11
5. Jalal, F.E., Xu, Y., Jamhiri, B., Memon, S.A., Graziani, A.: On the recent trends in expansive soil stabilization using calcium-based stabilizer materials (CSMs): a comprehensive review. Adv. Mater. Sci. Eng. (2020). https://doi.org/10.1155/2020/1510969
6. Saride, S., Puppala, A.J., Chikyala, S.R.: Swell-shrink and strength behaviors of lime and cement stabilized expansive organic clays. Appl. Clay Sci. **85**(1), 39–45 (2013). https://doi.org/10.1016/j.clay.2013.09.008
7. Taylor, M., Tam, C., Gielen, D.: Energy efficiency and CO_2 emissions from the global cement industry. In: IEA-WBCSD Workshop. Energy Efficiency and CO_2 Emission Reduction Potentials and Policies in the Cement, pp. 4–5 (2006)
8. Gartner, E.: Industrially interesting approaches to "low-CO_2" cements. Cem. Concr. Res. **34**(9), 1489–1498 (2004). https://doi.org/10.1016/j.cemconres.2004.01.021
9. Pourakbar, S.: Application of alkali-activated agro-waste reinforced with wollastonite fibers in soil stabilization. J. Mater. Civ. Eng. **29**(2), 04016206 (2016). https://doi.org/10.1061/(asce)mt.1943-5533.0001735
10. Bordoloi, S., Garg, A., Sekharan, S.: A review of physio-biochemical properties of natural fibers and their application in soil reinforcement. Adv. Civ. Eng. Mater. **6**(1), 20160076 (2017). https://doi.org/10.1520/acem20160076
11. Consoli, N.C., Arcari Bassani, M.A., Festugato, L.: Effect of fiber-reinforcement on the strength of cemented soils. Geotext. Geomembr. **28**(4), 344–351 (2010)
12. Park, S.S.: Unconfined compressive strength and ductility of fiber-reinforced cemented sand. Constr. Build. Mater. **25**(2), 1134–1138 (2011). https://doi.org/10.1016/j.conbuildmat.2010.07.017
13. Mazhar, S., GuhaRay, A.: Effect of fiber reinforcement on mechanical behaviour of alkali activated binder treated expansive soil: a reliability-based approach. Int. J. Geomech. **20**(12), 04020225-1-14 (2020). https://doi.org/10.1061/(ASCE)GM.1943-5622.0001871

14. Mazhar, S., GuhaRay, A.: Effect of natural fiber reinforcement on strength response of alkali activated binder treated expansive soil: experimental investigation and reliability analysis. Constr. Build. Mater. **273**(1), 121743 (2021). https://doi.org/10.1016/j.conbuildmat.2020.121743
15. Mazhar, S., GuhaRay, A., Goel, D.: Strength characterisation of fiber reinforced expansive subgrade soil stabilised with alkali activated binder. Road Mater. Pavement Des. (2021). https://doi.org/10.1080/14680629.2020.1869062
16. Tang, C., Shi, B., Gao, W., Chen, F., Cai, Y.: Strength and mechanical behavior of short polypropylene fiber reinforced and cement stabilized clayey soil. Geotext. Geomembr. **25**(3), 194–202 (2007). https://doi.org/10.1016/j.geotexmem.2006.11.002
17. Anggraini, V., Asadi, A., Farzadnia, N., Jahangirian, H., Huat, B.B.K.: Reinforcement benefits of nanomodified coir fiber in lime-treated marine clay. J. Mater. Civ. Eng. **28**(6), 1–8 (2016). https://doi.org/10.1061/(ASCE)MT.1943-5533.0001516
18. Sudhakaran, S.P., Sharma, A.K., Kolathayar, S.: Soil stabilization using bottom ash and areca fiber: experimental investigations and reliability analysis. J. Mater. Civ. Eng. **30**(8), 1–10 (2018). https://doi.org/10.1061/(ASCE)MT.1943-5533.0002326
19. Gupta, S., GuhaRay, A., Kar, A., Komaravolu, V.P.: Performance of alkali-activated binder-treated jute geotextile as reinforcement for subgrade stabilization. Int. J. Geotech. Eng. **15**(3), 299–312 (2018). https://doi.org/10.1080/19386362.2018.1464272
20. Sekhar, D.C., Nayak, S.: Utilization of granulated blast furnace slag and cement in the manufacture of compressed stabilized earth blocks. Constr. Build. Mater. **166**, 531–536 (2018). https://doi.org/10.1016/j.conbuildmat.2018.01.125
21. Miao, S., Wei, C., Huang, X., Shen, Z., Wang, X., Luo, F.: Stabilization of highly expansive black cotton soils by means of geopolymerization. J. Mater. Civ. Eng. **29**(10), 04017170 (2017). https://doi.org/10.1061/(asce)mt.1943-5533.0002023
22. Al-Mukhtar, M., Khattab, S., Alcover, J.F.: Microstructure and geotechnical properties of lime-treated expansive clayey soil. Eng. Geol. **139–140**, 17–27 (2012). https://doi.org/10.1016/j.enggeo.2012.04.004
23. Sivapullaiah, P.V., Prasad, B.G., Allam, M.M.: Effect of sulfuric acid on swelling behavior of an expansive soil. Soil Sedim. Contam. **18**(2), 121–135 (2009). https://doi.org/10.1080/15320380802660289
24. Mohammadinia, A., Arulrajah, A., Sanjayan, J., Disfani, M.M., Win Bo, M., Darmawan, S.: Stabilization of demolition materials for pavement base/subbase applications using fly ash and slag geopolymers: laboratory investigation. J. Mater. Civ. Eng. **28**(7) (2016). https://doi.org/10.1061/(asce)mt.1943-5533.0001526
25. Zhang, Y., Johnson, A.E., White, D.J.: Laboratory freeze-thaw assessment of cement, fly ash, and fiber stabilized pavement foundation materials. Cold Reg. Sci. Technol. **122**, 50–57 (2015). https://doi.org/10.1016/j.coldregions.2015.11.005
26. Rios, S., Cristelo, N., Viana da Fonseca, A., Ferreira, C.: Structural performance of alkali-activated soil ash versus soil cement. J. Mater. Civ. Eng. (2016). https://doi.org/10.1061/(ASCE)MT.1943-5533.0001398
27. Lekha, B.M., Goutham, S., Shankar, A.U.R.: Evaluation of lateritic soil stabilized with Arecanut coir for low volume pavements. Transp. Geotech. **2**, 20–29 (2015). https://doi.org/10.1016/j.trgeo.2014.09.001
28. Sharma, A.K., Sivapullaiah, P.V.: Swelling behaviour of expansive soil treated with fly ash–GGBS based binder. Geomech. Geoeng. **12**(3), 191–200 (2017). https://doi.org/10.1080/17486025.2016.1215548
29. Ouhadi, V.R., Yong, R.N., Amiri, M., Ouhadi, M.H.: Pozzolanic consolidation of stabilized soft clays. Appl. Clay Sci. **95**, 111–118 (2014). https://doi.org/10.1016/j.clay.2014.03.020

30. Moghal, A.A.B., Chittoori, B.C.S., Basha, B.M.: Effect of fibre reinforcement on CBR behaviour of lime-blended expansive soils: reliability approach. Road Mater. Pavement Des. **19**(3), 690–709 (2018). https://doi.org/10.1080/14680629.2016.1272479
31. Moghal, A.A.B., Chittoori, B.C.S., Basha, B.M., Al-Shamrani, M.A.: Target reliability approach to study the effect of fiber reinforcement on UCS behavior of lime treated semiarid soil. J. Mater. Civ. Eng. **29**(6), 04017014 (2017). https://doi.org/10.1061/(asce)mt.1943-5533.0001835

Bearing Capacity of Tubular Piles: Technological Improvements and Model Testing

Michael Doubrovsky(✉), Liudmyla Kusik, and Vladyslava Dubravina

Odessa National Maritime University, 34 Mechnikov Street, Odessa 65029, Ukraine
m.doubrovsky@gmail.com

Abstract. Some offshore and marine structures often include steel tubular piles of essential length (80–100 m and more) that should provide rather high bearing capacity regarding essential axial load. To increase piles bearing capacity under static pressing load, such an additional element as the internal diaphragm has been applied in some practical cases. Presented research aimed to study two connected processes during steel tubular pile driving: soil plug formation at the tip of the open-end pile and soil behavior under the internal diaphragm fixed inside the tubular pile's shaft. Results of physical modelling in laboratory conditions and their numerical analysis are discussed. Numerical analysis of the gained experimental data gave the possibility to apply approximating function with good correlation indexes. Obtained information of internal diaphragm application may be useful to provide an increase of pile's bearing capacity (in case of bearing capacity deficit) or to justify pile length reduction. Gained conclusions may facilitate design and construction of deep-water piled structures.

Keywords: Bearing capacity · Tubular pile · Soil plague · Internal diaphragm · Laboratory modelling

1 Introduction

Modern marine transportation and offshore structures such as deep-water port's berths, oil and gas platforms, raid and offshore fixed single point moorings, submerged stores and others often include steel tubular piles of essential length (80–100 m and more) as main bearing elements. Some examples of these structures are presented in Figs. 1, 2 and 3 [1].

Such tubular piles should provide high bearing capacity in case of external axial loads application [1–4].

One of the interesting peculiarities of long tubular open-end piles behavior is the formation of soil plugs at the piles' tip [5, 6].

From this point of view, we support a known opinion that it is important to study the influence of the soil plug not only on the pile's tip bearing capacity but also on soil behavior inside the tubular shaft [4].

H.-H. Zhu et al. (Eds.): BRWSG 2021, LNCE 230, pp. 137–154, 2022.
https://doi.org/10.1007/978-981-16-9963-4_11

Fig. 1. Single point mooring fixed to the sea bottom by long steel tubular piles (LNG carrier service).

Fig. 2. Piled cluster applied to fix mooring device of the tanker.

Fig. 3. Use of large-diameter bearing monopile as fixed single point mooring for the offshore industry.

In case of necessity, the bearing capacity of long tubular piles may be increased by different methods:

- by driving a pile with a closed end to develop increased end-bearing resistance (but it requires application of too powerful hammers);
- by installing the pile at larger depth in order to reach the bearing soil strata (but such penetrations are often much greater than those required for fixity against lateral loading);
- by grouting beneath the pile toe (but the operations of cleaning-out the pile and grouting are slow and relatively costly);
- by welding a steel plate diaphragm across the interior of the pile in order to increase bearing capacity by use of soil reaction under the diaphragm (such method demonstrated good results on some marine projects and got positive references [4]).

Regarding that, in many cases, large diameter tubular piles of shelf structures are installed without plugging effect (so called "fully coring mode" [7]) or with partial plugging, the last approach (closure of the pile's shaft) looks rather attractive for deep water port, marine and offshore engineering but needs detailed consideration and study aiming to determine method's peculiarity, an appropriate sphere of application, details of diaphragm construction and proper location along the pile's shaft.

2 Application of Internal Diaphragm in Steel Tubular Piles

Recommended technology to install the internal diaphragm as described in [4] (Fig. 4).

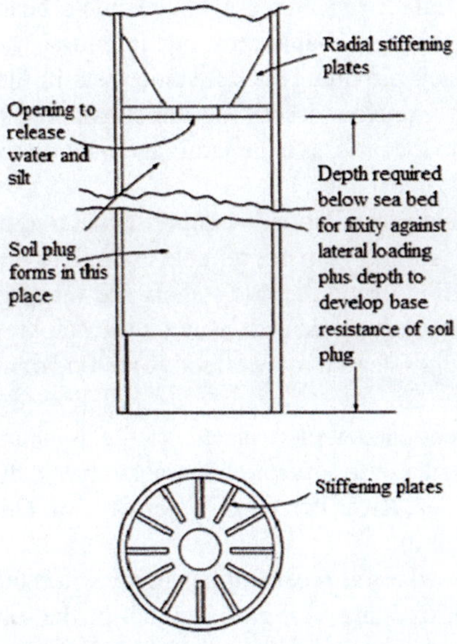

Fig. 4. Steel tubular pile with a diaphragm.

A hole is necessary for the diaphragm for the release of water pressure in the soil plug and to allow expulsion of silt. Stresses on the underside of the diaphragm are high during driving, and radial stiffeners are needed.

According to [4], the minimum depth above the pile toe for locating the diaphragm is the penetration below the sea bed required for fixity against lateral loading. There are formulas in some norms allowing determination of the fixity's depth depending on soil properties and pile's bending rigidity; roughly, this depth may be determined in the interval of (5–7) d. However, further penetration is necessary to form the soil plug under the diaphragm by compacting the soil within the plug and to develop the necessary base resistance. Thus, mentioned authors considered two locations for two soil plugs formation during the tubular pile driving: at the open end of the pile and under the internal diaphragm.

As an example of the diaphragm's practical application, we may refer to the piling works at the Hadera coal unloading terminal near Haifa [4]. Open-end piles 1424- and 1524-mm OD were proposed, but initial trial driving showed that very deep penetrations, as much as 70 m below sea bed in calcareous sands, would be needed to develop the required axial resistance. The blow count diagram showed quite low resistance at 36 m below the sea bed. Trials were then made of the diaphragm method. A diaphragm with a 600 mm hole giving 83% closure of the cross-section was inserted 20 m above the toe. This increased the driving resistance at 39 m below sea bed and another trial with a 300 mm hole (95% closure) gave a higher resistance at 37 m. It was supposed that such improvement of piles bearing capacity was stipulated by soil plug formation below the mentioned diaphragms.

3 Laboratory Model Testing

Regarding that obvious effect (an increase of the pile's bearing capacity) has been achieved by the use of the rigid diaphragm, our intention was to study peculiarities of the considered approach providing model static tests in the laboratory conditions. Our aim was to obtain parameters describing the considered pile driving processes – both qualitative (related to the process in general) and quantitative (characteristic for the applied model pile-soil system) ones.

As to the method of pile's installation, we suppose that traditional approaches (use of impact hammer or vibro hammer) are not reliable enough to provide safety of the rigid diaphragm fixed by welding inside the pile's shaft and interacting with soil under the diaphragm. In order to avoid dynamic actions upon the diaphragm during pile penetration, we prefer to consider a safer but more effective method of pressing load application [6].

To clarify above mentioned items related to the tubular pile with the internal diaphragm, we have started a series of experimental studies in the Geotechnical Laboratory of the Department "Sea, River Ports and Waterways" at Odessa National Maritime University (Odessa, Ukraine).

For pile testing, we used a soil box of dimensions: width 600 mm, length 750 mm, depth 1100 mm (Figs. 5, 6, 7 and 8). For the model of the tubular open-end pile, we apply steel pipe d = 50 mm external diameter, pipe wall thickness 1 mm, $l = 800$ mm length. To drive the pipe into fine sand mechanical jack has been applied.

For experimental studies, we used fine sand with the following characteristics: internal friction angle 33°; density 14.5 kN/m^3; void ration 0.71; moisture 0.07%; Young modulus 16 MPa, Poisson's ratio 0.3.

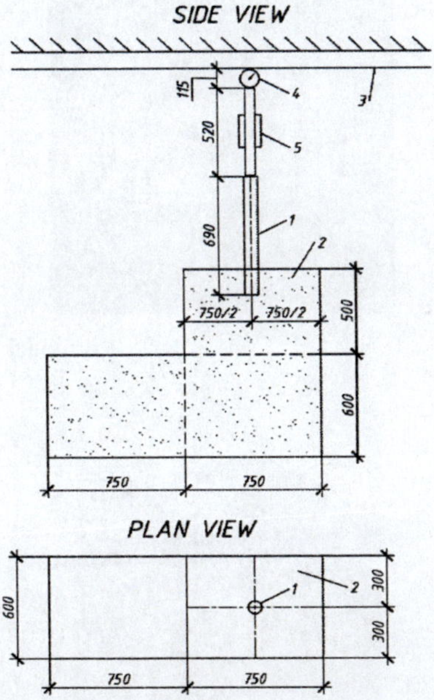

Fig. 5. Scheme of the experiment: 1 – pile model; 2 – sand box; 3 – bearing beam; 4 – force gauge (dynamometer); 5 – jack loading system (all sizes in millimeters).

Fig. 6. Experimental system: side view. 1 – soil box; 2 – model pile; 3 – mechanical jack; 4 – dynamometer; 5 – displacement gauge.

Fig. 7. Experimental system: front view. 1 – soil box; 2 – model pile; 3 – mechanical jack; 4 – displacement gauge.

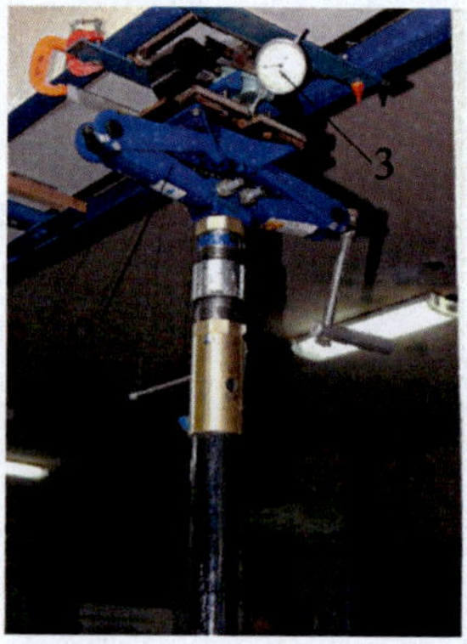

Fig. 8. Loading system. 1 – model pile; 2 – mechanical jack; 3 – dynamometer.

The first series of experiment was aimed to determine the conditions of the soil plug formation at the tip of the open-end pile model.

Measured parameters (at each stage of load application) were:

- applied vertical axial pressing force (named *Load, F* on the diagrams)
- pile's penetration depth (named *Displacement, t* on the diagrams)
- soil level inside the pile's model (the initial position was fixed when soil levels inside and outside the pile were equal).

In order to describe the process of the model tubular pile plugging, we applied:

- earlier proposed IFR (Incremental Filling Ratio) and PLR (Plug Length Ratio) characteristics (for example, recommendations [7–10] and others) presented in Figs. 9 and 10;

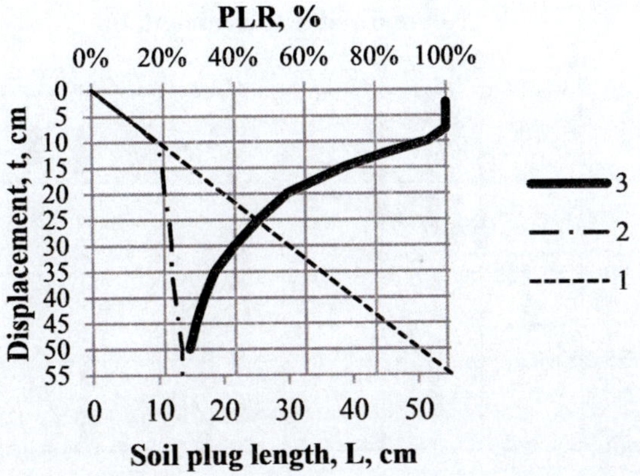

Fig. 9. Typical dependencies for the tests of the first series: 1 – fully coring mode; 2 – L = L(d); 3 – PLR.

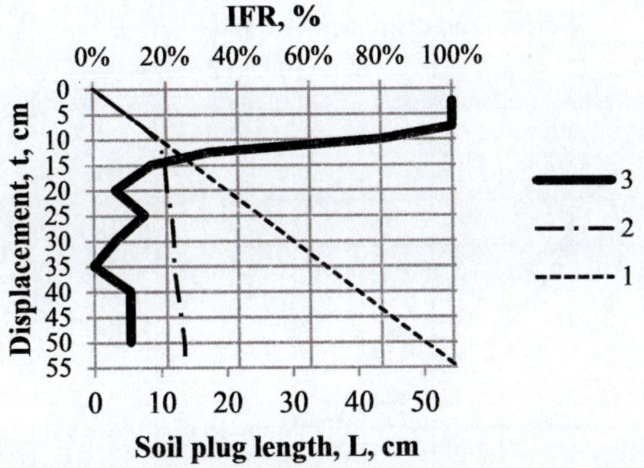

Fig. 10. Typical dependencies for the tests of the first series: 1 – fully coring mode; 2 – L = L(d); 3 – IFR.

- Assumption that as evidence of completing the process of soil plug formation, we may consider equal pile's vertical displacements and related settlements of the sand surface inside the shaft at each stage of further axial load application (related diagrams are presented in Figs. 11 and 12).

According to the PLR and IFR diagrams (Figs. 9 and 10) at the initial stage of pile installation [till t = (1.5–2) d] PLR = IFR = 100%, i.e., the pile is driven according to the "fully coring mode". From the penetration depth of some (4–5) d, soil plug length is almost unchanged; average IFR value is stable too.

Some IFR fluctuations below the penetration depth = 20 cm may be explained by technical reasons: resetting of the jack each 20 cm of the penetration process and corresponding installation of extension tubes.

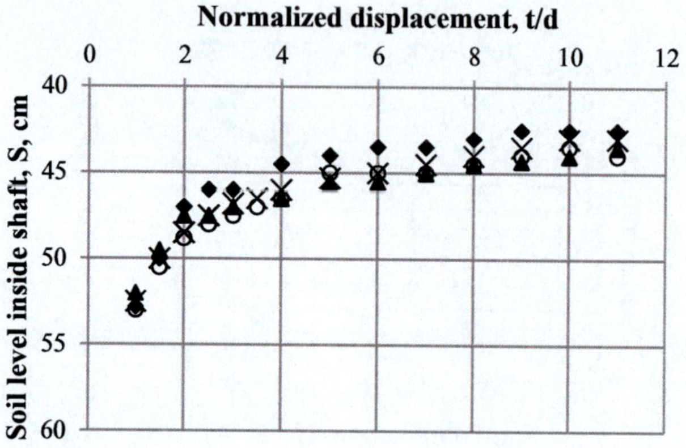

Fig. 11. Results of open-ended model pile jacking (4 similar tests): S – the distance between pile's top and soil in the shaft.

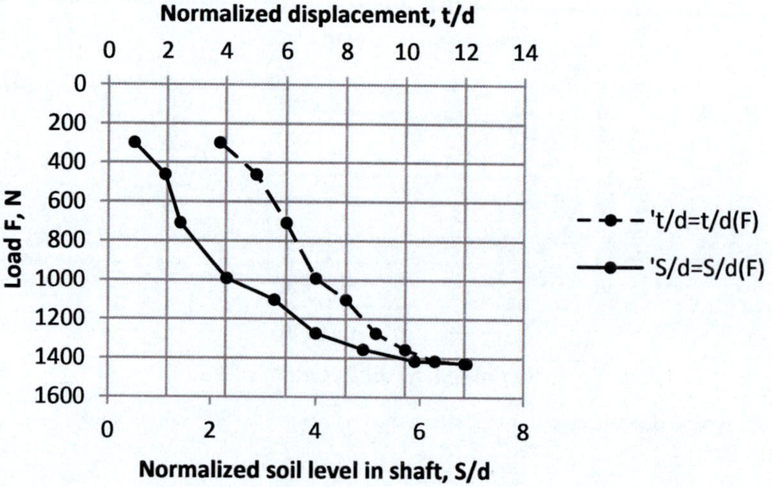

Fig. 12. Results of open-ended model pile jacking (average points of 4 similar tests).

A similar conclusion may be made on the basis of diagrams in Figs. 11 and 12. Soil level inside the pile's shaft becomes stable at the relative penetration depth (normalized pile displacement) approx. t/d = 5–6 (Fig. 11). Dependencies for t/d and S/d describing pile and soil displacements also become parallel starting from the t/d = 5–6 and till reaching the pile bearing capacity at t/d = 8–9 (Fig. 12).

Some important results of the first series of the laboratory experiments may be formulated in the following way:

- in fine sandy soil, the plug is formed at the comparatively early stage of pile's driving (in the considered case – at the penetration depth of around 4–5 pile's diameters);

- if to locate the internal diaphragm at the recommended depth required for fixity against lateral loading as described above (approx. at the penetration depth of around 5–7 pile's diameters), we may meet the situation of no contact between the diaphragm and the soil inside the shaft (clearance space); i.e., the diaphragm does not catch up with soil.

The second series of the experiment was devoted to clarification of the role and contribution of the internal diaphragm. For the model pile, the diaphragm was produced as a circular steel plate (4 mm thickness), with its diameter corresponding to the inner diameter of the pile.

By use of the rigid steel bar (located in the pile's shaft) the diaphragm was connected with the pile head. Varying the length of the mentioned rigid bar, we had the possibility to locate the internal diaphragm at different places along the model pile's shaft.

At the second series of the experiment internal diaphragm was fixed at several positions by changing the distance from the tip of the model pile: 0 (closed end); 3d; 6d; 9d (total length of the pile was equal to 16d).

It has been discovered that due to sand settlements inside the pile shaft during pile installation, there is an empty space under the diaphragm and, correspondingly, no contact between soil and diaphragm.

In order to avoid clearance space under the diaphragm and to provide constant contact of the sand with the underside of the diaphragm, we applied the diaphragm with several small holes allowing sand to fill into the space under the diaphragm (Fig. 13). For above mentioned options of the diaphragm location, the sand was filled after the pile driving on the depth 3d, 6d and 9d correspondingly (i.e., at the moment of the first potential contact of the soil surface with the underside of the diaphragm).

Fig. 13. Underside of the internal diaphragm with peripheral holes for sand filling.

The volume of the filled sand was calculated to provide the required diaphragm-sand contact during the whole process of model pile jacking.

The volume of the filled sand was calculated to provide the required diaphragm-sand contact during the whole process of model pile jacking.

As it is demonstrated by diagrams presented in Figs. 14, 15, 16 and 17, application of the internal diaphragm provides increasing open-end pile bearing capacity. The degree

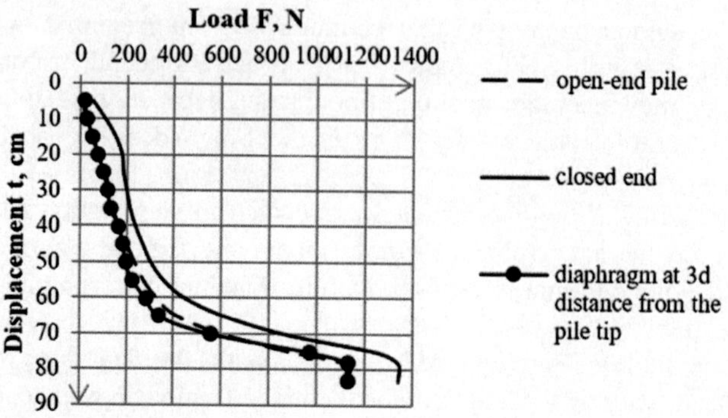

Fig. 14. Dependencies between vertical axial load upon the model pile and its displacements for the diaphragm at 3d distance from the pile tip.

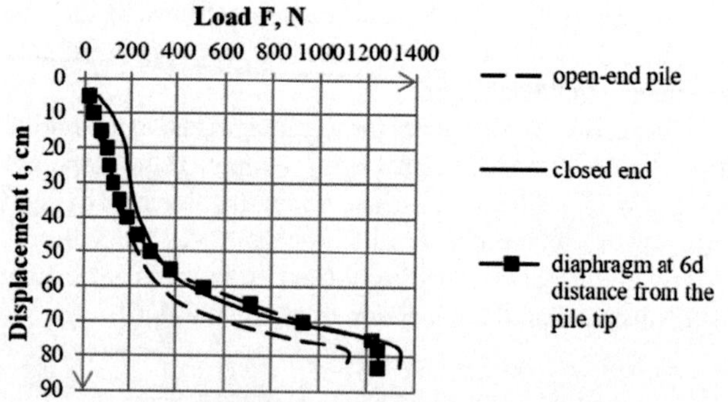

Fig. 15. Dependencies between vertical axial load upon the model pile and its displacements for the diaphragm at 6d distance from the pile tip.

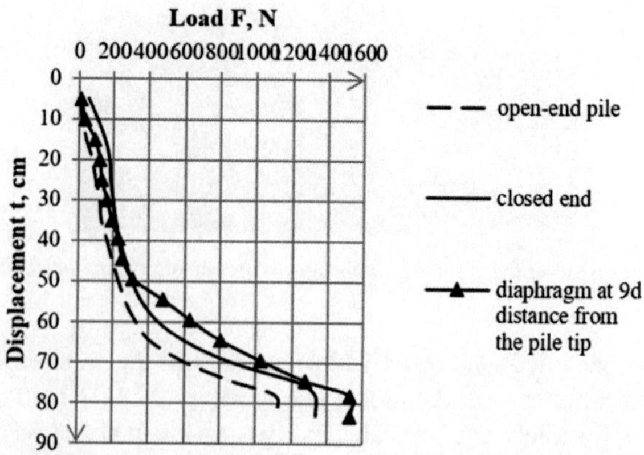

Fig. 16. Dependencies between vertical axial load upon the model pile and its displacements for the diaphragm at 9d distance from the pile tip.

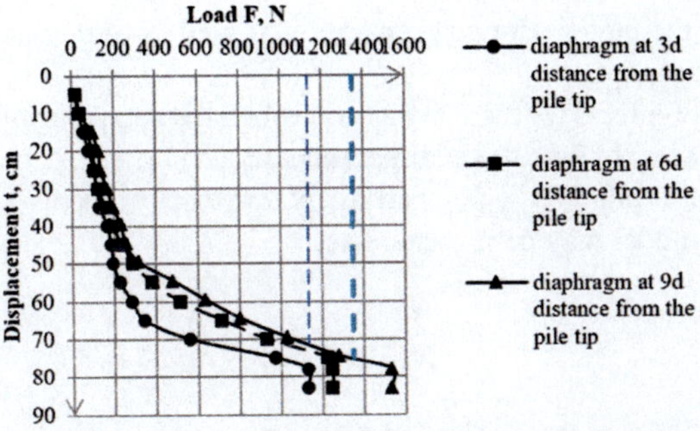

Fig. 17. Dependencies between vertical axial load upon the model open-end pile and its displacements.

of such increase depends on the diaphragm location. For the considered options of the diaphragm fixing point, the minimum increment of the open-end pile bearing capacity relates to the 3d distance between the diaphragm and the pile tip (Fig. 14 and 17) and the maximal increment is measured at the 9d distance (Fig. 16 and 17).

Perhaps mentioned circumstances may be commented by the following way. The upper plug under the diaphragm may be formed if there is the proper base reaction developed inside the shaft. Such a situation may occur if the upper plug (being in the process of formation) meets the already formed lower plug. The last transfers additional pressure to the soil under the toe and provokes an additional base reaction. Thus, an additional external force acts on the plug and increases soil density in it. In fact, after that stage, two plugs are combined and work as one large plug between the diaphragm and pile's toe. Obviously, the creation of the mentioned large plug and its effective contribution to the pile bearing capacity may be provided only in case of the "right" location of the diaphragm (not too low and not too high). For our model tests, the maximal bearing capacity of the open-ended pile was measured in the case of 9d distance of the diaphragm from the pile tip.

It may be explained, particularly, by the fact that for the considered test conditions, approximate driving depth t = (4–5) d at the initial stage of pile installation is needed to dense a soil due to the development of the friction forces inside the pile's shaft and to form a lower soil plug at the pile tip. If then to apply similar consideration for the follow-on stage of the driving process – compaction of the soil under the diaphragm due to the similar friction forces, required penetration depth for this stage to form the upper plug may be of similar value (4–5) d. So the total distance between pile toe and the diaphragm may be considered as the sum of these two parts of the penetration depth, i.e., approx. (8–10) d. Such location of the diaphragm may be optimal to form two plugs consecutively and to combine them in one large plug.

Regarding quantitate parameters of open-end pile bearing capacity (Fig. 17), we would like to note that due to the diaphragm's contribution, pile bearing capacity may be increased (in our tests up to 15–20%). Another effect consists in the possibility of decrease pile driving depth (10–15%). Obviously, mentioned figures should be considered with regard to possible experimental errors stipulated by differences in the reproducibility of the model ground preparation as well as to measurement inaccuracy (perhaps up to 10% in total).

Regarding scale-effects for the considered problem, it should be noted the following. From the point of view of so-called "direct modelling" [11, 12], dependencies between limit axial force in a pile $N_{lim,p}$ (as well as related displacements $U_{lim,p}$) and similar parameters of the model may be presented as

$$\frac{N_{lim,p}}{N_{lim,m}} = C_L^n \tag{1}$$

$$\frac{U_{lim,p}}{U_{lim,m}} = C_L^m \tag{2}$$

where C_L is the sizes (scales) correlation between prototype and the model; n and m are parameters depending on soil properties and pile dimensions.

For the conditions of our laboratory model testing (skipping the details of intermediate conversions and calculations), it was determined that the related prototype is a tubular pile of diameter 1.0 m driven up to 10 m into similar sandy soil. Its bearing capacity (sum of the toe and shaft bearing capacities) is 1723 kN. For comparison: the calculated value of the prototype bearing capacity according to the recommendation of the related Ukrainian code occurred to be 2020 kN (some 15% difference).

Also, for plugging effect assessment, we have to consider scale effects stipulated by the influence of internal pile diameter. This aspect is subject to a study for further investigations.

4 Numerical Analysis of Experimental Data

The graphical analysis of the experimental data led to the assumption that the relationship between the displacement of the pile (t) and the load (F) exerted on the pile can be described by power dependence. Assuming that the constraint equation is a function $F = a \cdot t^k$, by the least squares method after solving the linear (with respect to $ln\, a$ and k) system,

$$\begin{cases} n \cdot ln\, a + k \cdot \sum_{i=1}^{n} ln\, t_i = \sum_{i=1}^{n} ln\, F_i \\ ln\, a \cdot \sum_{i=1}^{n} ln\, t_i + k \cdot \sum_{i=1}^{n} ln^2 t_i = \sum_{i=1}^{n} ln\, F_i \cdot ln\, t_i \end{cases} \tag{3}$$

we have received the following values:

1) $a = 1{,}224$, $k = 1{,}485$ for pile with diaphragm at 9d distance from the pile tip;
2) $a = 0{,}272$, $k = 1{,}822$ for the open-end pile;
3) $a = 5{,}21$, $k = 1{,}1$ for the closed end pile.

The approximation error in all cases turned out to be greater than 7%:

1) $\overline{A} = 21\%$,
2) $\overline{A} = 22{,}2\%$,
3) $\overline{A} = 28{,}2\%$,

which suggests the need to select a more accurate model.

Empirical correlation relations are:

1) $\eta = 0{,}66$,
2) $\eta = 0{,}642$,
3) $\eta = 0{,}563$.

According to the Chaddock scale these relations indicate a salient correlation between t and F.

Correlation indexes for the considered cases

1) $R = 0{,}882$,
2) $R = 0{,}882$,
3) $R = 0{,}806$

are high enough, which means the tightness of the relationship of the considered values.

If we assume that the relationship between the considered values is described by exponential laws ($F = a - b \cdot e^{-mt}$, $F = n \cdot e^{kt}$), then the least squares method allows determination of the parameters b, m, n, k; the value of the parameter a may be obtained from the experiment. However, the discrepancy between the actual and calculated values of F is so great that further comparison becomes meaningless. A similar situation is observed with the inverse (hyperbolic) dependence.

However, the graphic interpretation of the "displacement – load" function (Figs. 14, 15, 16 and 17) prompted the idea to look for a model in the form

$$t = a \cdot \sqrt{F} - \frac{1}{F - b} + c \tag{4}$$

It turned out that in all cases it is possible to select the values of the coefficients a, b, c (see below) so that the approximation error lies in the range of 5%–7%:

1) $\overline{A} = 5{,}1\%$ for pile with diaphragm at 9d distance from the pile tip,
2) $\overline{A} = 4{,}6\%$ for the open-end pile,
3) $\overline{A} = 4{,}4\%$ for the closed end pile.

It confirms the correctness of the choice of the model. In such a case, the relationship between the quantities under consideration is very high:

1) $\eta = 0{,}985$,
2) $\eta = 1$,
3) $\eta = 0{,}986$.

Accordingly, the correlation indexes in all cases are practically equal to one.

Concretization of the function (4):

1) $t = 2{,}141\sqrt{F} - \frac{1}{F-1355} + \begin{cases} (-3)\ as\ F \in [12;\ 130), \\ 2,5\ as\ F \in [130;\ 240)\,, \\ 12\ as\ F \in [240;\ 800)\,, \\ 0\ as\ F \in [800;\ 1355) \end{cases}$

for pile with diaphragm,

2) $t = 2{,}243\sqrt{F} - \frac{1}{F-1249} + \begin{cases} (-3)\ as\ F \in [7;\ 13),\ F \in [1000;\ 1249)\,, \\ 0\ as\ F \in [13;\ 40), \\ 3\ as\ F \in [40;\ 70), \\ 6\ as\ F \in [70;\ 127),\ F \in [600;\ 1000), \\ 10\ as\ P \in [127;\ 240), \\ 9\ as\ F \in [240;\ 320),\ F \in [450;\ 600), \\ 15\ as\ F \in [320;\ 600) \end{cases}$

for the open-end pile,

3) $t = 2{,}152\sqrt{F} - \frac{1}{F-1255} + \begin{cases} (-8)\ as\ F \in [40;\ 120), \\ (-3)\ as\ F \in [120;\ 160), \\ 5\ as\ F \in [160;\ 180),\ F \in [1000;\ 1255), \\ 6\ as\ F \in [180;\ 230), \\ 9\ as\ F \in [230;\ 270),\ F \in [800;\ 1150), \\ 13\ as\ F \in [270;\ 300),\ F \in [500;\ 800) \\ 16\ as\ F \in [300;\ 500) \end{cases}$

for the closed end pile.

Comparison of the experimental values and calculated graphs for the considered cases is presented below (Figs. 18, 19 and 20).

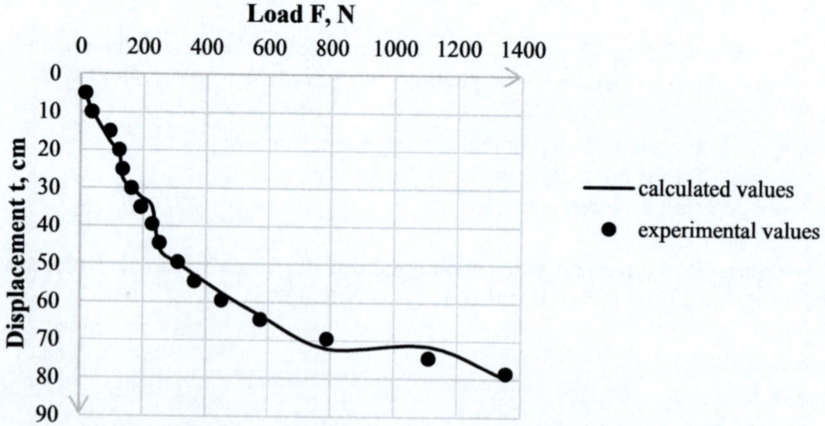

Fig. 18. Comparison of the experimental and calculated by formula (5) values for pile with diaphragm at 9d distance from the pile tip.

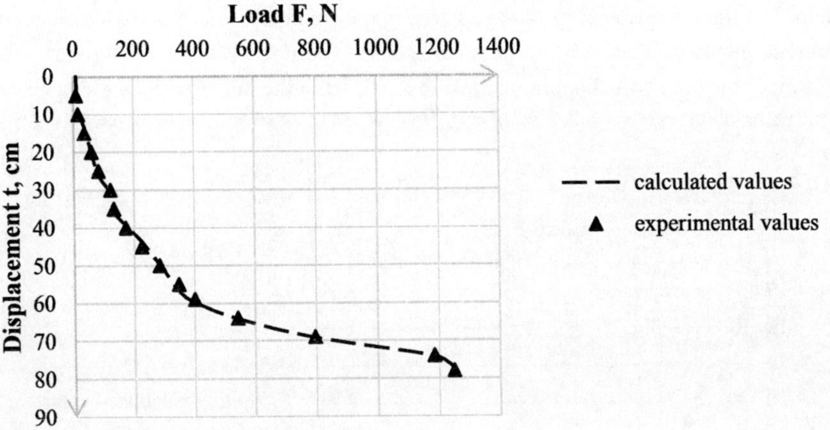

Fig. 19. Comparison of the experimental and calculated by formula (5) values for the open-end pile.

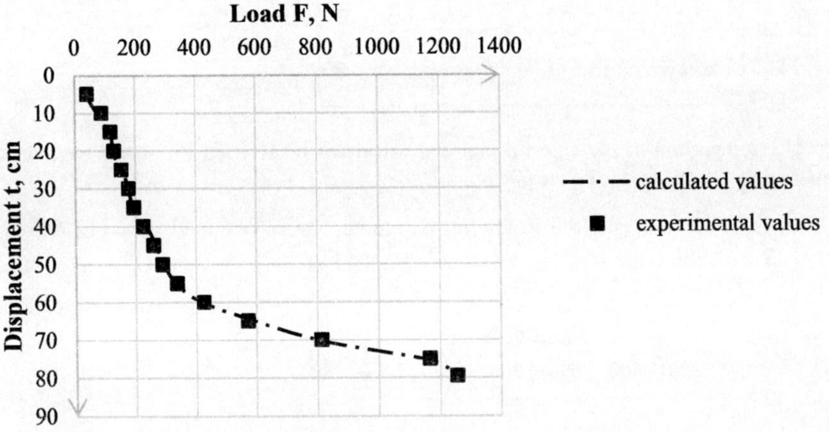

Fig. 20. Comparison of the experimental and calculated values for the closed end pile.

For the further development of the model, it will be useful to refine the coefficients a, c or to assume instead of them some functions depending on F, i.e., to look for a function in the form

$$t = a(F) \cdot \sqrt{F} - \frac{1}{F - b} + c(F) \tag{5}$$

In particular, for some considered experimental cases it has already been possible to obtain an approximation in the form:

$$t = a\sqrt{F} - \frac{1}{F - F_n - 1} + 3\sin(F - \overline{F}) + \begin{cases} \frac{\overline{t}}{\sqrt{n}} \ as \ F \in \left[\overline{F} - \frac{F_n - F_1}{k\sqrt{n}}; \overline{F} + \frac{F_n - F_1}{m\sqrt{n}}\right], \\ 0 \ as \ F \in [F_1; F_n + 1) \backslash \left[\overline{F} - \frac{F_n - F_1}{k\sqrt{n}}; \overline{F} + \frac{F_n - F_1}{m\sqrt{n}}\right] \end{cases} \tag{6}$$

where $\overline{F}$ is the sample average of the applied (experimental) loads, $\overline{t}$ is the average of the observed displacements, $a = \frac{\overline{t}}{\sqrt{\overline{F}}}$, k, m are some experimental parameters (constants).

So, for the pipe with diaphragm at 9d distance from the pile tip the following parameters were obtained: $\overline{A} = 8{,}2\%$, $\eta = 0{,}957$, $R = 0{,}993$ (at $k = 2, m = 1/\sqrt{2}$); see Fig. 21.

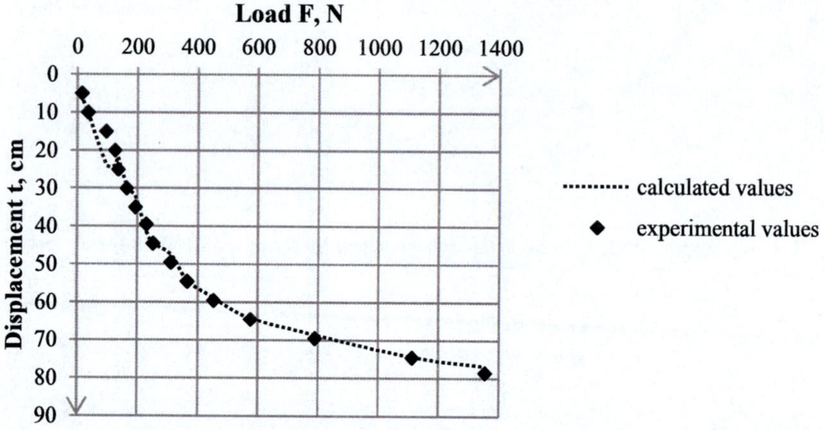

Fig. 21. Comparison of the experimental and calculated by formula (6) values for pile with diaphragm at 9d distance from the pile tip.

For the open-end pile the following parameters were obtained: $\overline{A} = 12{,}6\%$, $\eta = 0{,}997$, $R = 0{,}988$ (at $k = \sqrt{3/2}$, $m = \sqrt{2/3}$); see Fig. 22.

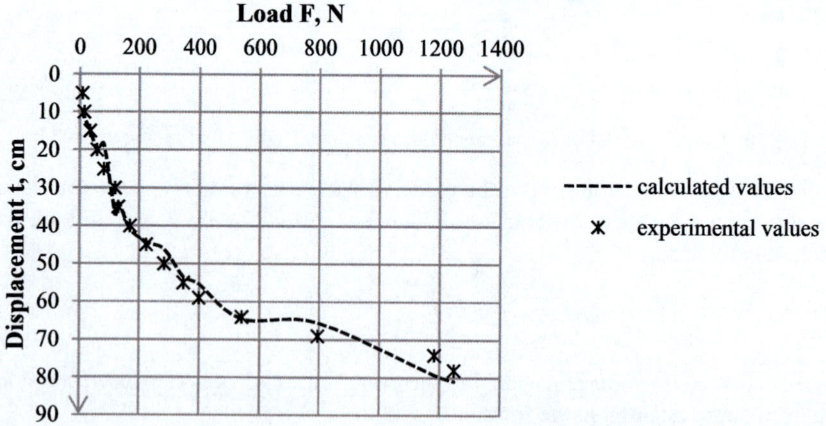

Fig. 22. Comparison of the experimental and calculated by formula (6) values for the open-end pile.

It is possible to note that there may be a model with another type of the function $c(F)$, but it is obvious that such a function should include some periodic component.

5 Conclusion

As obtained from the presented initial series of our experimental studies, a rigid diaphragm inside the tubular open-end pile may be a useful element for increasing the pile's bearing capacity.

Consecutive formation of two soil plugs (lower one formed just at the pile tip and then upper one formed under the diaphragm) leading to their partial or full integration is most effective when the optimal location of the diaphragm inside the pile shaft is provided. From the point of view of pile bearing capacity under axial compressive load and for the considered experimental conditions, such proper distance between the pile tip and internal diaphragm occurred to be around 9d (d – pile diameter).

As demonstrated by our tests, during the pipe jacking process, there is a possibility of clearance space between the diaphragm and soil inside the pile's shaft (no contact situation). For real construction site conditions and inhomogeneous soil base, it is a complex task to check proper diaphragm-soil contact and their interaction or to determine clearance space formation under the diaphragm. That's why it is proposed (and checked by our tests) the technological improvement based on sand filling into space under the internal diaphragm to provide constant diaphragm-soil contact and related soil resistance. Such an approach guarantees force interaction between the internal diaphragm and the soil inside the shaft via filled sand and may simplify the calculation scheme of such interaction (the last is a task of future development of the considered problem).

Improvement of the pile effectiveness determined by experimental modeling for the above-mentioned pile-soil conditions provides 15–20% increase of the bearing capacity or 10–15% reduction of the driving depth.

Numerical analysis of the gained experimental data gave the possibility to apply approximating function with good correlation indexes.

Presented experimental studies should be continued and developed in order to study installation peculiarities of piles with diaphragm by pressing technologies. Also, it looks prospective to investigate influence of internal diaphragm's design (different from the considered flat plate option).

Acknowledgement. The authors would like to express gratitude to colleagues from the Geotechnical Laboratory of the Department "Sea, River Ports and Waterways" at Odessa National Maritime University for their valuable support and assistance in model tests arrangement and fulfillment.

References

1. Doubrovsky, M.P., Dubravina, V.O.: Study of bearing capacity of tubular piles with diaphragm under pressing loads. In: Matsumoto T., et al.: Proceedings of the Second International Conference on Press-in Engineering 2021, Kochi, Japan, pp. 116–123 Taylor & Francis, UK (2021)
2. Doubrovsky, M., Gerashchenko, A., Dobrov, I., Dubrovska, O.: Piled structures for marine transportation facilities: innovative structures and technologies. In: Proceedings of the Second International Conference «Challenges in Geotechnical Engineering 2017», Kiyv, Ukraine, pp. 104–105 (2017)

3. Doubrovsky, M., Gerashchenko, A., Dobrov, I., Dubrovska, O.: Innovative design and technology solutions for development of port and offshore pressed-in piled structures. In: Proceedings of the First International Conference on Press-in Engineering, Kochi, Japan, pp. 91–99 (2018)
4. Tomlinson, M., Woodward, J.: Pile Design and Construction Practice, 5th edn. Taylor & Francis, London (2008)
5. Randolph, M.F., Leong, E.C., Houlsby, G.T.: One-dimensional analysis of soil plugs in pipe piles. Geotechnique **41**(4), 587–598 (1991)
6. White, D.J., Deeks, A.D., Ishihara, Y.: Novel piling: axial and rotary jacking. In: Proceedings of the 11th International Conference on Geotechnical Challenges in Urban Regeneration, London, UK, CD (2010). 24 p.
7. Gudavalli, S.R., Safaqah, O., Seo, H.: Effect of soil plugging on axial capacity of open-ended pipe piles in sands. In: Proceedings of the 18-th International Conference on Soil Mechanics and Geotechnical Engineering, Paris, France, pp. 1487–1490 (2013)
8. Brucy, F., Meunier, J., Nauroy, J.F.: Behaviour of pile plug in sandy soils during and after driving. In: Proceedings of 23rd Annual Offshore Technology Conference, Houston, vol. 1, pp. 145–154 (1991)
9. Paik, K., Salgado, R.: Determination of bearing capacity of open-ended piles in sand. J. Geotech. Geoenviron. Eng. **129**(1), 46–57 (2003)
10. Lehane, B.M., Gavin, K.G.: Discussion of "Determination of bearing capacity of open-ended piles in sand." J. Geotech. Geoenviron. Eng. **130**(6), 656–658 (2004)
11. Florin, V.A.: Fundamentals of Soil Mechanics: Volume 1 General Relationships and State of Stress Caused by Foundation Loads. National Technical Information Service, Moscow (1961). 357 p.
12. Ivanov, P.L.: Soils and Bases of the Hydraulic Structures. Textbook. High School, Moscow (1985). 352 p.
13. Lehmann, E.L., Romano, J.P.: Testing Statistical Hypotheses. Springer, New York (2005). https://doi.org/10.1007/0-387-27605-X. 211 p.
14. Larry, W.: All of Statistics. Springer, New York (2006). https://doi.org/10.1007/978-0-387-21736-9. 458 p.
15. Sobolev, I., Babichenko, S.: Application of the wavelet transform for feature extraction in the analysis of hyperspectral laser-induced fluorescence data. Int. J. Remote Sens. **34**, 7218–7235 (2013)

Features of the Rock Massif in the Influence Zones of Tectonic Disturbances in the Gold Deposits of Kyrgyzstan

G. A. Kadyralieva(✉) and K. Ch. Kozhogulov

Institute of Geomechanics and Development of Subsoil, National Academy of Sciences of the Kyrgyz Republic, Bishkek 720055, Kyrgyzstan
gulzat_7@list.ru

Abstract. Kyrgyzstan is a mountainous country where more than 90% of the surface is occupied by mountains and almost all gold deposits are located in the highlands above 2700 m from sea level. The mountainous terrain, high seismicity and active tectonics form a special discrete anisotropic medium with an initial stress state. Assessment and management of the stability of the sides of quarries, underground workings and dumps in the zones of influence of tectonic disturbances is one of the urgent tasks in the development of high-altitude deposits. This article presents the results of studies of the properties of the rock mass and the determination of the physical and mechanical properties of the main enclosing rocks of deposits by assessing the effect of metamorphism on their stability.

Keywords: Rock mass · Metamorphism · Stability · Properties of rocks

1 Introduction

Kyrgyzstan is rich in minerals, has large deposits of gold, coal, antimony, mercury, uranium, zinc, rare earth metals and nepheline syenites, and the mining industry is one of the priority sectors of economic development.

Both primary and placer deposits were developed on the territory of Kyrgyzstan. To date, about 2,500 indigenous gold deposits have been explored on the territory of Kyrgyzstan. Of these, deposits with ore reserves of more than 30 tons have been allocated for industrial development according to gold reserves [1].

The gold deposits of Kyrgyzstan are mountainous, located in zones of tectonic disturbances, mainly faults and joints. It is revealed that the main host rocks in these zones are metamorphosed rocks of shear, thermal (volcanites) and thermal-shear origin. Active structure-forming elements are tectonic faults and cracks.

The development of such deposits already at the design stage requires the most reliable information about the structure of the massif, the degree of fracturing, and the properties of rocks. Assessment of the stability of an array of upland quarries in the zones of influence of tectonic disturbances is one of the urgent tasks in the development of upland deposits by the open method.

H.-H. Zhu et al. (Eds.): BRWSG 2021, LNCE 230, pp. 155–160, 2022.
https://doi.org/10.1007/978-981-16-9963-4_12

Mining operations in upland quarries are complicated by the relief of the surface and are characterized by complex mining-geological, tectonic and hydrogeological conditions.

Table 1. Reserves of the largest gold deposits of Kyrgyzstan

No.	Gold deposits	Gold reserves, t	Gold content, g/t
1	Kumtor	396,1	4,26
2	Jerui	80,9	7,03
3	Taldybulak	77,7	5,82
4	Chaarat	76,7	3,3
5	Unkurtash	45,0	3,1
6	Bozymchak	43,0	2,0
7	Ishtamberdy	35.6	7,8
8	Andash	32,0	1,2
9	Kichisandyk	27,0	6,6
10	Tohtazan	27,3	2,8

The host rocks are, as a rule, intrusive and effusive rocks subjected to hydrothermal-metasomatic changes.

Rocks are strongly fractured, and the intensity of fracturing increases at the contacts of rock varieties and tectonic disturbances. As an example, Fig. 1 shows a tectonic map of the Kuran-Jailoo deposit.

The deposit is located in the zone of influence of both disjunctive and plicative tectonic disturbances, in the hanging side of the Oktorkoysky right upswing.

The ore-containing structure is a series of separation cracks, feathering the Oktorkoi fault. The main host rocks of the deposit are siltstones, fine-grained sandstones and conglomerates. The rock mass is strongly cracked. Up to 45 differently oriented cracks were detected on one linear meter. Characteristic violations of the rock mass of rocks in outcrops are shown in Fig. 2.

The geological environment has a number of properties that are directly related to the safety and safety of the operation of upland quarries. The geological environment, in turn, is viewed from an evolutionary point of view as a complex self-organizing nonlinear system in which physical processes in the earth's crust are of decisive importance.

The geological environment includes such elements as terrain, rock mass, rocks and their properties, groundwater, permafrost, natural gravitational processes, and during technogenic activity it experiences an impact that significantly changes the properties of its individual elements, the speed and direction of the processes occurring in it.

The main factors of the geological environment that can significantly affect the stable condition of the sides of the quarry and the dump during the development of upland deposits, the stability of the instrument array and the safety of mining operations include:

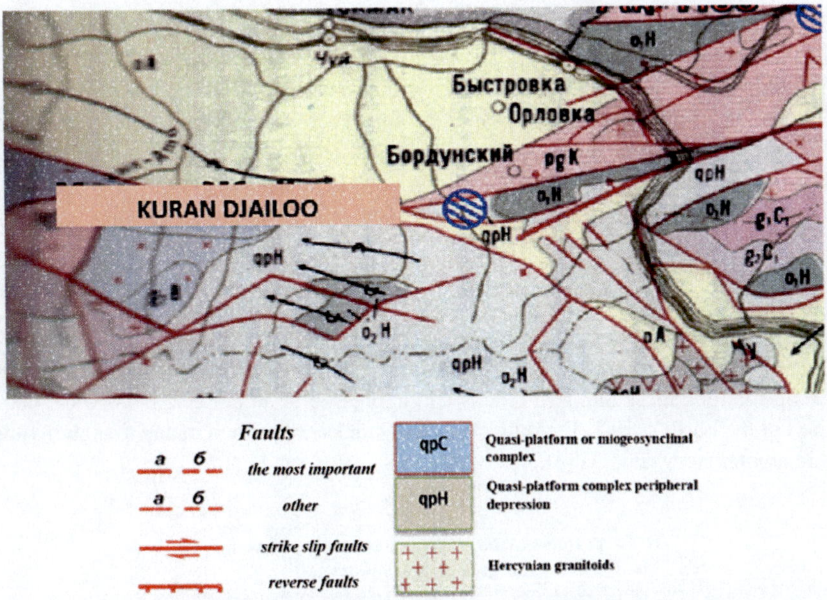

Fig. 1. Tectonic map of the Kuran-Jailoo gold deposit

A - Depth of selection H=75-78м, heavily crushed sandstones

B - Quartic sandstones in crushing zones

Fig. 2. General view of core rocks from various wells «Kuranjailoo» deposit

- Type of rocks, soils and soils, their structure, texture, physico-mechanical and geochemical properties.
- Tectonic disturbance and heterogeneity of the rock mass.
- Parameters of hydrogeological properties and processes
- Endogenous and exogenous processes, thermal field.
- Features of geological structures.
- Relief of the Earth's surface.
- Geodynamic activity, seismicity.
- Geophysical fields.

Active structure-forming elements of the geological environment are tectonic faults and cracks, which represent a zone of disruption of the continuity of the Earth's crust, dividing the rock mass into two blocks and expressed in a complex and diverse deformation of the Earth's crust. Tectonic faults and cracks are present in any mountain range and in any territory (Fig. 3) [3].

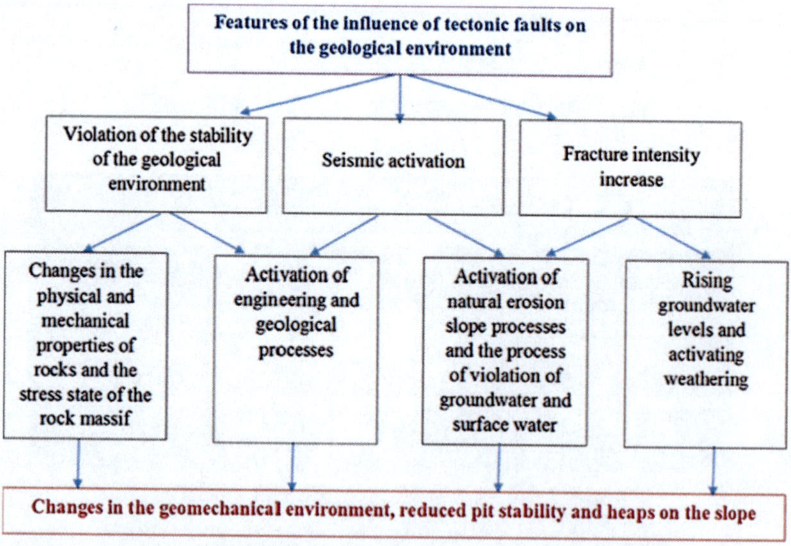

Fig. 3. The impact of faults on the geological environment and changes in the geomechanical environment

The gold deposits of Kyrgyzstan are confined to large regional and regional faults, various discontinuous faults, discharges and thrusts. The host rocks in these zones are represented by crushed metasomatites, metamorphosed shales, tectonic breccias, quartz metosamotites, quartz-carbonate metosamotites, strongly fractured metasomatites, represented by quartz tourmalines, quartz carbonates, quartz sericites, gneiss, amphibole shales, quartz sandstones, diorites and others.

The metamorphism of rocks is understood as significant changes in their mineralogical composition, structure and texture that occur under the influence of internal endogenous processes of the earth's crust with the preservation of the solid state of the rock.

As a result of metamorphism, igneous and sedimentary rocks change their physical and mechanical properties; the degree of these changes depends on the intensity and nature of metamorphic processes. It is customary to distinguish several types of metamorphism: dislocation, thermal and dynamothermal. Each of them has its own typical rocks with characteristic structural and textural features and physico-mechanical properties (Fig. 4).

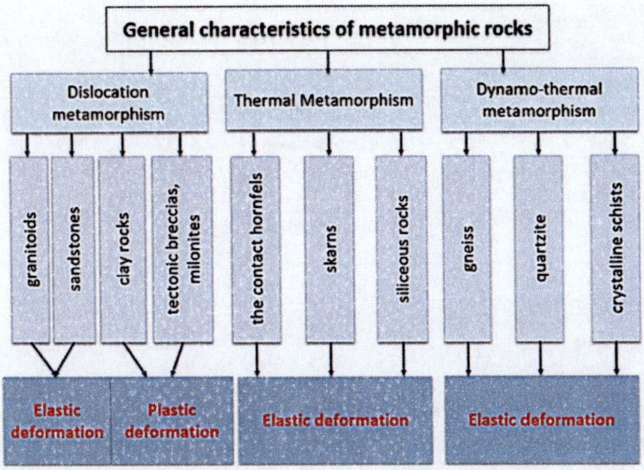

Fig. 4. Types of metamorphic rocks and their general characteristics

Based on the analysis of the obtained results of the properties of rocks of various genesis and long-term studies, we compiled a table of the values of the physical and mechanical properties of rocks lying mainly in the zones of influence of tectonic disturbances. This table shows the main properties necessary for calculating the stability of the sides and slopes of the quarry (Table 1).

Thus, the features of the properties of the main host rocks in the zones of influence of tectonic faults include:

1. strong fracturing and shale
2. low porosity on average, which is 0.37–7.14% and low water absorption, which does not exceed 0.12–4.70%
3. high compressive strength values and low tensile strength values;
4. the fragility of rocks which averages 24–25.

Table 2. Average values of physical and mechanical properties of rocks in the zones of influence of tectonic disturbances

Types of metamorphism	Properties / Rocks	Density γ, kg/m3	Water absorption W, %	Porosity P, %	Compressive strength σcs, MPa	Tensile strength σt, MPa	Shear strength τss, MPa	Angle of internal friction ϕ, deg	Cohesion C, MPA
Dislocation	Granitoids	2595	2,17	2,40	50,7–292,5	2,02–11,70	24,34–140,4	67	5–29,2
	Sandstones	2575	2,39	4,30	59,0–292,5	2,36–1170	32,44–145,74	66–70	5,0–29,0
	Tectonic breccia	2560	1,98	4,74	68,1–192,5	2,72–7,70	36,87–97,22	67	1,9–19,5
Thermal	The contact hornfels	2705	0,49	8,35	48,1–68,6	9,1–18,3	24,5–31,36	60	2,5–2,7
	Skarns	3150	2,62	9,40	63,5–259,9	7,8–23,4	32,68–125,13	46–56	17,5–34,0
Dynamo-Thermal	Gneiss	2665	2,31	1,80	46,6–167,5	4,5–15,3	25,4–80,9	39–40	4,4–5,8
	Quartz tourmaline	2792	0,46	12,01	75,8–231,6	3,03–9,26	40,61–116,22	67	8,15–23,31
	Crystal schists	2610	0,22	5,42	75,0–155,0	3,0–6,2	36,0–74,4	43	4,2–4,62

2 Conclusions

- The gold deposits of Kyrgyzstan are located in zones of tectonic disturbances of various genesis, and the properties of rocks are characterized by almost all types of metamorphism.
- It has been established that rocks in the zone of shear metamorphism have high values of compressive strength and low values when stretched, which is typical of fragile rocks.
- The properties of rocks in the zones of thermal metamorphism substantially depend on humidity.
- The separation resistance of rock blocks depends on the type and adhesion of the aggregate, while the Substantiated by that use of the indicator of quality of the rock mass of the field for the assessment of the marginal parameters of open pits, to ensure long-term sustainability.

References

1. Nikonorov, V.V., Karaev, Yu.V., Borisov F.I.: Gold of Kyrgyzstan Book 2. Description of Deposits B. "Nasi" (2004)
2. Sergeev, E.M.: Problems of engineering geology in connection with the protection and rational use of the geological environment. Vestn. MSU Ser. 4. Geol. **47**(5), 77–86 (1987)
3. Nikolskaya, O.V., Kadyralieva, G.A.: Features of rock properties in zones of influence of tectonic disturbances. Modern Probl. Cont. Mech. **17**, 136–143 (2013)

Disaster Risk Reduction in Italy: A Case History of a High-Risk Landslide

Alessandro Pasuto[1,2](✉) and Luca Schenato[1,2]

[1] CNR-IRPI, National Research Council-Research Institute for Geo-Hydrological Protection, C.so Stati Uniti, 4, 35127 Padova, Italy
alessandro.pasuto@irpi.cnr.it

[2] Sino-Italian Joint Laboratory on Geological and Hydrological Hazards, No. 9, Block 4, South Renmin Road, Chengdu, China

Abstract. The aim of the paper is to provide a short commentary on the risk reduction approach in Italy with special emphasis on landslides, floods and extreme rainfall events. In dealing with this topic, it's very important to preliminary defined a clear conceptual framework as well as a commonly accepted terminology to adopt in order to avoid misinterpretation and promote a culture of sharing data and experiences among researchers, professionals, stakeholders and any other actors involved in decision making. Therefore, the most important documents on risk reduction and adaptation strategies so far accepted at the international level will be introduced along with some basic concepts on hazard and risk assessment and management. The Italian system of Civil Protection will be then discussed, providing some data on the national meteorological monitoring system, moreover, an outline of the administrative and management structure which copes with risk situations and emergency procedures will be illustrated. The Italian law provides for risk reduction to be achieved by means of the Civil Protection Plan at the municipality level. This document considers all different types of risk affecting the territory and defines the main actions to be taken in order to reduce their impact on the natural, economic and social environment. One of the first attempts to adopt and test this plan in a very high-risk condition due to landslide hazard was carried out in the north-eastern Italian Alps several years ago; therefore, a short description of such experience will be reported in the final section of the paper.

Keywords: Landslide · Disaster risk reduction · Civil protection plan · Tessina landslide

1 Introduction

"Disaster risk (DR) is the likelihood of loss of life, injury or destruction and damage from a disaster in a given period of time" [1]. Such a simple definition, which is now worldwide accepted, implies a galaxy of keywords and concepts, among which it is quite difficult to see our way. In recent decades extreme natural events such as a rainstorm, windstorms, heat waves, drought etc., significantly increased their impact on anthropic and natural "landscape" due to climate changes. The frequency and intensity of these

H.-H. Zhu et al. (Eds.): BRWSG 2021, LNCE 230, pp. 161–174, 2022.
https://doi.org/10.1007/978-981-16-9963-4_13

phenomena show an unsafe trend inducing serious risks in many areas of the planet. Moreover, the growing interconnectivity and interdependence across socio-economic systems, as well as physical, biological, environmental and social systems, make the conditions even more complex with a consequent increase in vulnerability and risk. For this reason, the scientific community has for some time been questioning about the limits of development and is now elaborating important concepts such as planetary boundaries and tipping points [2]. Climate change, life below water and on land are also among the Sustainable Development Goals that were adopted by UN in 2015 as a universal call to action to protect the planet [3].

The growing attention that climate change and its effects are having on public opinion also derives from a series of important initiatives implemented by the United Nations and other important international organizations that have wanted to bring to the attention of governments the dangerous situation induced by a rapid development devoid of any attention to environmental issues. Starting from the International Decade for Disaster Risk Reduction (DRR) (1990–1999) three milestones can be mentioned that are the outcomes of the UN World Conferences on DRR held in Yokohama in 1995, Kobe in 2005 and Sendai in 2015 [4–6]. These documents solicited the governments to implement adequate policies to face with increased frequency of extreme events and develop adaptation strategies to mitigate their effects.

In this framework, each country tries to promote legislative initiatives that could help the public administrators and stakeholders in managing the risk situations affecting their territory and to find the best solutions to provide effective mitigation measures and a reliable civil protection system to cope with possible disaster risks.

As concerns of natural hazards, Italy can be considered one of the most affected countries in Europe, being its territory almost 80% mountainous and very active from a tectonic point of view. Catastrophic earthquakes frequently occur as well as floods and landslides and this situation is even worse due to the numerousness of cultural heritage sites present both in large historic cities and small villages. For such reasons, the regulatory situation and the structuring of the main offices for planning and emergency management are among the most advanced in Europe. Nevertheless, every year Italy pays a great tribute in terms of victims, damages, social and economic losses to hazard phenomena related to both endogenous and exogenous causes, namely floods, landslides and earthquakes.

The main aim of this contribution is to illustrate the experience carried out in Italy in DRR and emergency management related to natural hazards and the initiatives so far adopted in order to build up a reliable system of civil protection. A case of successful application and test of DRR policies in a high-risk situation induced by a large-scale landslide will then be presented.

2 Basic Concepts

The sharing of knowledge and experiences in DRR is of paramount importance in developing a new culture of prevention and preparedness, and therefore, the use of common terminology as well as the adoption of common basic concepts in describing natural hazards and their effects would be desirable. This issue is even more important if we consider the need to assist the authorities, practitioners and public in DRR efforts.

For such reason, the United Nation Office for Disaster Risk Reduction (UNDRR, formerly UNISDR, UN International Strategy for Disaster Reduction) promoted the publication of guidelines and recommendations to encourage the usage of common DRR concepts and terminology (e.g., [7, 8]), moreover the General Assembly of United Nations of December 1st 2016 transmitted a report of an expert group on indicators and terminology relating to DRR which stressed once again the importance of such issue, integrating the previous documents.

2.1 Terminology and Risk Equation

It can happen quite often while reading the newspaper or listening to TV or radio programs to hear terms like "natural disaster" or "natural catastrophe". Such terms are incorrect, and their usage has to be strongly discouraged. Thus, a key question is the term "natural". Disasters or catastrophes are never "natural" but disasters often follow natural hazards. Therefore, we can have natural hazardous phenomena that can turn into a catastrophe while affecting anthropic elements. In such perspective, urban and socio-economic development planning become strategic to be implemented according to current knowledge about hazards affecting the territory. Hence the importance of disaster risk knowledge based on the systematic collection of data on hazard processes and vulnerability of the potential affected areas.

Considering the DR definition quoted at the beginning, it implies a good understanding of the prevailing hazards and the pattern of population and socio-economic development. Therefore, in term of assessment of DR the role of academics and research is of paramount importance in order to provide exhaustive information and knowledge about the dynamic of natural and socio-economic processes characterizing each country. So, it would be urgent to bridge the gap between academy and decision makers and implement the synergy among all the actors operating in DRR activities with the aim of breaking the link between "hazards", which are generally unavoidable and "disasters" which can be avoidable or mitigated with good planning and policies.

Going into the detail of both terms UNIDRR [7] defines "disaster" as a "serious disruption of the functioning of a community or a society [*omissis*] which exceed the ability of the affected community or society to cope using its own resources" while "risk" is "the combination of the probability of an event and its negative consequences".

In order to develop the best strategies to reduce the impact of such dangerous events a correct quantification of risk have to be carried out by means of the following (apparently) simple equation (Eq. 1):

$$R = H \times E \times V \tag{1}$$

where: H is "hazard", and it is defined as a "dangerous phenomenon… that may cause loss of life, injury and other health impacts, property damage, loss of livelihoods and services, social and economic disruption, or environmental damage" E is "exposure" that is "people, property… or other elements present in hazard zones that are thereby subject to potential losses". V is "vulnerability" which is defined as "the characteristics… of a community, system or asset that make it susceptible to the damaging effects of a hazard".

A correct natural risk assessment that is the first step in DRR process, therefore needs different expertise that is not easy to combine, e.g., earth scientists or engineers in charge of analysing different aspects of hazard phenomena and their impacts; social scientists, to evaluate the direct and indirect consequences on life and activities of people; economists, that should estimate the total losses in terms of direct and indirect costs.

As concern landslides and all related topics (e.g., type, dimension, velocity) it must be stated that a great variety of classifications with different names indicating the same type of movement or different types of movements termed with the same name are still in use worldwide. This can create a confusion, especially in discussing at the international level. The adoption of a common language could favour the exchange of experiences, and a better comprehension of the description of phenomena that occurred in different countries. The most accepted landslide classification in western countries is the one proposed by Cruden e Varnes in 1996 [9] and integrated by Hungr et al. in 2014 [10].

In Asia, there is no accepted classification, and every country tends to use its own methods to describe landslides. At the moment, no attempt to solve such a problem by adopting a common classification proposed by an international organization has been done, but this would be more than welcome to facilitate the transfer of knowledge worldwide.

2.2 Milestones in DRR

As stated before, several initiatives have taken place since the nineties to promote a culture of DRR worldwide and to stimulate the governments to take actions in order to reduce the losses and damages due to natural hazards and climatic changes. It has been a long way marked by several important steps (Fig. 1).

In such a perspective the United Nations organized three World Conferences with the purpose of providing the Member States with concrete actions to protect development gains from the risk of disaster. Each conference was preceded by a preparatory agenda characterized by stakeholder consultations and inter-governmental negotiations to prearrange the main outcomes, which now represent the pillars on which the national policies and platforms on DRR should be based.

Fig. 1. Major steps to Sendai framework for DRR. Squared in red the 3 main outcomes of the UN World Conferences on DRR which represent the pillars at the base of the current strategy.

Yokohama Strategy. It was the first attempt to delineate a common strategy at global level to prevent disasters and take actions to be prepared and mitigate their impact. Yokohama conference took place in the mid of the International Decade for Natural Disaster Reduction (IDNDR) with the aim of reviewing "the accomplishments of the

Decade at national, regional and international levels" [4] and boost for the adoption of concrete actions to be implemented with particular emphasis to developing countries.

The first part describes the principles on which a disaster reduction strategy should be based, then a plan of action agreed upon by all member states of the United Nations is reported. The last section illustrates some guidelines concerning the follow-up of actions.

The used terminology is still not entirely appropriate, but a very important concept is introduced that is "living with risk" to press the civil society to recognize and strengthen traditional methods and explore new ways to prevent and reduce the effects of natural hazards.

Hyogo Framework for Action (HFA). In 2005 the World Conference on Disaster Reduction held in Kobe, adopted this outcome "to promote a strategic and systematic approach to reducing vulnerabilities and risks to hazards" [5]. The HFA was built starting from a review of the implementation of the Yokohama Strategy and the awareness that a more effective involvement of people in DRR was needed. Specific gaps have been identified and, on the basis of deliberations of the World Conference, the following five priorities for action have been adopted:

- ensure that DRR is a national and local priority with a strong institutional basis for implementation;
- identify, assess and monitor disaster risks and enhance early warning;
- use knowledge, innovation and education to build a culture of safety and resilience at all levels;
- reduce the underlying risk factors;
- strengthen disaster preparedness for effective response at all levels.

The innovative approach of involving many different "actors" in DRR is clearly contained in these priorities: end-users, stakeholders, public administrations, professionals, research and education institutions, volunteers etc. must operate all together to substantially reduce disaster losses by means of prevention, preparedness and emergency management.

Sendai Framework for DRR (SFDRR). As a follow-up of the HFA the Sendai Framework of DRR represents the main outcome of the 3rd World Conference on DRR held in Japan in 2015; in a time span of 15 years, this document "aims to guide the multi-hazard management of disaster risk" [6] reducing the impacts and consequent losses in all sectors (e.g., economic, physical, social, cultural, environmental). With respect of HFA the Sendai Framework introduced an important perspective shift: from managing events to managing processes creating risk [11]. This means to prevent or reduce the risk occurrence strengthening the resilience of the affected communities instead of focusing on preparedness and response.

The SFDRR outlines 7 global targets to guide the progress in substantially reducing existing disaster risk through the implementation of integrated countermeasures (Fig. 2).

In order to achieve the seven Global Targets, SFDRR defines four priority areas where special efforts and resources must be allocated at both national and local levels,

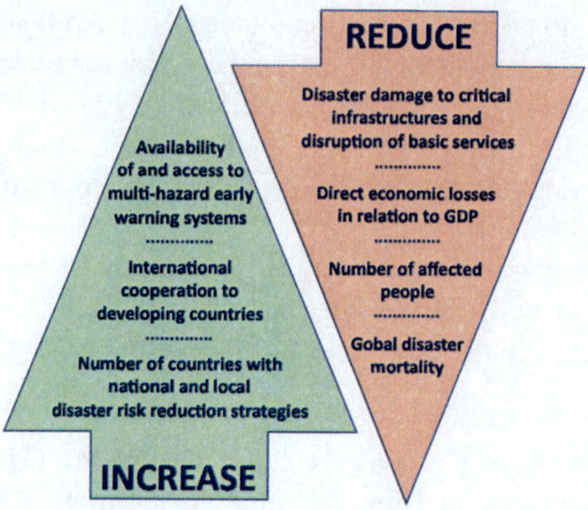

Fig. 2. Seven global targets outlined by SFDRR are to be achieved by 2030.

thus emphasizing the role of national platforms for DRR and the strong involvement of the local communities.

The four priority areas for action are:

- Understanding disaster risk. To better manage the disaster risk, the comprehension of the hazardous processes is mandatory. Therefore a closer connection between policy and decision makers and the so-called "knowledge providers" that are scientists, engineers, economists, sociologists etc., who investigate different aspects of hazard, is needed.
- Strengthening disaster risk governance to manage disaster risk. The governance is a key issue at all levels. It is important to clearly define roles and responsibilities, to enhance the coordination among involved organizations and administrations, to adopt effective and reliable strategies and plans, to foster collaboration and partnership among institutions, private sectors, civil society, including volunteers.
- Investing in disaster risk reduction for resilience. This is a sort of "call for investments"; the public and private have to allocate resources to reach the common goal of reducing losses due to natural hazards and protecting strategic infrastructures and assets. But it is not a mere economic issue. Further strengthening and reinforcements of building codes, a culture of maintenance services and infrastructures, protection of cultural sites, improvement of resilience of workplaces and health system, adoption of non-structural countermeasure etc., all these actions represent a fruitful investment for our future.
- Enhancing disaster preparedness for effective response and to "Build Back Better" in recovery, rehabilitation and reconstruction. This area puts stress on the concept of "preparing for recovery, rehabilitation and reconstruction before the disaster happens" [11], moreover it underlines the importance of the multi-hazard and multi-sectoral approach as well as the concept of "cascading disasters" that is, a chain of sequential hazard processes which magnitude and, therefore, their negative effects are amplified by their interconnection both in time and space. Planning can be seen as a keyword

in this priority area. Therefore the integration of DRR into development plans and measures can help the governments in making their communities safer.

For each of the priority areas just described, SFDRR lists a series of actions to be taken at global, regional, national and local levels in order to achieve the main objective of the area.

All these initiatives stimulated the various countries to invest and improve their own organizational structures devoted to DRR, and Italy also made great efforts in building a more reliable and effective civil protection system that could be able to cope with a growing risk situation related to climate change and fast urban development.

3 Outline of the Italian Civil Protection System

The Italian Civil Protection system is a permanent structure coordinated by Prime Minister involving central and peripheral State administrations, municipalities, provinces, regions, national and territorial bodies. It is a quite complex system and one of the most effective and well-oiled at European level. This is obviously due to the wide range of hazards affecting the national territory (e.g., earthquakes and tsunamis, volcanic eruptions, extreme meteorological events, floods and landslides, wildfires, chemical and industrial accidents). This chapter is intended to give a short outline on the Italian experience not claiming to be exhaustive, and treat all the aspects of such a complex issue.

Figure 3 shows the main "actors" of the system as well as an outline of the activities they are asked to do. Its functioning is guarantee by a huge number (more than 1.5 million) of trained volunteers that are grouped into more than 5000 associations scattered throughout the national territory. Therefore, each municipality has its own group of volunteers that can operate locally or, in case of large-scale catastrophes, on the entire territory and abroad.

Therefore, the system is strictly related to the administrative organization of Italy that is composed, from largest to smallest unit, by 20 regions, 107 provinces (in recent years these administrative units are subjected to a significant reorganization process) and 7915 municipalities; the mayor is the first authority of civil protection.

The scientific and academic community is strongly involved in supporting activities to decision makers and consultancy at each administrative level. A network of Centre of Competence, represented by universities, research institutions, national agencies and administrations and River Basin Authorities provides services, data, technical and scientific contributes to the National Civil Protection Department (NCPD) and regional governments in case of disaster occurrence as well as operative support in emergency and post-emergency phases.

Moreover, the Major Risk National Committee, composed of outstanding scientists, operates at central level with the aim of providing technical and scientific advice to the Head of NCPD on a different type of risks, i.e., seismic, volcanic, geo-hydrological, chemical, industrial, environmental etc.

The NCPD coordinates the emergency activities along with 21 Regional Functional Centers (one per region apart Trentino-Alto Adige, which has two), which are focused on forecasting, monitoring and surveillance of weather phenomena in real-time and managing the alerting system located in their own territory. Each center has to collect and share data gathered from a different technological platforms (e.g., ground-based, remotely sensed, satellite) and elaborate probabilistic scenarios on the base of which emitting warning bulletins to alert the exposed population.

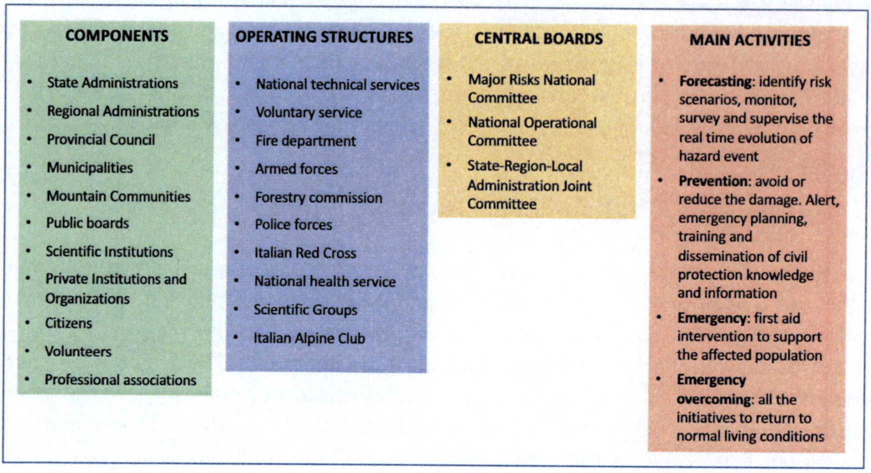

Fig. 3. Main structure and activities of the Italian civil protection system.

As for the geo-hydrological hazards mainly induced by rainfall (e.g., floods and landslides), the forecasting activity is based on the National Alerting System constituted by 23 meteorological radars, more than 4500 monitoring stations (Fig. 4a, b) equipped with different sensors to measure hydro-meteorological parameters. Roughly 1200 hydrometers, 3000 rain gauges and 3500 sensors measuring different weather variables (e.g., wind, temperature, snow depth, solar radiation) are active to integrate the data provided

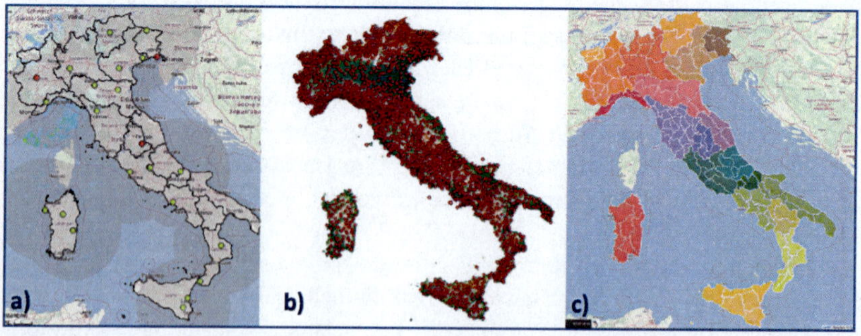

Fig. 4. National Alerting System is based on: a) meteorological radar network for rainfall estimation (https://radar.protezionecivile.it); b) spatial distribution of rain gauges network (from Peruccacci et al. [12]); c) 134 hazard zones on the base of which the alert bulletins are issued (www.polaris.irpi.cnr.it).

by satellites, radars and other remote sensing platforms. In addition, several local systems monitoring specific high-risk phenomena (namely landslides or debris flows) provide useful data at a regional level.

In order to better manage the forecasting and prevention activities, the national territory has been subdivided into 134 "hazard zones" (Fig. 4c), which represent homogeneous areas as concern type, intensity and effects of the expected meteo-hydrological events. On such basis, the Central Functional Centre (CFC) issues a daily national warning bulletin which provides information on possible critical situations as concern landslides and floods.

Whenever severe damages to population, infrastructures and property are forecasted, the regional government, through the Regional Functional Centre, issues a warning signal (yellow, orange or red with respect to the severeness of the expected effects). In the meantime, the CFC continues to monitor the meteorological situation and its impact on the territory. Depending on the level of warning, the region and/or municipality involved decides which type of operational phase to activate (warning, pre-alarm, alarm) and takes actions in accordance with the civil protection plan (CPP). The mayor is in charge to inform the population and activate evacuation procedures as well as declaring the "all clear" allowing the people to get back to normality.

4 DRR Main Tool: Municipality Civil Protection Plan

In May 1998, following extreme rainfall some very rapid earth/mudflows, mainly involving volcanic aches coverage, took place destroying several villages and claiming more than 150 lives in Campania (southern Italy). This shocking event induces Italian government to adopt new legislative tools to guarantee accurate land and urban planning, which carefully recognize and zone the different types of risk and implement reliable measures to mitigate their impact on the territory and manage the emergencies which might occur. The smallest administrative units selected to implement the new strategy have been the municipalities that have been asked to adopt the CPP that is an instrument that allows the public authorities to prepare and coordinate the relief interventions to protect the population and the heritage in a risk area.

The Plan consists of three fundamental parts [13]:

- General part: it includes all the information on the features and structure of the territory, including a detailed analysis of hazards;
- Lines of planning: they establish the objectives to achieve in order to provide a suitable civil protection response to any emergency situation, and the skills of the various operators;
- Intervention model: assigns the decision-making responsibilities to the various levels of command and control, uses the resources in a rational manner, defines a communication system that allows a constant exchange of information.

The Plan is aimed to:

- assign the responsibility to involved organisations and individuals to perform specific actions, planned in time and in places, in case of an emergency which exceeds the capacity of response or the competence of a single organisation;
- describe how these actions and the relationships among organisations are coordinated;
- describe how to protect the people and property in emergency and disaster situations;
- identify personnel, equipment, skills, funds and other resources available to be used during the response operations;
- identify the initiatives to be activated in order to improve the living conditions of any evacuated people.

Great emphasis has been done on the recognition of the main hazards affecting the territory and the development of different risk scenarios related to the magnitude of the events that might occur. This represents the fundamental information on which a reliable CPP has to be based.

The plan includes a set of emergency operational maps among which risk maps related to the different types of hazard, map of vulnerable strategic elements and infrastructures (e.g., hospitals, schools, public offices, roads), logistic map to facilitate the rescue operations and the evacuation procedures, map of the safe areas to mount tent camps etc. Moreover, CPP is linked with the municipal register of births and all the people who need any kind of assistance to be evacuated or moved are reported. It follows that the CPP needs to be continuously updated not only as a concern of the risk affecting the territory but also in terms of resident people, strategic facilities and infrastructures, tools and equipment and so on.

Once the plan is adopted, it must be periodically tested through public drills involving general public, volunteers, decision makers and all the actors to be involved in coping with a possible catastrophic event. This is a crucial issue since awareness and preparedness are key factors to reduce vulnerability and the potential losses. Periodical drills are useful to test protocols and procedures and, in case, revise them; at the same time, they are useful to keep the people prepared to promptly react in case of emergency.

5 A Case History

The approach briefly introduced before has been developed and tested in case of a high-risk landslide affecting a couple of villages in eastern Italian Alps: namely the Tessina landslide (Fig. 5). This event therefore represented a sort of laboratory where almost thirty years ago the modern approach to DRR has been applied.

The Tessina landslide, which was first triggered in October 1960, is a complex movement with a source area characterized by rotational and translational slides turning downhill into a mudflow through a narrow steep channel. In April 1992 a relevant reactivation took place involving almost 1 million cubic meters of material forming a thick mudflow which skimmed the village of Funes and stretched downhill as far as the village of Lamosano 2.5 km far from the source area and completely filling the valley which was some 50 m deep. The serious possibility of a sudden collapse and a consequent mud overflow involving the two villages induced the public authorities to take urgent actions to protect houses and resident people.

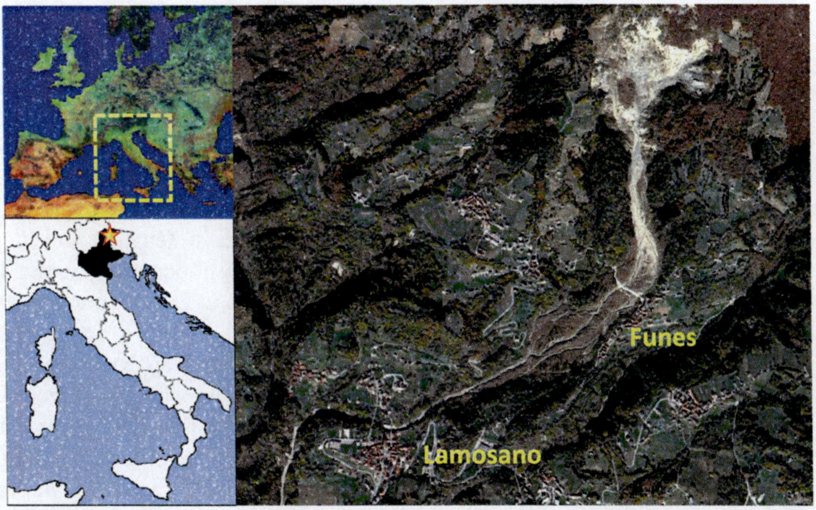

Fig. 5. Geographical location of the Tessina Landslide in northeaster Italian Alps (photo from Google Earth).

Following this event, a monitoring and alarm system was designed and installed, along with some countermeasures to temporary protect urban settlements. At the same time the complex mechanism of civil protection was activated with the involvement of all actors at national, regional and local levels as well as volunteers, army and a scientific group to support the decision makers on the basis of data gathered by a monitoring system and different risk scenarios derived from numerical modelling. In this framework three different hypothesis were formulated i.e., a further detachment of 0.5, 1 and 5 million cubic meters of material. On such base maps of the elements at risk in both the villages

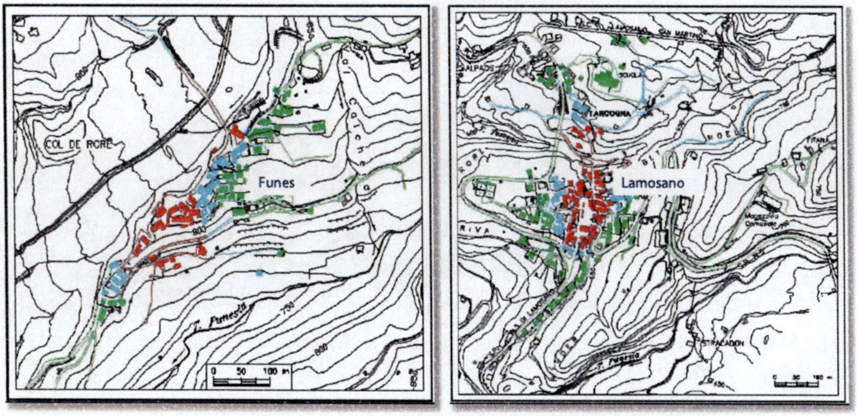

Fig. 6. Elements at risk (red: high; cyan: moderate; green: low) related to the worse scenario derived from numerical modelling.

were done (Fig. 6) thus allowing the definition of the number of people to be evacuated and the structures to be protected.

The first action of the mayor, who was the main authority of civil protection, is the opening of the Municipality Operational Centre that is composed by local people supporting the mayor's decision and coordinating the activities of the various sectors involved in the emergency (e.g., technical, scientific, health, logistic, services, telecommunication). At the same time, he appoints the Safety Commission, which plans all the activities to be carried out and, in case of need, trigger the evacuation procedures and Surveillance Commission that is in charge of manage the monitoring system and organize the control activities, alerting the interested bodies in case of flow velocity increase.

Following the Civil Protection Plan some safe areas where to arrange the tent camps and safe buildings to host elder and in need of care people were identified. Once activated all the emergency structures at municipality level with the support of regional and national civil protection departments, the mayor, on the base of three different evolutionary scenarios, ordered the evacuation of the population at risk and the activation of alarm procedures. These three scenarios, defined with the support of the involved scientists and based on the data provided by the monitoring system, were the follows:

Quasi-static evolution: in this case Surveillance Commission keeps in touch with the Safety Commission and organizes activities checking up on the efficiency of the monitoring system and validate the acquired data. All the interested bodies are informed.

Slow dynamic evolution: The checkup frequency increases as well as the contact with the Safety Commission. The Surveillance Commission is authorized to activate alarm procedures in case of need.

Fast dynamic evolution: on the base of information provide by monitoring system and Surveillance Commission a complex procedure which can lead to evacuation is activated. The Fig. 7 illustrates the main steps of such procedure that represents the final phase of the emergency.

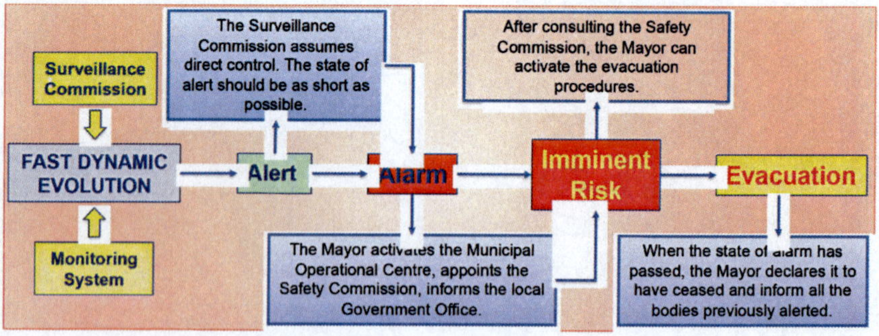

Fig. 7. Main phases of the evacuation procedure to be activated in case of significant reactivation of the Tessina Landslide.

Fortunately, the emergency phase related to the relevant reactivation of the Tessina Landslide in 1992 was over in a couple of months but such experience was very useful to better set up the whole system and give valuable indications at all levels to improve the DRR policies. The monitoring system activated at the time is still active and in the last 30 years has been improved and updated several times. It represented an open-air laboratory for many research groups that were able to test innovative monitoring instrumentation and techniques validating their performance.

The Civil Protection plan and all the related activities were tested several times afterward with contribution of the local people and all the bodies that have lent themselves with great willingness to join the drills to make the community more resilient.

6 Final Remarks

The DRR assumed in recent years a strategic role in making the people more aware of the risks they can experience and, at the same time, more prepared to face them. As stressed during the UNDRR World Conferences, we must invest in knowledge and education because people aware are generally less vulnerable. Therefore the focus shifted from disaster management to "integrated and anticipatory disaster risk management based on trend and losses" [11], promoting research activities and information sharing to maximize the understanding of natural and anthropic processes that can turn into disasters. This represents a fundamental change of strategy in facing disaster risk, thus recognizing the importance of investments in economic, social, and cultural resilience through structural and non-structural measures.

Following this new perspective, the Sendai Framework for DRR established three main goals to be achieved before 2030: (i) preventing the creation of risk, (ii) reduction of existing risk, and (iii) strengthening of the resilience of people and assets. Therefore the scientific community, practitioners, and public bodies should cooperate to promote a more effective interchange bridging the existing gaps among sectors allowing ideas and innovation to circulate quickly.

A small but significant example representing the Italian approach to DRR, has been provided by the tiny community affected by the Tessina Landslide. During the strong reactivations of 1992 and the following years, the local community has been subjected to a severe threat. Therefore, the implementation of the civil protection plan has been needed to face imminent risk. In that case, the solid intersectoral connection and the population's involvement made it possible to reduce damages and inconveniences significantly. Moreover, a monitoring and alarm system was designed and installed to evaluate the landslide displacements and support the decision makers in taking proper actions to mitigate the risk. Such a system has been improved in the following years and is still operating now, providing a reliable public safety measure to help people live with risk in the last 30 years.

The set up of a robust emergency structure with the contribution of local and national public institutions, scientific organizations, volunteers, and all the operational rescue bodies (e.g., fire brigade, red cross, armed forces) represented a reliable and effective example of how DRR issues can be tackled and solved. This experience was also used to promote a culture of prevention and self-protection that should be considered mandatory in the risk areas.

References

1. UNDRR: Global Assessment Report on Disaster Risk Reduction 2015, Making Development Sustainable: The Future of Disaster Risk Management. United Nations Office for Disaster Risk Reduction (2015). https://www.preventionweb.net/english/hyogo/gar/2015/en/gar-pdf/GAR2015_EN.pdf
2. Folke, C., et al.: Our future in the Anthropocene biosphere. Ambio **50**, 8234–8269 (2021)
3. UN Department of Economic and Social Affair: The 17 Sustainable Development Goals (2015). https://sdgs.un.org/goals
4. Various Authors: Yokohama Strategy and Plan of Action for a Safer World. Guidelines for Natural Disaster Prevention, Preparedness and Mitigation (1994). https://www.ifrc.org/Docs/idrl/I248EN.pdf
5. UNISDR: Hyogo Framework for Action 2005–2015, Building the Resilience of Nations and Communities to Disasters. United Nations International Strategy for Disaster Reduction. https://www.preventionweb.net/files/1037_hyogoframeworkforactionenglish.pdf
6. UNDRR: Sendai Framework for Disaster Risk reduction 2105–2030. United Nations Office for Disaster Risk Reduction (2015). https://www.preventionweb.net/files/43291_sendaiframeworkfordrren.pdf
7. UNISDR: Terminology on Disaster Risk reduction. United Nations International Strategy for Disaster Reduction (2009). https://www.undrr.org/publication/2009-unisdr-terminology-disaster-risk-reduction
8. UNDRR: Hazard Definition and Classification Review, Technical Report. United Nations Office for Disaster Risk Reduction (2020). https://www.undrr.org/publication/hazard-definition-and-classification-review
9. Cruden, D.M., Varnes, D.J.: Landslides types and processes. In: Turner, A.K., Schuster, R.L. (eds.) Landslides: Investigation and Mitigation. Special Report 247. Transportation Research Board, pp. 36–75. National Academy of Sciences, Washington, DC (1996)
10. Hungr, O., Leroueil, S., Picarelli, L.: The Varnes classification of landslide types, an update. Landslides **11**(2), 167–194 (2013). https://doi.org/10.1007/s10346-013-0436-y
11. UNISDR: Reading the Sendai Framework for Disaster Risk Reduction 2015–2030. United Nations International Strategy for Disaster Reduction (2015). https://www.preventionweb.net/files/46694_readingsendaiframeworkfordisasterri.pdf
12. Peruccacci, S., Brunetti, M.T., Gariano, S.L., Melillo, M., Rossi, M., Guzzetti, F.: Rainfall thresholds for possible landslide occurrence in Italy. Geomorphology **290**, 39–57 (2017)
13. Dipartimento della Protezione Civile website. https://servizio-nazionale.protezionecivile.gov.it/en/activities/prevention/emergency-planning. Accessed 23 Sept 2021

Comparison of Overlay Design in Between Lightweight Deflectometer and Benkelman Beam Deflection Test Results: A Case Study in India

Vinod Kumar Adigopula[1(✉)], Chandra Bogireddy[2], and Sunny Deol Guzzarlapudi[3]

[1] Department of Civil Engineering, Madanapalle Institute of Technology and Science, Angallu, Andhra Pradesh, India
vinodkumara@mits.ac.in

[2] Department of Civil Engineering, Vardhaman College of Engineering, Hyderabad, Telangana, India
hari2006chandra@vardhaman.org

[3] Department of Civil Engineering, National Institute of Technology, Raipur, India
sdguzzarlapudi.ce@nitrr.ac.in

Abstract. Strengthening is an essential aspect of the operation, maintenance and management of low volume roads. In India, maintenance of other districts roads and village roads is a very significant task. Benkelman Beam Deflection (BBD) method is laborious and time-consuming because of the state of practice. However, the state-of-art Lightweight Deflectometer (LWD) method was employed to overlay the thickness design of low volume roads (LVR), utilizing a rationale that accounts for the determination of dynamic deflections of thin paved surfaces. This paper aimed to look into the behavior of in-service low-volume roads at a project level and develop a simplified method of designing overlay thickness for low-volume roads using two NDT methods. The results were compared and correlated for low volume roads (LVR). In this paper, responsive stress and strain of thin asphalt pavement beneath layers were analyzed using KENPAVE software. The overlay thickness for different sections ranges from 27.00 mm to 184.39 mm by LWD and BBD. The correlation for deflection values between LWD and BBD was R2 (0.741), while the overlay thickness designed by LWD and BBD showed a high correlation with R2 value of 0.994. This paper also serves to assess the capability of LWD to serve as a substitute to traditional BBD on low volume roads (LVR). The practical limitation of BBD test procedure and analysis may be overcome by conducting more test points using LWD on Indian thin asphalt low volume roads.

Keywords: LVR · LWD · BBD · Overlay thickness

1 Introduction

Lightweight Deflectometer (LWD) was investigated as a tool to aid in determining when to rehabilitate low-volume roads (LVR) by the overlay [1]. LWD directly measures the stiffness of pavement systems and a compacted layer needed for mechanistic pavement

H.-H. Zhu et al. (Eds.): BRWSG 2021, LNCE 230, pp. 175–182, 2022.
https://doi.org/10.1007/978-981-16-9963-4_14

design [2]. LWD has been used over thin, flexible pavements over the past decades [3–18]. India has adopted the Benkelman Beam deflection (BBD) technique for overlay thickness design [19]. Strengthening existing pavement may provide additional thickness in one or more layers over the existing thin asphalt layer [20]. An analytical method of overlay design has a few advantages, such as considering the variation of loading types, which will give more exact and accurate results. The purpose of this study is to make an attempt to analyze the existing pavement structural condition and to calculate the residual life and determination of overlay thickness required based on the pavement deflection measurements using LWD lateral geophones. The deflection bowl was analyzed by Method of Equivalent Thickness (MET), firstly proposed by Sharif and Mustaffa [21], and the results were obtained as elastic modulus in each layer. Determining the pavement structural condition by considering the factors of fatigue and rutting life of a pavement and estimating the residual life of the pavement and the overlay thickness needed were obtained. The results of Resilient Moduli signify in such a way that the lesser the resilient moduli value, the lesser the strength of the pavement. The estimation of residual life for each section was carried out, and the results show that the residual life of pavement sections between 0–1 year needs immediate overlay and the residual life of pavement sections more than 20 years is strong enough with no need for pavement sections overlay.

The LWD and BBD test was performed on a given LVR section, and elastic properties and overlay thickness were correlated. The stress and strain were also estimated of each layer to understand the performance of each layer under dynamic wheel load using KENPAVE software.

2 Methodology

A Test section was selected based on the reconnaissance survey; subsequently, a traffic survey was carried out on the test section for the duration of eight months from June 2011 to March 2012 and the commercial vehicles per day (CVPD) was 398, which are less than 450 CVPD which satisfies the criteria of low volume road to determine overlay thickness [22]. The length of the test section was 550.00 m and was divided into 11 sub-sections, i.e. 50 m each [19].

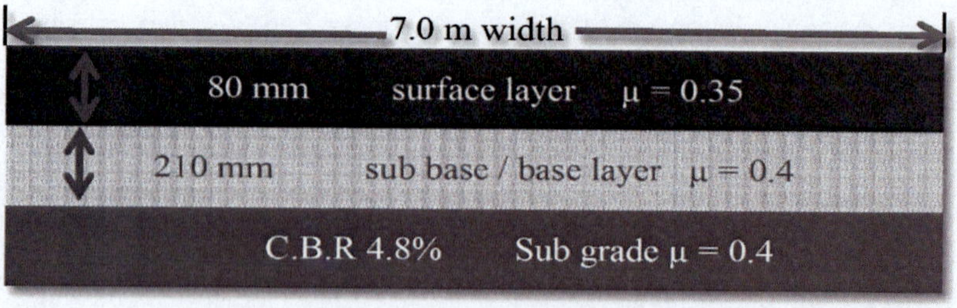

Fig. 1. Crust thickness of the test Sect. 0.0 to 550.00 m

The cross-section details of the pavement, which notifies the thickness, Poisson's ratio of various layers and subgrade CBR, is shown in Fig. 1. Further, a Pavement condition survey was carried out, and it was found that functional distress over existing pavement was more than 40 to 60%. KENLAYER is a sub-program of KENPAVE and it is used to determine the vehicle wheel load responses in terms of stresses, strains, and displacements in flexible pavement structures. Main reason for selecting this software is having an advantage of considering different axle load configurations such as, single, dual, tandem and tridem etc. Eventually, flexibility in selecting different material properties of pavement behavior models like linear elastic, nonlinear elastic and viscoelastic. The primary failure criteria of flexible pavements are fatigue cracking and permanent deformation or rutting. In this paper, based on the estimated stress and strain values, the allowable number of load repetitions was estimated.

2.1 LWD and BBD Experiments

The LWD test was conducted on the selected test location per ASTM standards with a sand pad on the wheel path [2, 23–2]. This LWD study 150 mm diameter plate was considered for the surface course, and 300 mm diameter plate was used over sub-base and subgrade layers. The deflections observed from LWD are recorded in Personal Digital

Fig. 2. **(a).** LWD tests points, **(b).** Testing of LWD, **(c).** BBD test on the study location.

Assistant (PDA) of Trimble make, and a mass of 20 kg produce approximate impact load values of 16 kN, and test procedure was shown in Fig. 2 (a) and (b). Similarly, the BBD test was conducted on the test points of LWD, as shown in Fig. 2 (c).

3 Results and Discussion

Total eleven sections were selected for LWD and BBD test. LVR is 2 lane (7.00 m) two way traffic Road. Tests were conducted on the wheel path of both directions in staggered manner. LWD data was collected at the same location where BBD was conducted at every 50.0 m section. LWD observed the maximum deflection values at Chainage 150.0–200.0 m, and the maximum deflection was observed at 150.00 – 250.00 m by BBD, as shown in comparisons of LWD and BBD deflections is shown in Fig. 3. LWD and BBD data were correlated and was found with R2 (0.740), as shown in Fig. 4. The modulus of sub-grade and sub-base elasticity was estimated, i.e. 48 MPa and 108 MPa, respectively [25].

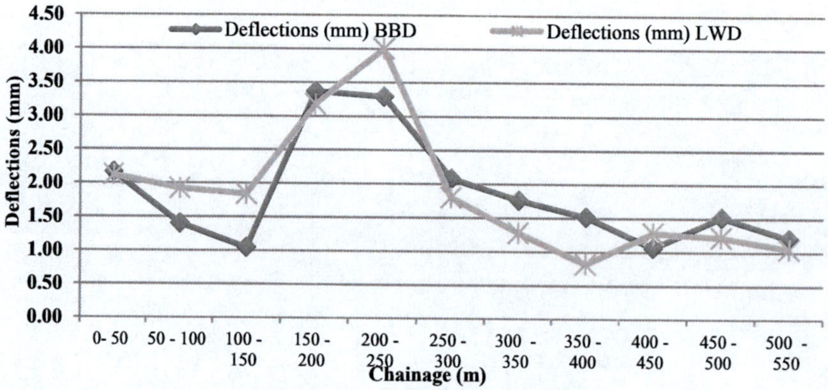

Fig. 3. Deflections by LWD & BBD

The modulus of elasticity for the surface layer was estimated through back-calculation by LWD mod software, as shown in Table 1 [26]. Results illustrate that the higher modulus of elasticity value, the higher the stiffness and structural property of the pavement. The maximum modulus of elasticity value was observed at 400.00–450.00 m indicates that the existing pavement condition is good. Bituminous Macadam was considered for overlay, and hence the thickness of overlay was estimated for distinct surface moduli 500, 550, 600, 650, 680, 700 and 760 MPa [25].

The in-situ moisture content and the plasticity index were 20.8% and 28%. The recorded annual rainfall is more than 1300, and the same was considered while designing the overlay thickness for 0.5 msa, 1.0 msa, 2.0 msa, 5.0 msa, 10.0 msa, 20.0 msa and 100.0 msa. The overlay thickness estimated from BBD analysis for 0.5, 1.0, 2.0, 5.0, 10.0, 20.0 and 100.0 msa [25] and overlay thickness estimated from the LWD analysis for the required elastic modulus for an overlay material of 500, 550, 600, 650, 680,

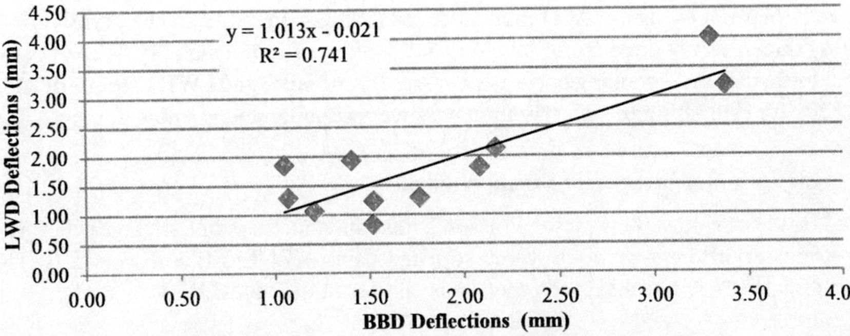

Fig. 4. Correlation between LWD and BBD deflections

Table 1. Modulus of elasticity values.

Sl. no.	Chainage (m)	Modulus of elasticity (MPa) asphalt layer
1	0–50	1381.7
2	50–100	3862.6
3	100–150	4590.0
4	150–200	494.7
5	200–250	2968.4
6	250–300	4621.5
7	300–350	7832.2
8	350–400	7366.0
9	400–450	9611.4
10	450–500	6531.9
11	500–550	3597.5

Table 2. Overlay thickness by BBD and LWD

Sl. no.	BBD (MSA)	LWD (MPa)	Overlay thickness (mm)	
			BBD	LWD
1	0.5	500	24.36	27.00
2	1.0	550	34.43	42.90
3	2.0	600	56.00	67.20
4	5.0	650	87.02	106.30
5	10	680	105.30	131.10
6	20	700	119.19	153.80
7	100	760	171.60	202.10

700 and 760 MPa by using LWD mod software is shown in Table 2. The correlation of overlay thickness obtained from the LWD and BBD was found suitable (R2 = 0.994). Henceforth, the corresponding equivalent values for the BBD and LWD in terms of MSA and MPa for estimating the overlay thickness were summarized in Table 2.

3.1 Estimation of Stress and Strain Analysis

Based on the modulus of elasticity, Poisson's ratio and thickness of each layer, the stress and strains at different sections were estimated using KENPAVE software [27]. This also estimates the allowable load repetitions, as shown in Table 3.

Table 3. Example analysis of structural evaluation at Chainage 400.00–450.00 m.

Chainage (m)	point	Vertical Coordinate (cm)	Vertical Displacement (cm)	Vertical Stain	Vertical Stress (kPa)	Tensile Strain	Tensile Stress (kPa)
400-450	1	0	0.09105	-7.549E -05	549.170	6.773E -05	8136.647
		8	0.09141	8.835E -05	53.959	-9.535E -05	-8472.126
		8.1	0.09138	3.470E -04	53.861	-9.617E -05	21.153
		29	0.08343	4.249E -04	35.411	-2.220E -04	-10.832
		29.1	0.08336	6.126E -04	35.359	-2.218E -04	7.585
	2	0	0.09341	-8.231E -05	0.000	7.257E -05	8793.888
		8	0.09373	9.053E -05	56.490	-1.028E -04	-8402.559
		8.1	0.09369	3.608E -04	56.387	-1.035E -04	24.125
		29	0.08526	4.539E -04	37.354	-2.322E -04	-12.295
		29.1	0.08519	6.519E -04	37.297	-2.321E -04	8.028

The maximum tensile strain and vertical compressive strain values observed on the bitumen layer were −1.028E-04 and 6.519E-04. Based on the tensile strain value, allowable load repetitions to prevent fatigue failure were estimated, i.e. 50.24 msa and based on the vertical compressive strain values, allowable load repetitions for rutting criteria were estimated, i.e. 11.50 msa [25]. NDT and assessment of deflection behavior of flexible pavements for low volume roads concerning the in-situ material properties is a promising procedure for evaluating the structural capacity of pavements.

4 Conclusions

In this paper, the overlay thickness of LVR was studied with conventional BBD and LWD techniques. The study was conducted on a 550 m stretch of 2-lane undivided road of 7.00 m carriageway width. The data was collected at every 50.0 m interval at both sides of the wheel path pavement. The deflections by BBD and LWD were correlated and found in poor relation with R2 (0.741). However, the correlation between overlay thickness estimated by LWD and BBD is good (R2 = 0.994). The elasticity modulus of the surface layer was estimated by LWD mod software.

Further, the equivalency values of MSA of BBD and MPa of LWD is obtained. Analysis of stress and strain analysis was estimated using KENPAVE software. The allowable load repetitions for fatigue and rutting failure criteria were estimated. The allowable load repetitions were estimated for fatigue and rutting criteria based on the estimated layer elastic properties. Finally, it was concluded that more data points are required to get the reliable correlation between LWD and BBD for overlay thickness design. However, this study proves that LWD may substitute conventional BBD in estimating the overlay thickness for low volume roads.

References

1. Owner's Manual: Dynatest 3031 LWD test system. Dynatest (2006)
2. E2583–07: Standard test method for measuring deflections with a light weight deflectometer. Published by American Society of Testing Materials (ASTM) (2007)
3. Amarnath, A., Srinivasamurthy, B.R.: Characterization of polymer stabilized soils for pavements. Indian Highways, 40(3) (2012)
4. Hoffman, M.S.: A direct method for evaluating the structural needs of flexible pavements based on FWD deflections. In: Proceedings of TRB 82nd Annual Meeting, Transportation Research Board, Washington, DC (2003)
5. Kessler, K.: Use of DCP (Dynamic Cone Penetrometer) and LWD (Light Weight Deflectometer) for QC/QA on subgrade and aggregate base. In: GeoHunan International Conference, ASCE (2009)
6. Abaza, K.A.: Performance –based models for Flexible pavement structural overlay design. J. Transp. Eng. **131**(2), 149–159 (2005)
7. Bertulienė, L., Laurinavičius, A.: Research and evaluation of methods for determining deformation modulus of road subgrade and frost blanket course. Baltic J. Road Bridge Eng. **32**(1), 71–76 (2008)
8. Mooney, M.A., Miller, P.K.: Analysis of light weight deflectometer test based on in situ stress and strain response. J. Geotech. Geoenviron. Eng. **135**(2), 199–208 (2009)
9. Saltan, M., Saltan, S., Şahiner, A.: Fuzzy logic modeling of deflection behavior against dynamic loading in flexible pavements. J. Constr. Build. Material. **21**(7), 1406–1414 (2006)
10. Senseney, C.T., Mooney, M.A.: Characterization of a two-layer soil system using a lightweight deflectometer with radial sensor. Transp. Res. Record **2186**(1), 21–28 (2010)
11. Steinert, B.C., Humphrey, D.N., Kestler M.A.: Portable falling weight deflectometers for tracking seasonal stiffness variations in asphalt surfaced roads. In: presented at the 85th Transportation Research Board meeting, National Research Council, CD-ROM, Washington DC, USA (2006)
12. Vinod, K.A.: Overlay Design of Low Volume Road Using Light Weight Deflectometer. M.Tech dissertation, SVNIT, Surat (2012)
13. Guzzarlapudi, S.D., Adigopula, V.K., Kumar, R.: Comparative studies of lightweight deflectometer and Benkelman beam deflectometer in low volume roads. J. Traffic Transp. Eng. **3**(5), 438–447 (2016)

14. Kumar, R., Adigopula, V.K., Jr., Guzzarlapudi, S.D.: Stiffness-based quality control evaluation of modified subgrade soil using lightweight deflectometer. J. Mater. Civil Eng. **29**(9), 04017137 (2017)
15. Kumar, V., Deol, S., Kumar, R.: Structural evaluation of flexible pavement using non-destructive techniques in low volume road. In: Mohammad, L. (eds) Advancement in the Design and Performance of Sustainable Asphalt Pavements. GeoMEast 2017, Sustainable Civil Infrastructures. Springer, Cham (2017). https://doi.org/10.1007/978-3-319-61908-8_13
16. Adigopula, V.K., Bogireddy, C., Kumar, R.: A study on the application of lightweight deflectometer during the construction of low volume road in India. In: Garg, A., Solanki, C.H., Bogireddy, C., Liu, J. (eds.) Proceedings of the 1st Indo-China Research Series in Geotechnical and Geoenvironmental Engineering. LNCE, vol. 123, pp. 113–127. Springer, Singapore (2021). https://doi.org/10.1007/978-981-33-4324-5_8
17. Adigopula, V.K.: A simplified empirical approach for prediction of pavement layer moduli values using lightweight Deflectometer data. Int. J. Pavement Res. Technol. 1–13 (2021)
18. Kumar, R., Adigopula, V.K.: A correlation between LWD Backcalculated moduli with Dynamic Cone Penetrometer test results for subgrade layer. No. 17–04169 2017. Transportation Research Board, Annual meeting (2017)
19. IRC: 81: Guidelines for strengthening of flexible road pavements using Benkelman Beam Deflection Technique (1997)
20. Zhou, L., Wu, Q., Ling, J.: Comparison of FWD and Benkelman beam in evaluation of pavement structure capacity. In: Geo Shanghai 2010, International Conference ACSE Journal (2010)
21. El-Badawy, S.M., Kamel, M.A.: Assessment and improvement of the accuracy of the Odemark transformation method. Int. J. Adv. Eng. Sci. Technol. **5**(2), 105–110 (2011)
22. IRC: SP: 72: Guide lines for design of low volume roads (2007)
23. D4695–03: Standard guide for general pavement deflection measurements. Published by American Society of Testing Materials, (ASTM) (2008)
24. D5858–96: Standard guide for calculating in situ equivalent Elastic Moduli of pavement materials using layered elastic theory. Published by American Society of Testing Materials (ASTM) (2008)
25. IRC: 37: Guide lines for design of Flexible pavements (2001)
26. Lee, Y.H., Ker, H.W., Lin, C.H., Wu, P.H.: Study of back calculated pavement layer moduli from the LTPP database. Tamkang J. Sci. Eng. **13**(2), 145–156 (2010)
27. Huang, Y.H.: Pavement Analysis and Design. Englewood Cliffs, NJ, USA (1993)

Physical and Mechanical Properties of Lime and Wood Ash Plastering Mortars

Ayat Gamal Ashour[1], Wafaa Mohamed Shaban[2], and Khalid Elbaz[3,4](✉)

[1] Department of Civil and Environmental Engineering, College of Engineering, University of Sharjah, Sharjah 27272, UAE

[2] Department of Civil Engineering, Misr Higher Institute of Engineering and Technology, El Mansoura, Egypt

[3] MOE Key Laboratory of Intelligent Manufacturing Technology, Department of Civil Engineering, College of Engineering, Shantou University, Shantou 515063, Guangdong, China
khalid@stu.edu.cn

[4] Department of Civil Engineering, Higher Future Institute of Engineering and Technology, El Mansoura, Egypt

Abstract. Cement industry is one of the causes of increasing carbon dioxide in the world. Using sustainable cementitious materials as a full or partial replacement of cement is become vital to reduce the massive amount of carbon dioxide emission and encourage the use of waste materials. This study aims to build a database for the characteristics of different types of plastering mortars using waste materials, mainly, lime and wood ash as a partial replacement for cement. Four different cement types (CEM I 32.5R, CEM I 42.5R, CEM II B-S 32.5N, and CEM II B-LL 42.5N) were used with two types of lime and wood ash with different proportions were studied. To investigate the different properties of mortar, flow table test, compressive strength, flexural strength, and drying shrinkage strain were carried out. Results showed that the mortars with CEM II B-LL 42.5N have the highest values in flexure and compressive strength. However, CEM II B-S 32.5N showed the least value of dry shrinkage strain. Wood ash decreased the flexure and compressive strength of all mortars except for CEM I 32.5R which had the same value when using lime and cement. Moreover, fly ash increased the drying shrinkage for all mortars while the dry shrinkage value of mortar with CEM I 42.5R decreased.

Keywords: Cement · Plastering mortar · Lime

1 Introduction

Cement is being a vital constituent in the construction industry which is the major binding material in plastering mortar. Cement production is one of the most energy-consuming products in Egypt. It also consumes more than 9% of the available primary energy [1, 2]. In recent years, the government raised the diesel price per ton every year, and because of increasing costs, manufacturers had no other choice but to increase the charge on customers [3]. Consequently, alternative types of cement come up with a technical option to reduce the cost and have the prospective to make a significant contribution towards affordable shelter [4]. In addition, increasing the global production of cement, which

H.-H. Zhu et al. (Eds.): BRWSG 2021, LNCE 230, pp. 183–199, 2022.
https://doi.org/10.1007/978-981-16-9963-4_15

reaches around 4.1 billion metric tons around the world in 2030 which is considered one of the most consumed materials worldwide [5]. This huge amount of fuel energy brings on high carbon dioxide, around 1 ton of carbon dioxide emission [6]. Therefore it is persuasive to partially replace cement. Lime is one of the binding materials which is used from ancient times for over 2000 years for the building's construction. Lime is a white powdery substance produced from heating limestone or similar material in a furnace at 982 °C. Heating converts the carbonate into calcium oxide or quicklime. Hydraulic lime is a variety of lime slaked [7].

Cement-lime mortars set slower and remain more pliable than cement-sand mortars. Hence, lime promotes the brickwork's ability to absorb stresses caused by building movement and periodic changes without too much cracking. Therefore, lime has been used with cement in local practice in Egypt as a partial replacement for cement. Numerous studies have been carried out to study the properties of cement mortar containing lime as partially replace of cement Itim et al. [8] studied a compressive strength and shrinkage of mortar containing different amounts of limestone powder. Results illustrated that using 15% of limestone powder achieved compressive strengths similar to those of ordinary cement at 28 days. Moreover, when it comes to shrinkage at an early age, the autogenous shrinkage increases when the cement is replaced by limestone powder. Limestone powder did not affect the overall shrinkage, however, with the increase in the proportion of limestone powder, the autogenous shrinkage of the mortar increased, and the drying shrinkage reduced with a high replacement rate. Another investigation studies the influence of cement on the properties of lime mortars. This study added CEM I 52.5R to HML-4 (mortar for restoration of the historical building) with a different amount. Results indicated that, as cement content increased, the density of mortar and consequently, the compressive strength increased. However, the increase in cement content decreased the air content [9]. David and Hassan [10] investigated the use of lime and bamboo ash as a partial replacement of cement in the mortar with a mix proportion 1:6. Results indicated that by increasing the percentage of bamboo ash in the mix, the compressive strength increased at 28 days. Kate and Murnal [11] showed that the effect of different ratios of fly ash as a partial replacement of cement to investigate the impact on shrinkage strain of high strength concert. In spite of the rate of shrinkage increase over time being uniform for low amount FA concrete samples, there was a considerable increase in shrinkage for large amount FA concrete samples at 28 days. Furthermore, the large amount of FA concrete displayed a slow rate of growth in strength at an early age of curing. Few studies have investigated the effect of lime and wood ash which is a type of fly ash According to ASTM C 618–19 specification [14].

This study has been performed to evaluate the physical and mechanical properties of some available cement types in Cairo and concentrated to build a database of the characteristics of plastering mortars according to local practice by using lime which is locally produced in Egypt, and wood ash which is available from burned waste wood on the properties of mortar. These materials are utilized as a partial replacement for cement.

2 Materials and Methods

In this study, the mixtures of plastering mortar consist of cement, sand, lime, wood ash, and water. Four types of common and available cement in Cairo (CEM I 32.5R,

CEM I 42.5R, CEM II B-S 32.5N, CEM II B-LL 42.5N) were used in this study. The initial and final setting, Specific gravity, Blain fineness, and Autoclave expansion tests are performed to assess the physical properties of cement types. Iso sand was used with cement according to ES 4756-1/2009 to determine the compressive strength. The granular gradient curve of ISO sand is displayed in Fig. 1. The physical and mechanical properties of cement are illustrated in Tables 1 and 2. Two types of lime were used in this study, type I named by Sultany Hydrated Lime, and type II named by Balady lime both of them were manufactured by HAMCO for building materials factory, which is a local factory in Egypt. The specification of the two types of lime is summarized in Tables 3 and 4. The used wood ash is collected from some bakeries in the countryside that use wood as fuel. The wood ash with specific gravity 3.2 and bulk density 450 kg/m^3. Another type of sand used to produce cement mortar that locally produced in Cairo. This sand was nature silicon sand with specific gravity 2.5, bulk density 1500 kg/cm^2, and passing from sieve size 2.4 mm, grading of this sand is illustrated in Fig. 2.

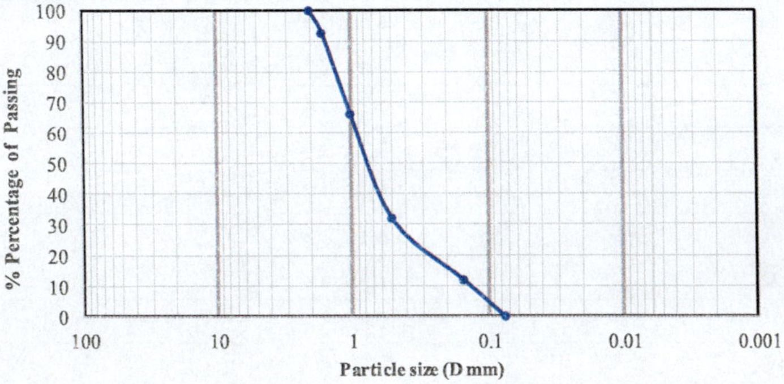

Fig. 1. The granular gradient curve of ISO sand according to ES 4756-1/2009

Table 1. Physical properties of cement types used

Cement type	Physical properties				
	Specific gravity	Fineness (cm^2/gm)	Consistency w/c ratio	Initial/Final setting time (min)	Autoclave expansion (mm)
CEM I 32.5R	3.1	3360	0.26	105/186	0.028
CEM I 42.5R	3.12	3367	0.27	110/190	0.027
CEM II B-LL 42.5N	3	4710	0.31	78/189	0.028
CEM II B-S 32.5N	3.02	3955	0.3	115/250	0.019

Table 2. Mechanical properties of cement types used

Cement type	Compressive strength (N/mm^2)			
	1 day	2 days	7 days	28 days
CEM I 32.5R	8	16.2	18	23
CEM I 42.5R	9.5	12	15	22.16
CEM II B-LL 42.5N	7	10.4	19.7	22.41
CEM II B-S 32.5N	8.5	13.5	21	23.5

Table 3. The specification of lime I (sultany lime)

Tested item	Specification
$Ca(OH)_2$	Min 80%
MgO	Max 0.1%
SO_3	Max 0.50%
CL	Max 0.52%
Fe_2O_3	Max 0.08%
Al_2O_3	Max 0.01%
Defects	Max 2.5%
Particle size	75 μm
Color	White

Table 4. The specification of lime II (balady lime)

Tested item	%CaO	Extinguishment time	%$CaCO_3$	Defects	Particle size	Color
Specification	Min 70%	More than 90 °c in 5 min	max 3%	10:15%	Up to 200 μ	Pure or dark white

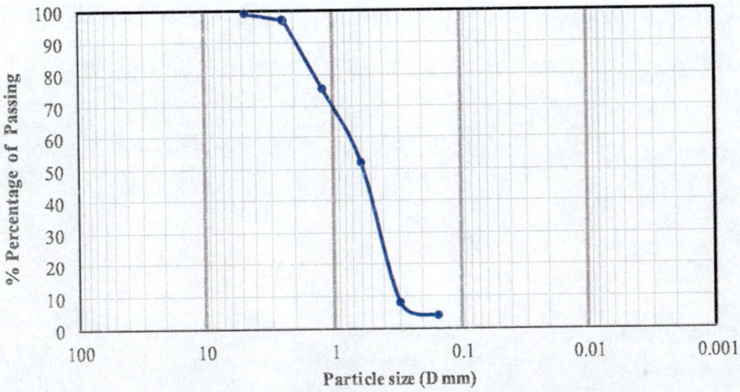

Fig. 2. The granular gradient curve of natural sand

3 Mix Proportions

A total of 16 mixtures are carried out and classified into four groups. Each group contains four mixes with 4 types of cement. The mixtures proportion are prepared according to the local experience.

Group 1: (cement + sand + water) with cement: sand: water ratio (1:4.5:1.56) by volume.

Group 2: ((cement + lime I) + sand + water) with cement: lime: sand ratio (1:1:5.6:1.95) by volume, lime used in this group is type I.

Group 3: ((cement + lime II) + sand + water) with cement: lime: sand ratio (1:1:5.6:1.95) by volume, lime used in this group is type II.

Group 4: ((cement + lime I + wood ash) + sand + water) use wood ash with 15% by weight of cementitious materials according to the percentage of groups 2 and 3.

4 Mortar Testes

4.1 Fresh Mortar Test

Flow table test was carried using two different cones. The first is the standard concrete cone with a diameter of 25 cm with 15 drops. The second is the standard mortar cone with a diameter of 10 cm with 7 drops and 5 cm height, which was manufactured in the lab. A flow table was used to measure the workability of the mortar mixes, as shown in Fig. 3.

4.2 Hardened Mortar Testes

A flexural strength test was performed on a standard prism (40 × 40 × 160 mm) according to ES 2421–7/2006 specification with l/d ratio = 4.0 [15]. 12 prisms were cast and cured in water to be tested at 3, 7, 28, and 56 days by using the compressive and flexural machine with 1000 kN capacity and 3 mm/s loading rate, as shown in Fig. 4. The

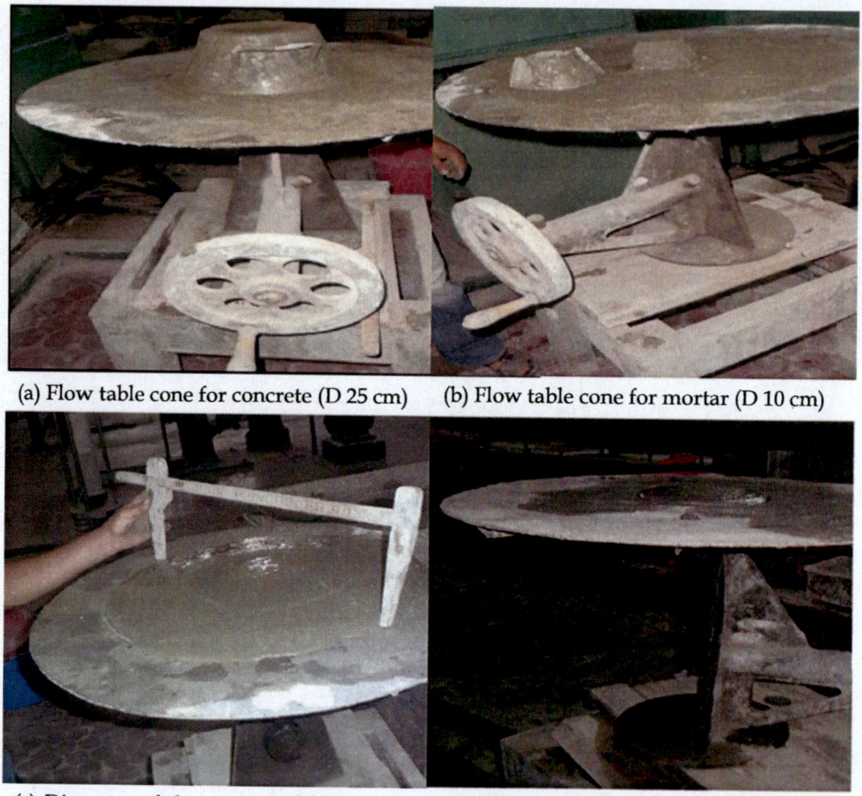

(a) Flow table cone for concrete (D 25 cm)

(b) Flow table cone for mortar (D 10 cm)

(c) Diameter of the mortar resulted from concrete cone

(d) Diameter of the mortar resulted from mortar cone

Fig. 3. Flow table test with two cones

Fig. 4. Flexure strength sample

pretested specimens used in the flexural strength test were broken into two parts. Each part was covered by a steel plate with an overall dimension (40 × 40 × 10 mm) on the top and bottom surfaces and exposed to the compressive strength test, as shown in Fig. 5. Dry shrinkage strain test was measured on the specimens with an overall dimension (25 × 25 × 270 mm) and cast in the mold according to the Egyptian Code specification. Measuring for samples was performed by micrometer which accuracy reached 0.01 mm by every day for the first week, every week for the first month, and every month a year, as shown in Fig. 6.

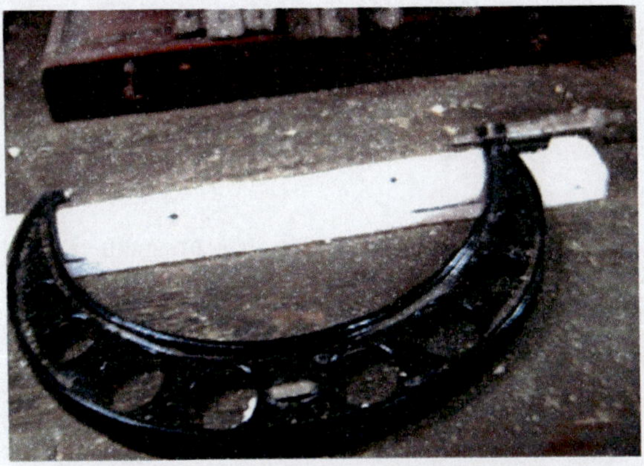

Fig. 5. The length change instrument

Fig. 6. The compressive strength sample

5 Results and Discussion

Flow table test was performed on mortars to determine the percentage of flow on all mixes; cement only, cement plus lime, and cement plus lime plus fly ash. The flow table test was performed by both the standard concrete cone and another one manufactured in the lab for mortar only. Results of the flow table test for mortars are shown in Table 5. The same proportion of the mixture (cement: lime) is (1:1) with lime I; the percentage of flow is higher than the percentage of flow when using lime II. In the case of adding fly ash with a percent equal to 15% by weight of cement and lime to the mixture containing cement plus lime I, the flow of mortar is higher than the flow of mortar without fly ash. In addition, the result of using standard concrete cones has the same trend as the result of the mortar cone. For group 1, which cement is only used as cementitious material with sand and water, CEM I 32.5R has the maximum percentage of flow followed by CEM I 42.5R. Both CEM II types achieve the same result. With the mix proportion in group 2, CEM I 32.5R had the least percentage of flow. However, lime (Type I) increased the flow for CEM II B-LL 42.5R and CEM II B-S 32.5N. Lime (Type II) type had a remarkable effect on decreasing the flow of mortar for all cement types used than lime I. In group 4 when wood ash was used with lime I and cement, wood ash had a little effect on decreasing the flow compared with group 2 when using lime I only with cement.

Table 5. Percentage of flow for all mortars using different cones

Group	Cement	%Flow using (mortar cone)	%Flow using (concrete cone)
Group 1	CEM I 32.5R	148	138
	CEM I 42.5R	125	112
	CEM II B-S 32.5N	105	105
	CEM II B-LL 42.5N	105	99
Group 2	CEM I 32.5R	95	100
	CEM I 42.5R	103	100
	CEM II B-S 32.5N	113	113
	CEM II B-LL 42.5N	113	117
Group 3	CEM I 32.5R	93	84
	CEM I 42.5R	68	71
	CEM II B-S 32.5 N	104	104
	CEM II B-LL 42.5N	73	61
Group 4	CEM I 32.5R	96	98
	CEM I 42.5R	98	95
	CEM II B-S 32.5N	100	100
	CEM II B-LL 42.5N	103	112

5.1 Flexural Strength Results

Flexural strength was measured through the average of three specimens (40 × 40 × 160 mm). The flexure strength was measured at 3, 7, 28, and 56 days of curing. The obtained results for all groups at 28 days were presented in Table 6. Results indicated that CEM I 43.5R had the best result when cement was used with sand without any other cementitious material. CEM II B-LL 42.5N had the maximum result in the lime I group. Using Lime II had a significant increase in flexure strength for all types of cement used. However, the CEM I type had flexure strength value more than CEM II types. CEM I 32.5R had a remarkable increase in flexure strength value when wood ash was added to lime I and cement.

Table 6. Comparison between flexural strength for all mortars at 28 days

Cement type	Flexural strength (kg/cm^2)			
	Group 1	Group 2	Group 3	Group 4
CEM I 32.5R	19.1	20.609744	16.92	29.8
CEM I 42.5R	28	20.712044	16.97	17.526
CEM II B-S 32.5R	18.2172	16.269986	13.35	14.3
CEM II B-LL 42.5N	26.5	22.839588	13.82	12.7

The flexure strength results are presented in Figs. 7, 8, 9, and 10. Figure 7 showed the effect of changing cement type on flexural strength at different curing times for group 1. At 28 days all types of mortar ranged from (18 to 28) kg/cm^2, mortar with CEM I 42.5R has the highest strength in this group, and CEM II B-S 32.5N has the lowest value of strength. After 3 days, the flexural strength of all types ranged from 42% to 52% of the strength at 28 days, except for CEM I 32.5R where its strength equal 29% of the strength of 28 days. After 7 days, the flexural strength of all types ranged from 71% to 80% of the strength at 28 days. After 56 days, the flexural strength of all types ranged from 110% to 130% of the strength at 28 days, CEM II 42.5N showed the largest value.

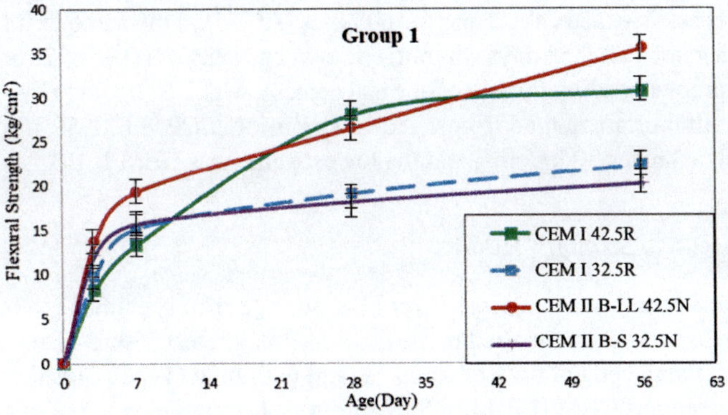

Fig. 7. Comparison between flexure strength values for group 1 mortars

Figure 8 showed the effect of adding lime I where cement: lime (1:1) by volume to various cement types on flexural strength at different curing ages for group 2. At 28 days all types of mortar ranged from 16 to 23 kg/cm^2, mortar with CEM II B-LL 42.5 N has the highest value of strength, but the mortar with CEM II B-S 32.5 N which decreases to 16 kg/cm^2. At 3 days, the flexural strength of all types ranged from 50% to 64% of the 28 days, flexural strength expects for CEM I 32.5 R where it represented 42% of the 28 days' strength. The strength gained after 7 days was ranged from 60% to 70% of 28 days' flexural strength. The strength gained after 56 days was ranged from 100% to 140% of 28 days' flexural strength and CEM I 42.5R showed the largest value of strength.

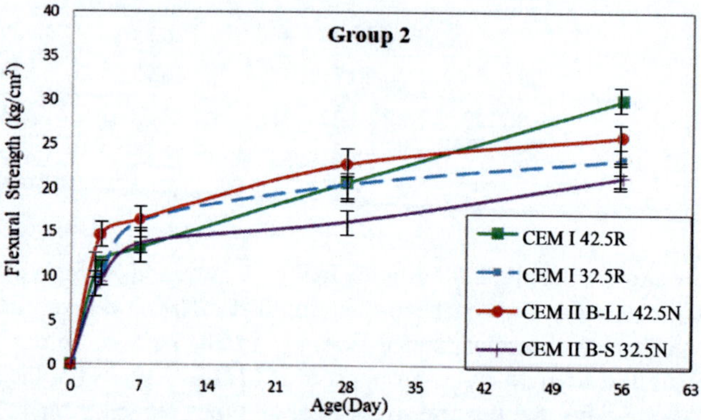

Fig. 8. Comparison between flexure strength values for group 2 mortars

Figure 9 showed the effect of adding lime II with percentage cement: lime 1:1 by volume all mortar's strength results at 28 days showed approximately the same strength around 17 kg/cm^2 except CEM II B-LL 42.5 R had the least value. At 3 days, the flexure strength of types is exactly the same for all types. At 7 days the same trend of 28 days. The strength gained after 56 days was ranged between (122%: 160%) of 28 days' flexural strength. Moreover, adding fly ash to lime and cement, which is shown in Fig. 10, flexural strength for all mortars ranged from 13 to 18 kg/cm^2, but CEM I 32.5R has the highest value, which achieves 30 kg/cm^2, and the lowest value was CEM II B-LL 42.5N.

5.2 Compressive Strength Results

Compressive strength was measured from the average of 6 specimens, which resulted from the pretested specimens used in the flexural strength test. Table 7 shows a comparison between using cement only or using cement with lime (Type I and Type II) and fly ash. Mortar contains CEM II B-LL 42.5N achieves the maximum value of compressive strength when used with sand without any additives, followed by CEM I 42.5R, CEM I 32.5R, and CEM II B-S 32.5R. However, in group 2 when lime type I was added to cement CEM II B-S 32.5R showed the maximum results followed by CEM I 32.5R, CEM I 42.5R, and CEM II B-LL 42.5R. Lime II increased the compressive strength for

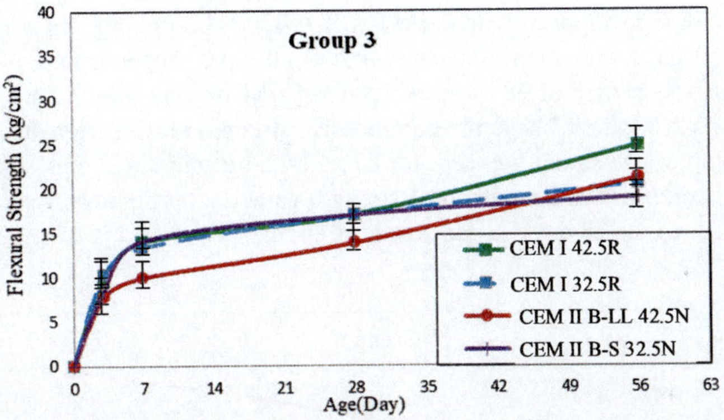

Fig. 9. Comparison between flexure strength values for group 3 mortars

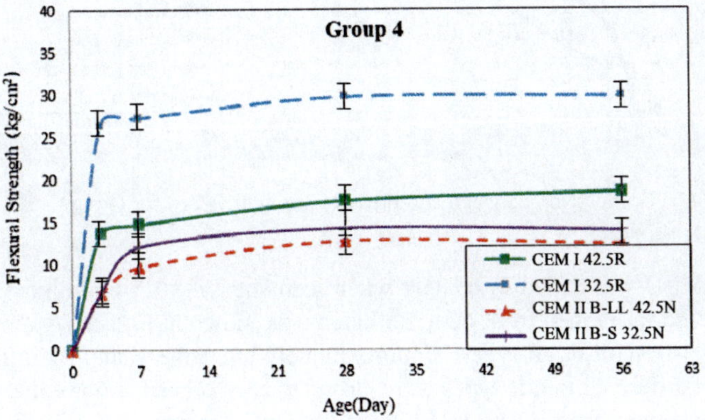

Fig. 10. Comparison between flexure strength values for group 4 mortars

Table 7. Comparison between compressive strength for all mortars at 28 days

Cement type	Compressive strength (kg/cm^2)			
	Group 1	Group 2	Group 3	Group 4
CEMI 32.5R	29	43.992	31.65	46.327
CEM I 42.5R	36	43.7	27.02	21.34
CEM II B-S 32.5 R	26.54	45.02	25.27	19.32
CEM II B-LL 42.5N	40.16	41.8	24.97	16.9

all mortars containing various cement types. Fly ash had a great effect on CEM I 32.5R, which achieved the greatest compressive strength in all mortars produced.

Figure 11 displays the impact of changing cement type on compressive strength at different curing ages. All types of cement mortar achieve compressive strength ranging from 26 to 41 kg/cm^2, mortar with CEM II B-LL 42.5N has the highest strength in this group. After 3 days, the compressive strength of all types ranged from 24% to 30% of the compressive strength at 28 days except for CEM II types which have the highest ratio reached 42%. After 7 days the compressive strength ranged from 40% to 62% of 28 days' compressive strength except for CEM I 42.5R where it is represented 34% of the 28 days' compressive strength. The strength gained after 56 days was ranged from 100% to 122% of the 28 days' compressive strength.

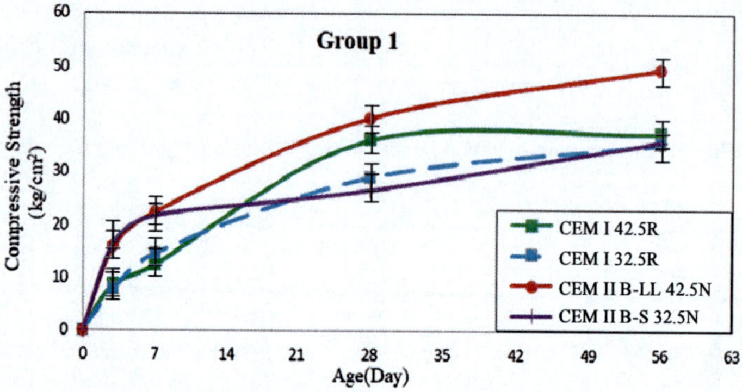

Fig. 11. Comparison between compressive strength values for Group 1 mortars.

When lime I was added to cement where cement: lime 1:1 by volume at 28 days all types of mortar ranged from 42 to 45 kg/cm^2 as shown in Fig. 12. After 3 days, the compressive strength of all types is approximately the same value. At different ages, all values of different mortars types were closed to each other. Compressive Strength at 56 days was ranged from 110% to 140% of 28 days' compressive strength. In Fig. 13 when lime type II is added to cement with ratio cement: lime (1:1) by volume, there was a remarkable reduction in all compressive strength for all types of mortar. The compressive strength for all cement results was in a ranged (25:32) kg/cm^2. After 3 days, the compressive strength of all types ranged from 48% to 55% of the 28 days' compressive strength. The strength gained after 7 days was around 60% of 28 days' compressive strength. The strength gained after 56 days was ranged from 101% to 119% of 28 days' compressive strength.

Adding fly ash to lime I and cement as shown in Fig. 14, the compression strength for all mortars after 28 days ranged from 17 to 24 kg/cm^2. However, CEM I 32.5R had the highest value, which achieved 46.355 kg/cm^2 after 28 days. At 3 days, the compressive strength of CEM I types ranged from 65% to 71%. Compressive strength after 7 days the compressive strength of all mortars types ranged from 70% to 75. Compressive strength after 56 days was ranged from 106% to 140% of 28 days' compressive strength.

5.3 Drying Shrinkage Results

Shrinkage test was performed to indicate the effect of cement type on the behavior of the mortars due to drying shrinkage and to study the influence of adding various types of

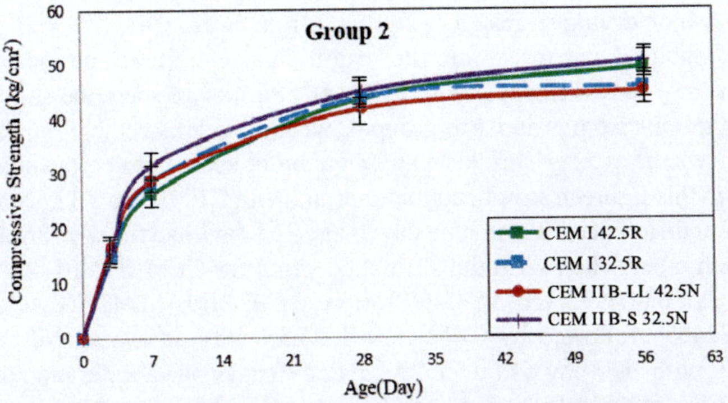

Fig. 12. Comparison between compressive strength values for group 2 mortars.

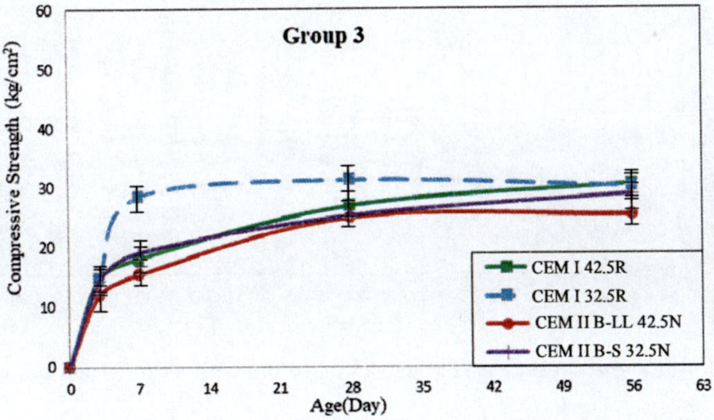

Fig. 13. Comparison between compressive strength values for group 3 mortars.

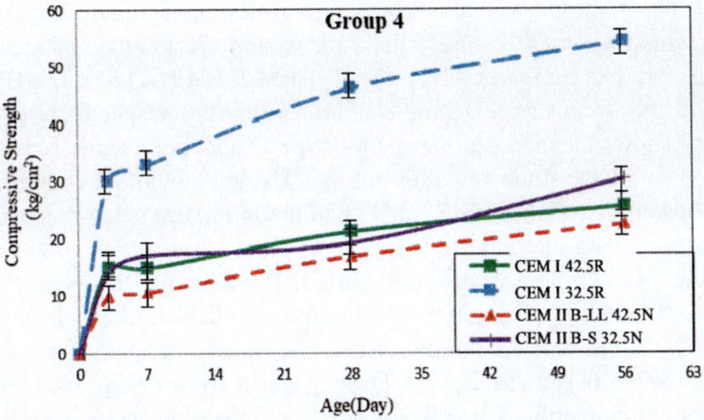

Fig. 14. Comparison between compressive strength values for group 4 mortars.

lime and fly ash on drying shrinkage. After 24 h from casting, the (25 × 25 × 250 mm) prism was unmolded and measuring the length change started. The measuring starts from age one day up to 315 days. Figures 15, 16, 17, and 18 show the shrinkage strain value for all groups from group 1 to group 4, which contains cement only as a binder. Figure 15 shows the average shrinkage strain values of specimens cast using all types of cement. From this figure, it is noticed that mortars with CEM II B-S 32.5N and CEM II B-LL 42.5N at different ages from one day to age 315 days had the dry shrinkage values closed to each other. The maximum shrinkage strain for CEM II B-LL 42.5N with an average value of 0.834993. CEM I 32.5R followed CEM II B-LL 42.5N with an average value of 0.824559 followed by CEM II B-S 32.5N with an average of 0.66471 and CEM I 42.5R with an average of 0.556293. After 90 days, the acceleration of shrinkage increased more slowly and remained nearly constant.

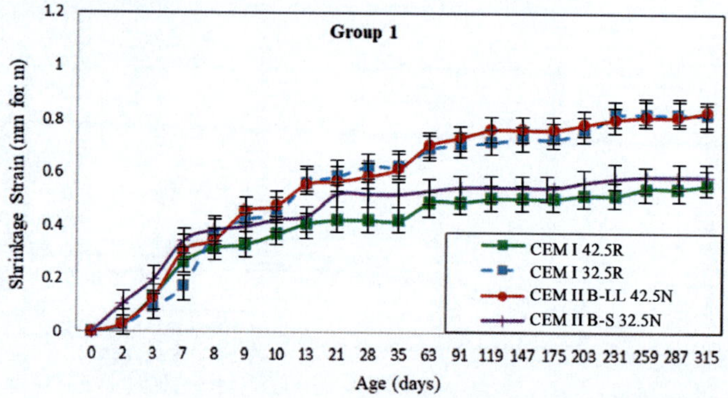

Fig. 15. Comparison between drying shrinkage strain values for group 1 mortars.

Figure 16 shows the effect of adding lime I where cement: lime (1:1) by volume to various cement types on shrinkage strain value with the proportion of group 2. Using lime type, I with different types of cement showed that CEM II B-S 32.5N had the minimum value of shrinkage strain which was around 0.45 mm/m after 315 days of observation which was approximately the same value from 150 days. The other mortars contain other types of cement; CEM I 42.5R, CEM II B-LL 42.5N, and CEM I 32.5R showed shrinkage strain values 0.83, 0.88, and 0.97 mm/m, respectively.

Figure 17 shows the effect of using lime type II with percentage cement: lime 1:1 by volume on shrinkage strain value for group 3. Using lime type II with different types of cement showed that CEM II B-S 32.5N had the minimum value of shrinkage strain which was around 0.45 mm/m after 315 days of observation which was approximately the same value from group 2 when using Lime I. The other mortars contain other types of cement; CEM I 42.5R, CEM II B-LL 42.5N, and CEM I 32.5R showed shrinkage strain values 0.72, 0.81, and 0.83 mm/m, respectively. As a comparison between group 2 when using lime I with cement: lime (1:1) and group 4 when using lime II with cement: lime (1:1), mortar containing CEM II B-S 32.5N has the same value of shrinkage strain using the two types of Lime. CEM II B-LL 42.5N with lime II has shrinkage strain, and lime I have approximately the same results. However, the shrinkage strain value for

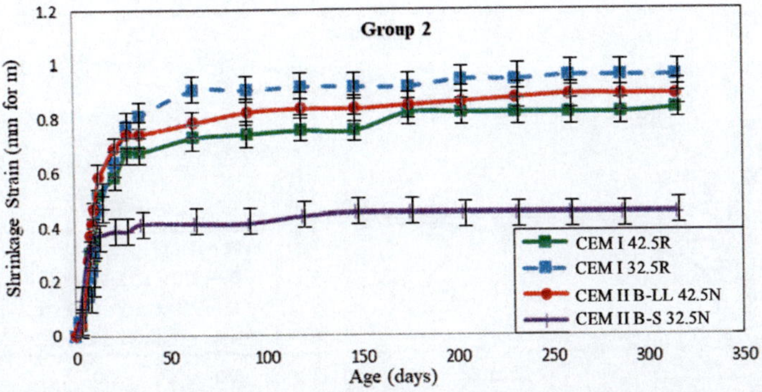

Fig. 16. Comparison between drying shrinkage strain values for group 2 mortars.

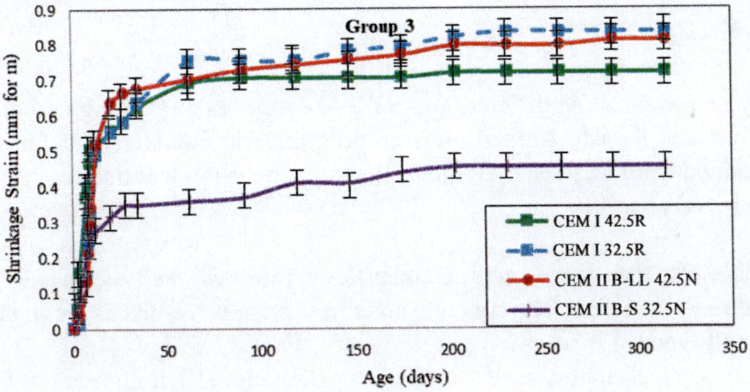

Fig. 17. Comparison between drying shrinkage strain values for group 3 mortars.

mortar when using lime II with CEM I 42.5R was more than using lime I, around 15%. CEM I 32.5R is more than mortar with lime I in group 2 by 16%.

Figure 18 shows the effect of adding fly ash to lime and cement (Fly ash 15%wt. of cementitious materials (lime + cement)) on shrinkage strain value to group 4. CEM I 42.5R, and CEM II B-S 32.5N showed the least values of shrinkage strain around 0.5 mm/m. However, the maximum values were CEM I 32.5R, about 1 mm/m followed by CEM II B-LL 42.5N that achieved 0.97 mm/m. comparing with using cement with lime I only which in group 2. This result illustrated that fly ash increased the shrinkage strain for all cement types while mortars with CEM II B-S 32.5N increased from 0.45 mm/m when lime was only used to 0.5 mm/m when adding fly ash with lime and cement. CEM I 42.5R decreased from 0.83 when using lime with cement to 0.5 when fly ash was added. CEM II B-LL 42.5N increased from 0.88 to 0.97. CEM I 32.5R increased from 0.97 to 1 mm/m.

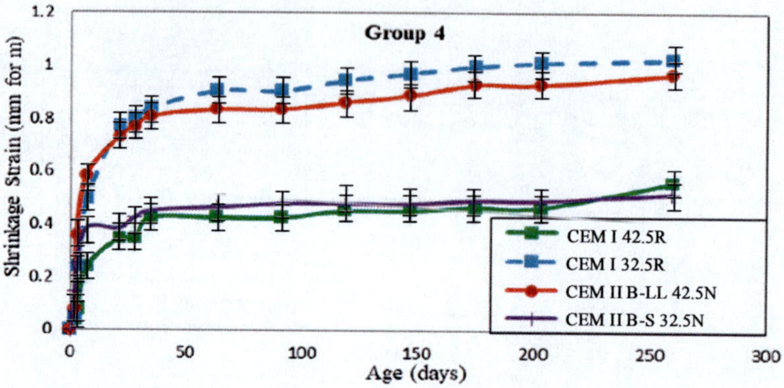

Fig. 18. Comparison between drying shrinkage strain values for group 4 mortars.

6 Conclusions

This study investigates the different properties of mortars containing various types of cement, lime, and fly ash. Abroad research program was conducted, and based on the results obtained from various tests; the following conclusion and finding were deduced and summarized as follow:

- Flow table test that carried with standard concrete cone and the mortar cone that manufactured in the lab. The concrete cone has the same results as the manufactured cone for all types of mortars.
- Sultana lime type I that is used with cement: lime ratio 1:1 improves the percentage of flow for all mortar that contains various types of cement used.
- Using lime I is more beneficial in flexural and compressive strengths than lime II. Fly ash decrease the compressive and flexure strength for all mortar except mortar which contains CEM I 32.5R.
- As a comparison between drying shrinkage strain value of mortars which contains cement only as a binder mortar with CEM I 42.5R has the lower value. For the mortars containing cement and lime I as a binder with cement: lime 1:1 mortar with CEM II B-S 32.5R has the lower value.
- All values of drying shrinkage strain of mortars used increase more than mortars containing cement only as a binder expect mortar contain CEM I 42.5N which decreased by 11% and mortar contain CEM II B-S 32.5N which decreased by 14%.
- Fly ash increases the drying shrinkage strain value for all mortar except mortar which contains CEM I 42.5R. In addition, after 90 days, the acceleration of shrinkage increased more slowly and remained nearly constant.

References

1. Egyptian Standard Specification. ES4756-1: Cement – Part 1: Composition, Specifications And Conformity Criteria For Common Cements, pp. 1–48 (2007)

2. Abdou, D.M.S., El Adaway, N.S.: Application on game theory: cement industry in Egypt - Arabian cement company. Am. J. Bus. Econ. Manag. **6**(3), 49–56 (2018)
3. Awed, A.M., Aboelela, A.E., El-Ashwah, A.S., Allam, M., El-Badawy, S.M.: Improvement of unbound granular pavement layers and subgrade with cement dust in Egypt. Int. J. Pavement Res. Technol. **13**(6): 621–629 (2020)
4. Akshay, S., Arun, S., Rushikesh, R., Neeraj, M., Ghadge, M.D.: Partial replacement of sand with sawdust in Cconcrete. Int. Res. J. Eng. Technol. **5**(3), 3098–3101 (2018)
5. Garside, M.: Cement production in the United States and worldwide from 2010 to 2020. https://www.statista.com/statistics/219343/cement-production-worldwide/
6. Assi, L.N., Majdi, A., Alhamadani, Y., Ziehl, P.: Early properties of concrete with alkali-activated fly ash as partial cement replacement. In: Proceedings of the Institution of Civil Engineers - Construction Materials, vol. 174, pp. 1–8 (2020)
7. Kashmire, S.S., Pawar, S.K., Patil, J.A.: Experimental study on lime mortar by using admixtures. Int. Res. J. Eng. Technol. **8**(1), 1443–1446 (2021)
8. Itim, A., Ezziane, K., Kadri, E.H.: Compressive strength and shrinkage of mortar containing various amounts of mineral additions. Constr. Build. Mater. **25**(8), 3603–3609 (2011)
9. Gulbe, L., Vitina, I., Setina, J.: The influence of cement on properties of lime mortars. Procedia Eng. **172**, 325–332 (2017)
10. David, N.L., Hassan, L.K.: Physical and mechanical properties of cement mortar using lime and bamboo-ash partical replacements. Malaysian J. Civ. Eng. **30**(2), 254–268 (2018)
11. Collepardi, M., Collepardi, S., Olagot, O., Simonelli, F.: The influence of slag and fly ash on the carbonation of concretes. Int. Concr. Abstr. Portal **221**, 483–494 (2004)
12. Kate, G.K., Murnal, P.B.: Effect of addition of fly ash on shrinkage characteristics in high strength concrete. Int. J. Adv. Technol. Civ. Eng. **1**, 2231–2721 (2013)
13. Pitroda, J., Umrigar, F.S.: Evaluation of sorptivity and water absorption of concrete with partial replacement of cement by thermal industry waste (Fly Ash). ISO Certif. Int. J. Eng. Innov. Technol. **2**(7), 245–249 (2008)
14. ASTM C 618-1: Standard specification for coal fly ash and raw or calcined natural pozzolan for use in concrete. (2019)
15. Egyptian Organization for Standards & Quality. E . S. E. 2421-7/2006. Cement-Physical and Mechanical Tests Part 7 – Determination of Strength Prism Method (2006)

Key-Block Theorem Application on Discontinuous Rock Slope Instabilities and Rock Mass Description

Mohammad Azarafza[1(✉)] and Hong-Hu Zhu[2]

[1] Department of Civil Engineering, University of Tabriz, Tabriz, Iran
m.azarafza.geotech@gmail.com
[2] School of Earth Sciences and Engineering, Nanjing University, Nanjing, China
zhh@nju.edu.cn

Abstract. Key-block theorem, also known as block theory or Goodman's approach, is considered the most accurate and flexible method for understanding structural instability conditions of discontinuous rock slopes. This procedure provides useful information on slope failure mechanism, rock blocks geometry, discontinuity networks and slope stability conditions. In this article, the application of the key-block analysis method was considered, which attempted to present a practical review of Goodman's theory in recent advancements.

Keywords: Slope stability · Discontinuous rock slope · Key-block method · Limit equilibrium analysis

1 Introduction

Slope stability is one of the essential issues in geoengineering for more than 300 years. Slope stability is involved in numerous geotechnical projects and covers different aspects of infrastructural developments [1–3]. The slope stability analysis can be categorized in various instabilities of earth or rock slopes, which are classified as primary and secondary groups. The primary group consists of wedge failure, toppling failure (rock slopes), planar failure, rotational failure (rock and earth slopes), and the secondary group consists of composite slips, exceptional cases, glacial slips, etc. [4]. The wedge failure is the most common instability type in discontinuous rock slopes with slip geometry in wedge form, leading to the unstable mass during the functional interface of two or more discontinuities in rock slope. The failure that contains several metres to hundreds of metres of geometric volume movable mass above the main slip surface can be triggered progressively [5]. The planar failure is considered a particular case of wedge failure due to its geometrical status of the discontinuity network of slope mass. This failure occurs along a planar surface, such as bedrock, main discontinuity surface, resistant layer, and dense clay lenses [6, 7]. The toppling failure is considered as anti-dip instability occurred in a discontinuous rock slope. The main discontinuity set and slope surface orientations are opposite, and rock columns are overturned at the geometric centre of the sliding mass [8]. Goodman and Bray (1976) categorized these types of instabilities, which called toppling into the main

H.-H. Zhu et al. (Eds.): BRWSG 2021, LNCE 230, pp. 200–208, 2022.
https://doi.org/10.1007/978-981-16-9963-4_16

and secondary groups [9]. The main group contains systematic toppling named flexural failure, block failure and block-flexural failure. The secondary group includes the non-systematic and complex form of toppling [10, 11]. Finally, the rotational or massive failure mainly occurs in homogeneous masses (i.e., soil, heavily jointed rock, weak geomaterials) where slip surfaces pass through the body of the mass in the most fragile state [7, 12, 13]. Generally, various approaches are used for slope stability assessments to investigate the instability conditions and quantify the failure mechanisms, classified as simple evaluations, planar failures, limit state criteria for limit equilibrium analysis, numerical methods, hybrid, and high-order approaches [2, 3, 14]. In the meantime, limit equilibrium methods (LEMs) are considered the most flexible approach due to their simplicity, continuous access, rapid implementation, accessible assumptions, closed-form analysis, and providing multiple answers, coupled with different methods [15]. This flexibility helps to develop different capable formulations like Goodman's block theory [16].

2 Key-Block Theorem

Block theory, also known as Key-block theory or Goodman's theorem, was introduced by Goodman and Shi in 1985 [16]. It is based on stability assessment of rock blocks using LEM and two primary mechanisms of failure called 'structural failure' because of discontinuities and 'stress-based failure' because of the presence of high stresses. The key-block theory uses the geometrical status of rock blocks and the discontinuity network on rock mass to claim the logical relationship between the block geometry and the critical failure surface of excavated or natural rock slopes [15]. Goodman's theorem aims to search and identify the finite key blocks involving local or progressive instability on discontinuous rock slopes [17]. Thus, the rock blocks' geometric position, discontinuity network, and emplacement on rock mass play essential roles instability assessments and are considered evaluation variables. This theory classified the blocks into finite and infinite groups, which the finite blocks are capable of movement, which causes structural failure. These blocks are divided into different blocks, as well as a presentation in Fig. 1.

Key-block theory uses "finiteness theorem" and "removability theorem" to identify, recognize, classify and analyze the discontinuity network based on spatial equations/inequalities of discontinuities which lead to recorded block information in polar coordinates (i.e., discontinuities' dip/dip direction) as rock mass which can easily convert to Cartesian coordinates [18]. Based on the removability theorem, the polyhedral geometry of blocks was convex convergence form to block's mobility. The non-convergent concave blocks are not movable. According to the finiteness theorem, the joint pyramid (the shared spaces between half-spaces of discontinuity planes that form part of the block pyramid), the excavation pyramid a group of extraction half-spaces (excavation half-spaces) that are displaced to create a block pyramid), space pyramid (complementary half-space of excavation pyramid), and block pyramid (the factor for determining the convexity and concavity of blocks) have to define and categorized [19]. These continuous equations give a mathematical definition for block geometry and their stability analysis leading to [16]:

$$\left.\begin{array}{l}\text{Block pyramid} \neq \emptyset \\ \text{Block pyramid} = \text{Joint pyramid} \cap \text{Excavation pyramid} \\ \text{Joint pyramid} \cap \text{Excavation pyramid} \neq \emptyset \\ \text{Joint pyramid} \not\subset \text{Space pyramid}\end{array}\right\} \quad \text{Infinite blocks}$$

$$\left.\begin{array}{l}\text{Block pyramid} = \emptyset \\ \text{Block pyramid} = \text{Joint pyramid} \cap \text{Excavation pyramid} \\ \text{Joint pyramid} \cap \text{Excavation pyramid} = \emptyset \\ \text{Joint pyramid} \subset \text{Space Pyramid}\end{array}\right\} \quad \text{Finite blocks}$$

$$\left.\begin{array}{l}\text{Block pyramid} = \emptyset \\ \text{Block pyramid} = \text{Joint pyramid} \cap \text{Excavation pyramid} \\ \text{Joint pyramid} \cap \text{Excavation pyramid} = \emptyset \\ \text{Joint pyramid} = \emptyset\end{array}\right\} \quad \text{Conical/trapped blocks}$$

$$\left.\begin{array}{l}\text{Block pyramid} = \emptyset \\ \text{Joint pyramid} \neq \emptyset \\ \text{Joint pyramid} \cap \text{Excavation pyramid} = \emptyset \\ \text{Joint pyramid} \subset \text{Space pyramid}\end{array}\right\} \quad \text{Key blocks}$$

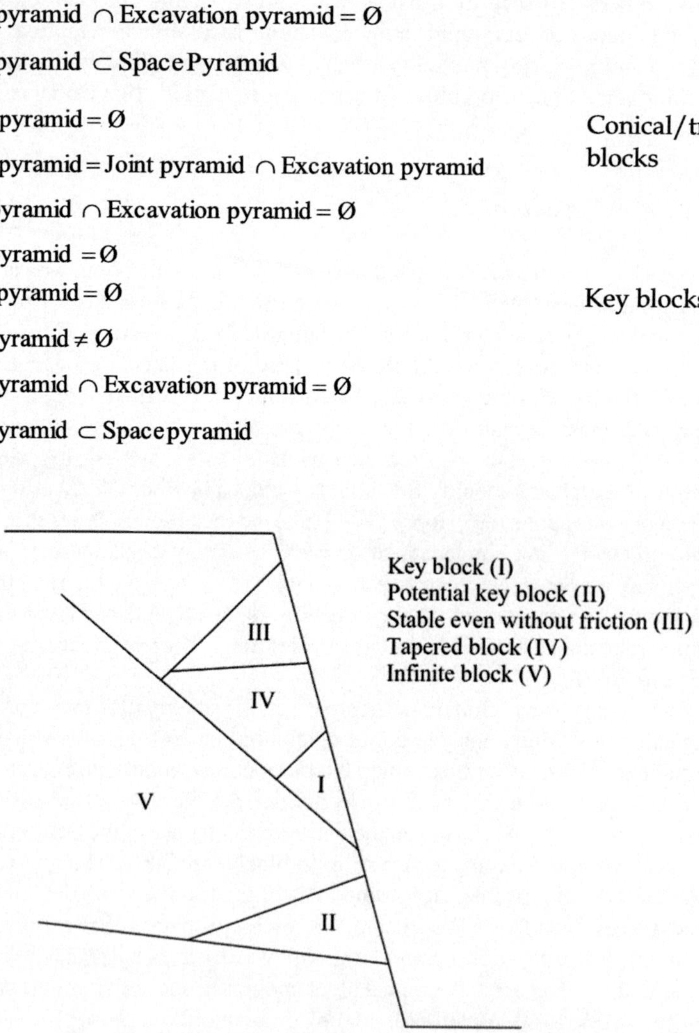

Fig. 1. Rock block classification based on block theory [13].

It is easy to restrict the trending evaluation loops to identify the active key-blocks based on the equations. This advantage has focused on key-block theory in stability analysis [20, 21]. Also, Goodman and Shi (1985) applied the removability theorem to express the tendency trend and removable blocks. They can displace in a particular direction without encountering the neighbour's actions. This is the definition of key-block as it is a key of instability [16, 22–24]. For the removable block [16]:

Block pyramid $= \emptyset$
Block pyramid $=$ Joint pyramid $\cap$ Excavation pyramid
Joint pyramid $\cap$ Excavation pyramid $= \emptyset$
Joint pyramid $\subset$ Space pyramid
Joint pyramid $\neq \emptyset$

They also defined non-removable blocks that blocks cannot move in any direction. The removable and non-removable status is presented in Fig. 2. So, the slope instability in discontinuous rock masses is limited to finite and removable blocks. This finiteness in the LEM-base stability analysis provides the evaluation much faster, more accurately, and less time than traditional methods like manual methods or commercial software.

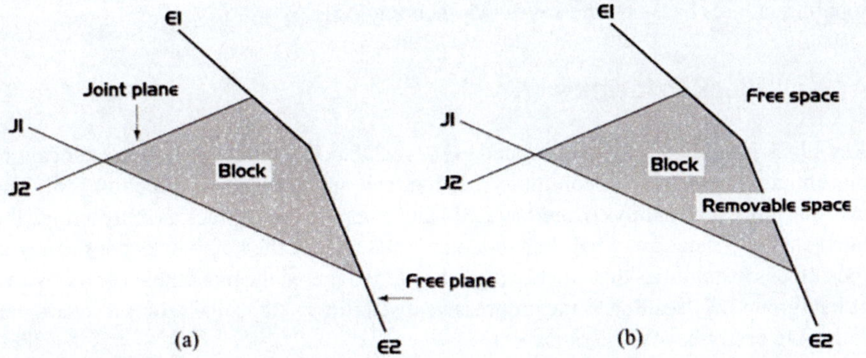

Fig. 2. Rock block in an excavation: (a) intersection of two joint surfaces and two free spaces, (b) removable space of the block [16].

3 Review of Literature on Key-Block Method

Um and Kulatilake (2001) provide a kinematic and LEM analysis based on block theory for stability evaluations on rock slope in Gorges dam site in China [25]. Huang et al. (2000) used key-block basics to describe discontinuity networks in rock slopes [26]. Yarahmadi-Bafghi and Verdel (2003) developed a key-group method based on Goodman's theory [27]. Jimenez-Rodriguez et al. (2006) utilized the systematic quantitative procedure for rock slope stability assessments based on key-block theory [28]. Haswanto and Abd-Ghani (2008) performed a rock slope stability analysis based on key-block theory in Fraser's Hill Pahang in Malaysia [29]. Kulatilake et al. (2011) applied the key-block method to estimate the stability condition of rock slopes at the

Yujian River dam site [15]. Greif and Vlčko (2013) used key-block analyses to evaluate the LEM-based stability status for 45 rock slopes in Slovakia [30]. Azarafza et al. (2013) applied key-block theory to stability analysis of discontinuous rock slopes in the southwest of Iran [31]. Sun et al. (2014) utilized key-block procedure for LEM-based stability analysis of rock slopes in the Jinping-I hydropower station in China [32]. Li et al. (2016) introduced a method based on block-group, and Sarma approaches stability assessment for rock slope mass instabilities [33]. Azarafza et al. (2017) provided the algorithm for three-dimensional simulation of discontinuities emplacements in jointed rock slopes named 3DDGM (three-dimensional discontinuity geometrical modelling) [18]. Also, the key-block based stability analysis was performed on a discontinuous rock slope in South Pars Gas Complex, southwest of Iran [17]. Liu et al. (2017) present the novel semi-deterministic method based on block theory named NSDBT and successfully applied it to Changhe dam in Sichuan, China. They mentioned that NSDBT is capable of identifying block geometry based on key-block instructions [23]. Wang et al. (2018) utilized a multi-level framework based on block theory and analytic hierarchy process, named GeoSMA-3D, to characterize rock block dimension in rock mass [21]. Azarafza et al. (2020) provided the key-block-based fuzzy logical decision-making procedure on estimating the wedge and planar instability status in jointed rock slopes [34]. Later on, Azarafza et al. (2020) introduce the novel approach to investigate the main toppling failures based on the key-block theorem [35].

4 Key-Block Principles

Key-block is built on principles related to the rock blocks geometry, key-block location, discontinuity networks, discontinuity friction, and applied loading direction [36]. The analysis process is mainly covered by LEM and computed safety factors representing the movable body's stability [16]. As Goodman's statements, the key-blocks are the key to rock blocks instabilities that can be single-handed triggered the movements or motivated as key-group [27], leading to the progressive instability in slope mass. In this regard, the following basics have to be satisfied [37]:

- Discontinuity network was considered planar and extended to slope boundaries,
- Key-block have to identify by applying vectorial method,
- Active finite key-block considered as unstable single key when LEM-based safety factor less than 1.0,
- Recognition of finite neighbour blocks as potential triggering key-blocks,
- Building all possible groups related to active finite key-blocks,
- Stability analysis based on LEM as evaluation loops,
- Iteration of new geometry for final stable state slope condition.

5 Real Case Implementation

Several case studies were used to apply the concept of the key-block method to investigate the stability analysis of discontinuous rock slopes. Figure 3 picked the view of the studied slope from Tabriz -Tehran highway path near Tabriz city, located in East-Azerbaijan province, in the northwest of Iran. The studied slope is composed of jointed limestone units in which three joint sets have been identified by field survey. According to the recent excavation, the slope is geometrically modified, leading to local instability in slope mass. The mechanism of failure is obtained as wedge failure and block sliding. So, conducting the key-block method can provide a fast and accurate way to locate the key-block and perform suitable stabilization procedures. Figure 4 illustrates the key-block-based geometrical identification of studied rock slope. The results of experiments and field studies have been used to estimate the geomechanical properties of rock slopes. The Mohr-Coulomb criterion has been used for obtaining the strength properties and LEM analyses. Also, Table 1 shows the results of slope mass mechanical properties. Table 2 and Fig. 5 present the stability analysis of slope based on LEM, which can be stated that the domain is generally stable and the estimated safety factor is up than 1.0. But it is detected some local failures which are required to conducting stabilizations.

Fig. 3. Photograph of the selected slope.

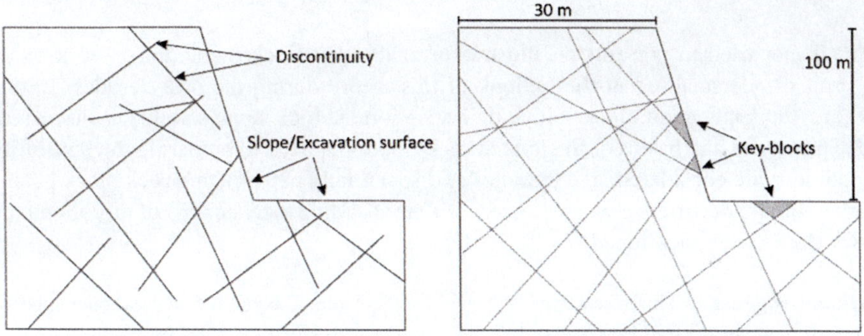

Fig. 4. Key-block-based geometrical identification of studied rock slope.

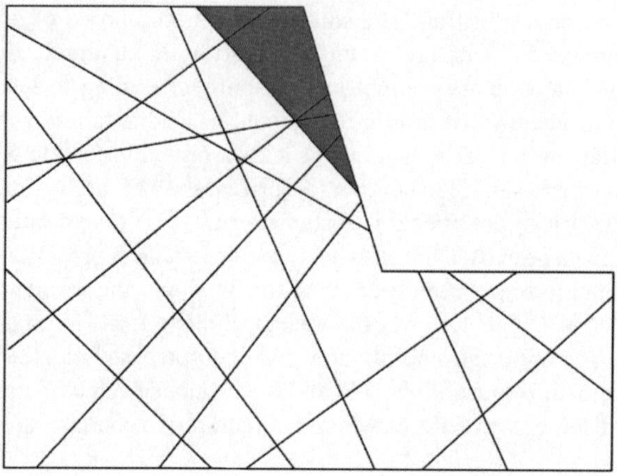

Fig. 5. LEM-based stability status of the rock slope under investigation.

Table 1. Mechanical properties of the slope mass

Parameter	c (kPa)	ϕ (°)	γ (kN/m^3)	E (GPa)	K_n (GPa)	K_s (GPa)	ν
Value	12	32	18.75	75	0.5	0.5	0.33

c: cohesion, ϕ: angle of internal friction, γ: unit weight, E: Young's modulus, ν: Poisson's ratio, K_n & K_s: Normal and shear joint stiffness coefficients

Table 2. Stability analysis by using the key-block method.

Stability	Weight of unstable stone (T_n)	Factor of safety	Analysis method
Stable	1.02	1.17	LEM

6 Conclusions

This paper tried to presents an illustration of the key-block methodology to provide a proper understanding of the outlook of this theory during the past decades. In this regard, the application of key-block theory on jointed rock slope stability is discussed. The principles and basics of the jointed rock slopes that lead to preparing the instability conditions are considered, and the roles of discontinuity network and rock block geometry in failure occurrence were stressed. Different sliding mechanisms of discontinuous rock slopes were mentioned.

Acknowledgments. This research was funded by the National Natural Science Foundation of China (Grant No. 42077235).

References

1. Frasheri, A.: Outlook on the possibility for slope stability evaluation according to petrophysical and integrated geophysical data. In: Proceedings of the 6th International Conference and Exposition on Petroleum Geophysics, Kolkata (2006)
2. Chen, X.-P., Zhu, H., Huang, J., Liu, D.: Stability analysis of an ancient landslide considering shear strength reduction behavior of slip zone soil. Landslides **13**(1), 173–181 (2015). https://doi.org/10.1007/s10346-015-0629-7
3. Zhu, H.H., Wang, Z.Y., Shi, B., Wong, J.K.W.: Feasibility study of strain based stability evaluation of locally loaded slopes: insights from physical and numerical modeling. Eng. Geol. **208**, 39–50 (2016)
4. Bromhead, E.N.: The Stability of Slopes. Spon Press, Milton Park (1998)
5. Zhou, J., Wang, J.: Lower bound limit analysis of wedge stability using block element method. Comput. Geotech. **86**, 120–128 (2017)
6. Wyllie, D.C., Mah, C.: Rock Slope Engineering, 4th edn. Spon Press, Milton Park (2004)
7. Huang, Y.H.: Slope Stability Analysis by the Limit Equilibrium Method. ASCE Publications, Virginia (2014)
8. Nikoobakht, S., Azarafza, M.: Stability analysis and numerical modelling of toppling failure of discontinuous rock slope (A case study). J. Geotech. Geol. **12**(2), 169–178 (2016)
9. Goodman, R.E., Bray, J.W.: Toppling of rock slopes. In: Proceedings of the ASCE Speciality Conference on Rock Engineering for Foundations and Slopes, Boulder, Colorado (1976)
10. Alejano, L.R., Gómez-Márquez, I., Martínez-Alegría, R.: Analysis of a complex toppling-circular slope failure. Eng. Geol. **114**(1–2), 93–104 (2010)
11. Amini, M., Ardestani, A., Khosravi, M.H.: Stability analysis of slide-toe-toppling failure. Eng. Geol. **228**, 82–96 (2017)
12. Abramson, L.W., Lee, T.S., Sharma, S., Boyce, G.M.: Slope Stability Concepts: Slope Stabilization and Stabilization Methods, 2nd edn. Wiley, New York (2001)
13. Albataineh, N.: Slope stability analysis using 2D and 3D methods. Master's Thesis, University of Akron, Ohio (2006)
14. Kliche, C.A.: Rock Slope Stability. 2nd edn. Society for Mining, Metallurgy, and Exploration Press, Englewood (2018)
15. Kulatilake, P.H.S.W., Wang, L., Tang, H., Liang, Y.: Evaluation of rock slope stability for Yujian river dam site by kinematic and block theory analyses. Comput. Geotech. **38**, 846–860 (2011)
16. Goodman, R.E., Shi, G.: Block Theory and its Application to Rock Engineering. Prentice-Hall, New York (1985)
17. Azarafza, M., Asghari-Kaljahi, E., Akgün, H.: Assessment of discontinuous rock slope stability with block theory and numerical modeling: a case study for the South Pars Gas Complex, Assalouyeh, Iran. Environ. Earth Sci. **76**(11), 397 (2017)
18. Azarafza, M., Asghari-Kaljahi, E., Akgün, H.: Numerical modeling of discontinuous rock slopes utilizing the 3DDGM (three-dimensional discontinuity geometrical modeling) method. Bull. Eng. Geol. Env. **76**(3), 989–1007 (2016). https://doi.org/10.1007/s10064-016-0879-1
19. Azarafza, M., Nikoobakht, S., Asghari-Kaljahi, E., Moshrefy-far, M.R.: Stability analysis of jointed rock slopes using block theory (case study: gas flare site in phase 7 of South Pars Gas Complex). In: Proceedings of the 32th National and the 1st International Geosciences Congress, Sari (2014). (in Persian)
20. Zhang, Y., Chen, G., Wu, J., Zheng, L., Zhuang, X.: Numerical simulation of seismic slope stability analysis based on tension-shear failure mechanism. SEAGS-AGSSEA J. **43**(2), 18–28 (2012)

21. Wang, F., Wang, S., Hashmi, M.Z., Xiu, Z.: The characterization of rock slope stability using key blocks within the framework of GeoSMA-3D. Bull. Eng. Geol. Env. **77**(4), 1405–1420 (2018). https://doi.org/10.1007/s10064-018-1291-9
22. Ling, H.I.: Recent applications of sliding block theory to geotechnical design. Soil Dyn. Earthq. Eng. **21**, 189–197 (2001)
23. Liu, T., Deng, J., Zheng, J., Zheng, L., Zhang, Z., Zheng, H.: A new semi-deterministic block theory method with digital photogrammetry for stability analysis of a high rock slope in China. Eng. Geol. **216**, 76–89 (2017)
24. Ma, Q., Li, G., Hong, Y.: The evaluation of stability of rock mass slope based on the block theory and DDA numerical. Appl. Mech. Mater. **353–356**, 1051–1056 (2013)
25. Um, J.G., Kulatilake, P.H.S.W.: Kinematic and Block theory analyses for shiplock slopes of the three gorges dam site in China. Geotech. Geol. Eng. **19**, 21–42 (2001)
26. Huang, C.C., Tsai, C.C., Chen, Y.H.: Generalized method for three-dimensional slope stability analysis. J. Geotech. Geoenviron. Eng. **128**, 836–848 (2000)
27. Yarahmadi-Bafghi, A.R., Verdel, T.: The key-group method. Int. J. Numer. Anal. Meth. Geomech. **27**, 495–511 (2003)
28. Jimenez-Rodriguez, R., Sitar, N., Chacón, J.: System reliability approach to rock slope stability. Int. J. Rock Mech. Min. Sci. **43**, 847–859 (2006)
29. Haswanto, W.A., Abd-Ghani, R.: Stability analysis of rock slopes using Block theory. In: Proceedings of the American Rock Mechanics Association Annual Conference (ARMA), Salt Lake (2008)
30. Greif, V., Vlčko, J.: Key block theory application for rock slope stability analysis in the foundations of medieval castles in Slovakia. J. Cult. Herit. **14**, 359–364 (2013)
31. Azarafza, M., Yarahmadi, A.R., Asghari-Kaljahi, E., Bahmannia, G.R., Moshrefy-far, M.R.: Stability analysis of jointed rock slopes using key block method (case study: gas flare site in 6, 7 and 8 phases of south pars gas complex). J. Geotech. Geol. **9**(3), 169–185 (2013). (in Persian)
32. Sun, G., Zheng, H., Huang, Y.: Stability analysis of statically indeterminate blocks in key block theory and application to rock slope in Jinping-I hydropower station. Eng. Geol. **186**, 57–67 (2014)
33. Li, M., Zhou, S., Wang, G.: 3D identification and stability analysis of key surface blocks of rock slope. Trans. Tianjin Univ. **22**(4), 317–323 (2016). https://doi.org/10.1007/s12209-016-2596-z
34. Azarafza, M., Akgün, H., Feizi-Derakhshi, M.R., Azarafza, M., Rahnamarad, J., Derakhshani, R.: Discontinuous rock slope stability analysis under blocky structural sliding by fuzzy key-block analysis method. Heliyon **6**(5), e03907 (2020)
35. Azarafza, M., Akgün, H., Ghazifard, A., Asghari-Kaljahi, E.: Key-block based analytical stability method for discontinuous rock slope subjected to toppling failure. Comput. Geotech. **124**, 103620 (2020)
36. Shi, G.: Discontinuous Deformation Analysis: A New Numerical Method for the Statics and Dynamics of Block Systems. Thesis, University of California Berkeley (1988)
37. Noroozi, M., Jalali, S.E., Yarahmadi-Bafghi, A.R.: 3D key-group method for slope stability analysis. Int. J. Numer. Anal. Meth. Geomech. **36**(16), 1780–1792 (2011)

Comparison of Accuracy in Prediction of Radial Strain in Stone Columns Using AI Based Models

Tanwee Mazumder[1] (✉) and Ankit Garg[2]

[1] Department of Civil Engineering, Indian Institute of Technology Madras, Chennai 600036, India

[2] Guangdong Engineering Center for Structure Safety and Health Monitoring, Shantou University, Shantou, China

ankit@stu.edu.cn

Abstract. Ground improvement of soft soil with construction of stone columns has been widely adopted. Lateral deformation of stone columns plays a significant role in behavior of columns. This study aims to explore the applicability of different AI techniques/mathematical models in predicting radial strain (ε) (change in radius/original radius of column) in stone columns as a function of significant input parameters viz. diameter (d) of stone column, *l/d* ratio, *s/d* (spacing/diameter) ratio, area ratio (A_r), λ (area of stone column/total area of loading), geosynthetic stiffness (k), β (clearance ratio). The radial strain (ε) in ordinary and encased columns is predicted with the help of linear regression, SVM, GPR and ANN models using Matlab software. The datasets of input parameters are obtained from already published literature. The values predicted by the models are compared to the corresponding true values of radial strain reported in the literature. A comparative analysis of the efficiency of all models is examined in terms of RMSE, R-squared, MSE and MAE values. It was observed that ANN models closely predicted the radial strain in columns with higher accuracy as compared to other models. ANN models may therefore be used to predict radial strain even in larger size columns in the field/in-situ conditions. However, these models are put forward as a complementary technique to evaluate the radial strain in columns and not as a substitute to field tests.

Keywords: AI techniques · ANN · Stone column · Radial strain

1 Introduction

Ground improvement of soft soil sites with construction of stone columns has been widely adopted across the globe. Construction of stone columns helps to increase the load bearing capacity and decrease the settlement of soft ground [1–5]. Numerous studies have been carried out to study the behavior of stone columns with different geometrical patterns and arrangement [4–9]. Various studies on design procedures and methodologies [2, 3, 7–12] have been established to enable a safe design of such columns. Diverse factors like diameter and length of stone columns, spacing between columns, area replacement

H.-H. Zhu et al. (Eds.): BRWSG 2021, LNCE 230, pp. 209–222, 2022.
https://doi.org/10.1007/978-981-16-9963-4_17

ratio [4, 5, 7, 9, 13, 14]; arrangement of columns and area of load application [6–9, 15]; length of encasement and encasement stiffness [3, 4, 9, 14, 16, 17] influence the behavioural characteristics of stone columns.

The bulging of stone columns significantly influence column behaviour and hence considered in design of stone columns. The load bearing capacity of the column is generated from lateral earth pressure mobilised by the surrounding soil against the bulging of columns [1, 2, 13]. In soft clay, the inability to resist bulging due to the extremely low shear strength of the clayey soil causes failure of such columns [4]. The confinement provided by encasements to stone columns constructed in very soft soil ($c_u \leq 15$ kPa) significantly cause reduction in column bulging [3, 7, 18]. Murugesan and Rajagopal [3] have found that with encasement the lateral deformation or bulging of columns reduced significantly and the trend of radial deformation is similar to the pattern of variation of hoop strain in columns. The study observed that maximum bulging occurred over a depth equal to twice the diameter of column and providing encasement over this range of depth would be sufficient to arrest column bulging. The geosynthetic encasement stiffness plays an important role in increasing the column stiffness by increasing the lateral confining stress. With increase in lateral confining stress in columns the lateral stress transferred to the surrounding soil decreased. It was noted that smaller diameter columns perform better than larger diameter columns due to generation of higher confining stress.

The depth of predominant bulging has been reported by various studies and it falls closely in the range of 1–4 times the diameter of columns. Irrespective of being an end-bearing or floating column, long unreinforced columns fail by bulging while short columns fail by punching [4]. The study also found that with an increase in diameter of columns, the failure stress increased with construction of reinforced or unreinforced stone columns. In layered ground it was observed that in case of column area loading the top soft soil stratum influenced significantly the limiting axial stress of the columns. On the contrary, for entire area loading, stiffness and load carrying capacity of column reinforced ground decreased as the thickness of top strata of weaker soil increased [19]. The length of columns invariably affects the column performance and failure. Malarvizhi and Ilamparuthi [15] reported that the load capacity of stone column increased with an increase in length of columns. From the review of literature it was found that stone columns having l/d ratio less than 4 fail by punching while bulging failure occurs in the columns having $l/d > 4$ [12]. Therefore, it is noted that a wide variety of factors like diameter and length of columns, arrangement of columns, area of load application, encasement stiffness significantly influence the lateral deformation/bulging of columns. Numerical simulations can be performed to evaluate the influence of each factor. However, for engineering purpose, it is more realistic to develop a simple model that can estimate radial strain as a function of all such factors (diameter and length of columns, spacing between columns, area replacement ratio, area of load application, type of columns i.e. floating or end-bearing nature of columns and encasement stiffness). Such modelling is less time consumable and gives an approximate idea of bulging characteristics. This study aims to predict the radial strain in ordinary and encased stone columns as a function of such factors using different numerical models/Artificial Intelligence (AI) techniques.

2 Methodology

2.1 Theory of Artificial Intelligence (AI)

The AI techniques attempts to solve the problems with training/learning from the input and output datasets presented. While training, it generates/devises the subtle functional relationships between the data/training examples without explaining the physical laws or determining the underlying relationships that govern the interdependence between the datasets [20]. Hence, AI models are entirely dependent on the training data to obtain the functional relationships as well as establish the parameters or structure of a system with minimum considerations of the physical laws. In case of models based on physical laws between different parameters, a prior knowledge of the governing relationships between various factors, as well as the assumptions and theories involved is necessary unlike AI techniques. This enhances the utility of such AI techniques as compared to other physically based empirical and statistical models in the application of problems where the target output can be evaluated [20]. Also, it was found that the AI models performed comparably similar or better than the traditional models [20].

2.2 Applications of AI in Geotechnical Engineering

Various studies on applications of AI techniques in geotechnical engineering have been carried out over the past few decades. Goh [21] calculated the friction capacity of piles in soft clay with the help of ANN models. The model was trained with data obtained from field tests. Subsequent studies were carried on application of ANN models to develop a correlation between undrained shear strength of soil, effective overburden pressure, undrained side resistance factor in case of drilled shafts [22]; to predict the axial load carrying capacity of driven piles and drilled shafts based on insitu/field tests; prediction of bearing capacity of stone columns [23] and so on. Studies have also been conducted to calculate the settlement of stone columns with the help of ANN models [24, 25], predict the bearing capacity of unreinforced as well as geogrid-reinforced sand bed constructed on stone columns in soft clay using SVR models [26]. Hence, application of AI based approaches/techniques in geotechnical engineering is wide.

2.3 Consideration of Input Parameters

As mentioned in previous section, various factors play a significant role in influencing the lateral deformation/bulging of stone columns. Encasing of columns to significantly reduce column bulging has been well established. Also *l/d* ratio and *s/d* ratio has a remarkable influence on the lateral deformation of columns. To account for the end-bearing as well as floating columns, β (clearance ratio) (*depth between the bottom of the column and test tank/diameter of column*) is taken as an input parameter. Thus, for end-bearing columns, the clearance ratio is 0. Area ratio (A_r) (area of stone column/total area of loading) were considered as other significant input parameters to take into account the variation in stone column arrangement and area of load application in evaluating the radial strain. The stiffness of geosynthetic used as encasements for stone columns play a remarkable role in the radial strain developed in stone column. Hence, geosynthetic stiffness (k) was considered as an input parameter to evaluate the maximum radial strain and its corresponding depth.

Data collected from published literature [4, 5, 7, 8, 14–16, 19, 27–36] was extensively used to evaluate the ε of ordinary and encased stone columns in the present study. However, considerably fewer studies have reported the radial strain in columns. The parameters considered in the study include data from experimental/model tests as well as numerical studies. The range of input parameters was taken into consideration taking account of the usual ranges adopted in model studies/large scale tests. The range of variation of each of these parameters is presented in Table 1.

Table 1. Range of input parameters

Range	Diameter (mm)	l/d	s/d	A_r (%)	λ (%)	Geosynthetic stiffness (kN/m)	Clearance ratio (β)
Maximum	1000	20	12	44	100	10000	12
Minimum	20	1	1	0.7	0.1	0	0

In the present study, Linear Regression, Support Vector Machine (SVM), Gaussian Process Regression (GPR) and Artificial Neural Network (ANN) models were built with the help of deep learning tools/Regression Learner app in *Matlab* software. The different models are described briefly in the following sections.

3 AI Techniques/Mathematical Models

Linear Regression Models
Linear regression models comprise of predictors that are linear in the model parameters. These models are simple to understand and enable quick predictions. However, these usually have low accuracy of prediction. The different linear regression models incorporated in the study include Linear, Interactions linear, Robust Linear and Stepwise Linear model (Matlab [37]).

Support Vector Machine Models
Support vector machine (SVM) analysis is a popular machine learning tool employed for classification and regression analysis. SVM regression is based on kernel functions and hence considered to be a nonparametric technique. SVM models with *Linear*, *Quadratic*, *Cubic* and *Fine*, *Medium* and *Coarse Gaussian* kernel functions are incorporated in this study. Unlike nonlinear SVMs, linear SVMs are simple models and can be easily interpreted. However, nonlinear SVMs have higher predictive accuracy (Matlab [37]).

Gaussian Process Regression Models

In Gaussian process regression, the predictions are based on probability distribution over a range of functions. These models have flexibility in choice of presets in the Model Type gallery to enable a small training error and, protects against over-fitting. These models are mostly non-parametric kernel based probabilistic models. Squared exponential GPR, Matern 5/2 GPR, Exponential GPR, Rational Quadratic GPR (Matlab [37]) models have been included in this study to predict the radial strain.

Artificial Neural Network Models

Artificial neural network is one of the widely adopted Artificial Intelligence techniques. It is a computational technique and tries to imitate the cognitive abilities of a human brain to solve complex problems with a simplistic approach [20]. ANN model architecture consist of three types of layers, i.e. input layer, hidden layers and output layer in each case. The independent variables are fed into the input layer which further transmits information through the hidden layer and the predictions are obtained from the output layer. The system develops a learning algorithm to determine the weights to the input signals as well as adjusts/sets the weights and biases and employs an activation function to develop a learning rule for input-output mapping that yields the smallest possible error [20]. Narrow, medium, wide, Bilayered and Trilayered Neural Network models are incorporated in the present study.

4 Determination of Accuracy of Prediction by Different Models

To determine the best model that make prediction with the closest accuracy, the RMSE, R-squared, MSE and MAE values are compared for each model. Also, the plot of predicted vs true value of radial strain for each model demonstrates the correlation between radial strain reported in various studies and that estimated using different models for the corresponding case.

RMSE (Root Mean Square Error): It is a measure of the difference between observed values and that predicted by a model/estimator. Lower values of RMSE are preferred [37].

R-square (Coefficient of determination): It is always smaller than 1 and mostly greater than 0. It enables comparison of the trained model to the model with a constant response and it equals the mean of the training response. R-squared values close to 1 is preferred [37].

MSE (Mean Squared Error): It is the square of the RMSE, smaller values of MSE are preferred [37].

MAE (Mean Absolute Error): This is similar to the RMSE values but it is insensitive of the outliers. Smaller values of MAE are preferred [37].

5 Results and Discussion

The plots of predicted vs true radial strain obtained from different models are shown in Figs. 1, 2, 3, 4, 5. It is observed from Figs. 1, 2, 3, 4, 5 and Table 2 that the predictions by Neural Networks (Fig. 5(a)–(e)) exhibit higher accuracy as compared to the predictions of Linear Regression (Fig. 1(a)–(d)), SVM (Fig. 2(a)–(c); Fig. 3(a)–(c)) and GPR (Fig. 4(a)–(d)) models. The predictions by neural networks demonstrate a significantly higher value of R-square (0.92–0.97) that shows a close correlation between the predicted and true radial strain reported. Also, RMSE, MSE and MAE values of the ANN models are lower than those of other models. Apart from ANN models, Interactions Linear Regressions model (Fig. 1(b)), Quadratic (Fig. 2(b)) and Cubic SVM (Fig. 2(c)), Squared Exponential GPR (Fig. 4(a)), Matern 5/2 GPR (Fig. 4(b)), Rotational Quadratic GPR (Fig. 4(c)) and Exponential GPR models (Fig. 4(d)) exhibit a high R-squared value in the range of 0.75–0.92. However, the RMSE, MSE and MAE values of these models are higher than

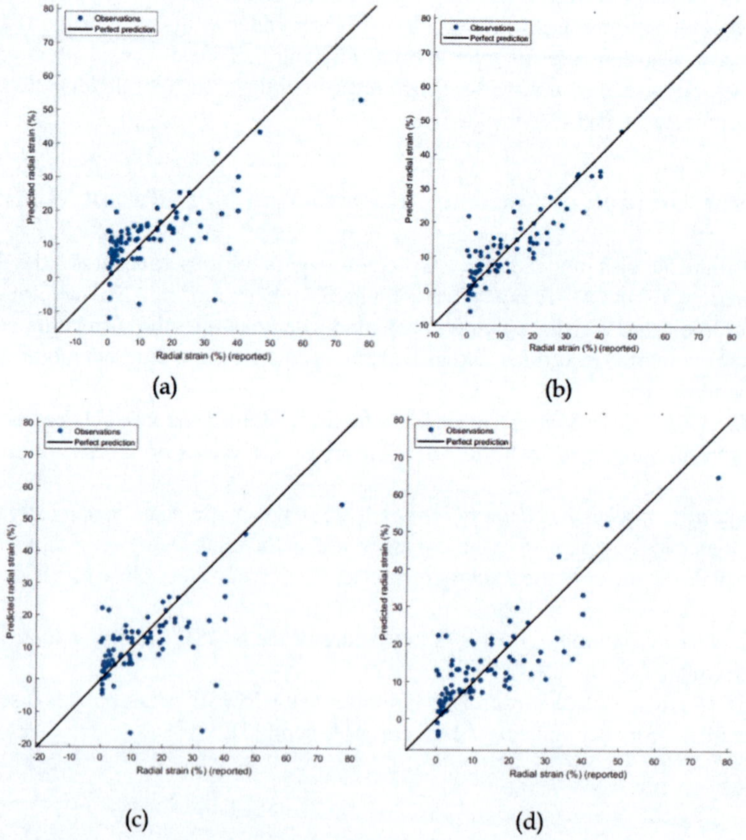

Fig. 1. Predicted vs Reported radial strain as obtained from **(a)** Linear **(b)** Interactions linear **(c)** Robust linear **(d)** stepwise linear regression model

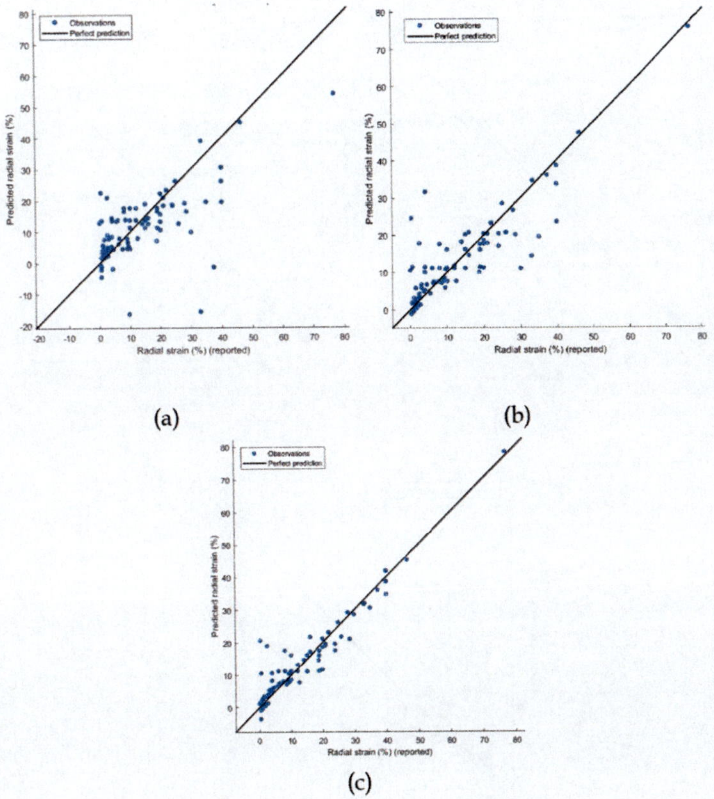

Fig. 2. Predicted vs Reported radial strain as obtained from **(a)** Linear **(b)** Quadratic **(c)** Cubic SVM model

that of ANN models. The values of RMSE, R-square, MSE and MAE of the different models are listed in Table 2.

This observation emphasizes the higher applicability of ANN models in predicting the radial strain.

Comparison of Performance of Different ANN Models

Of all the ANN models adopted in the study, predictions of Wide neural network model (Fig. 5(c)) exhibited the highest accuracy with maximum R-squared value of 0.97 and lowest values of RMSE, MSE and MAE of all the models. The R-squared value of predictions of medium, bilayered and trilayered neural network models were similar to wide ANN model unlike RMSE, MSE and MAE values. The R-squared value of correlation between predicted and reported values was the lowest at 0.92 for narrow neural network. However, the RMSE values of wide and trilayered neural network showed marginal

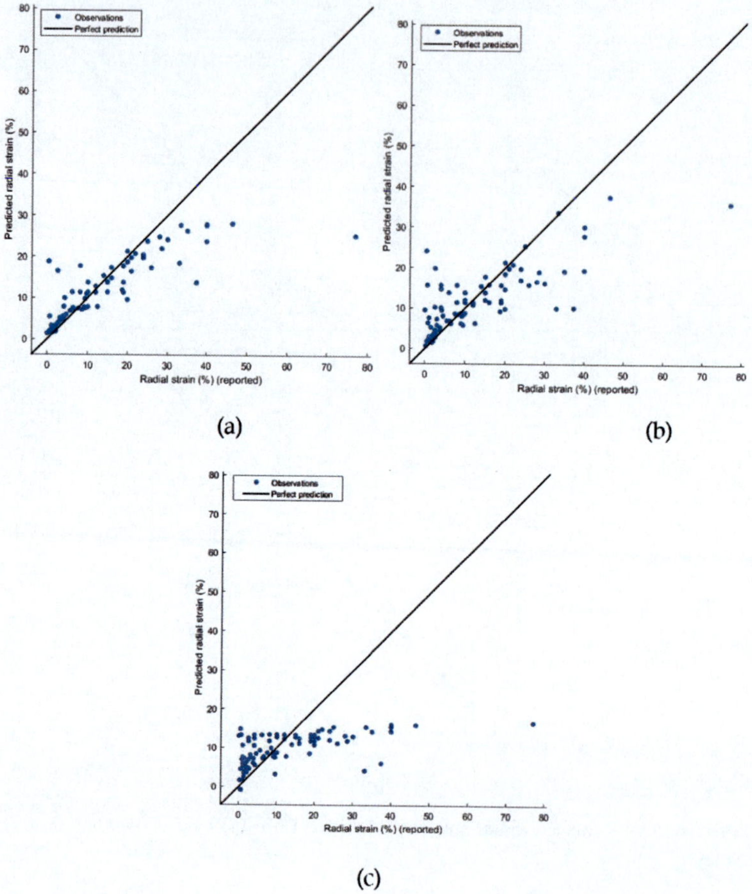

Fig. 3. Predicted vs Reported radial strain as obtained from **(a)** Fine **(b)** Medium **(c)** Coarse Gaussian SVM model

variation. It can therefore be concluded that of Linear Regression, SVM, GPR and ANN models, ANN models (Fig. 5(a)-(e)) give the best prediction of radial strain computed as a function of *d*, *l/d* ratio, *s/d* ratio, λ, *k*, *β*. Also, apart from the narrow ANN model, all the other ANN models yielded close predictions.

Limitations and Future Scope of Work

The models were built with limited number of datasets. Detailed parametric study will be carried out in future and data from such studies may help to better predict the lateral deformation trend/pattern using ANN models. Systematic experimental studies on stone

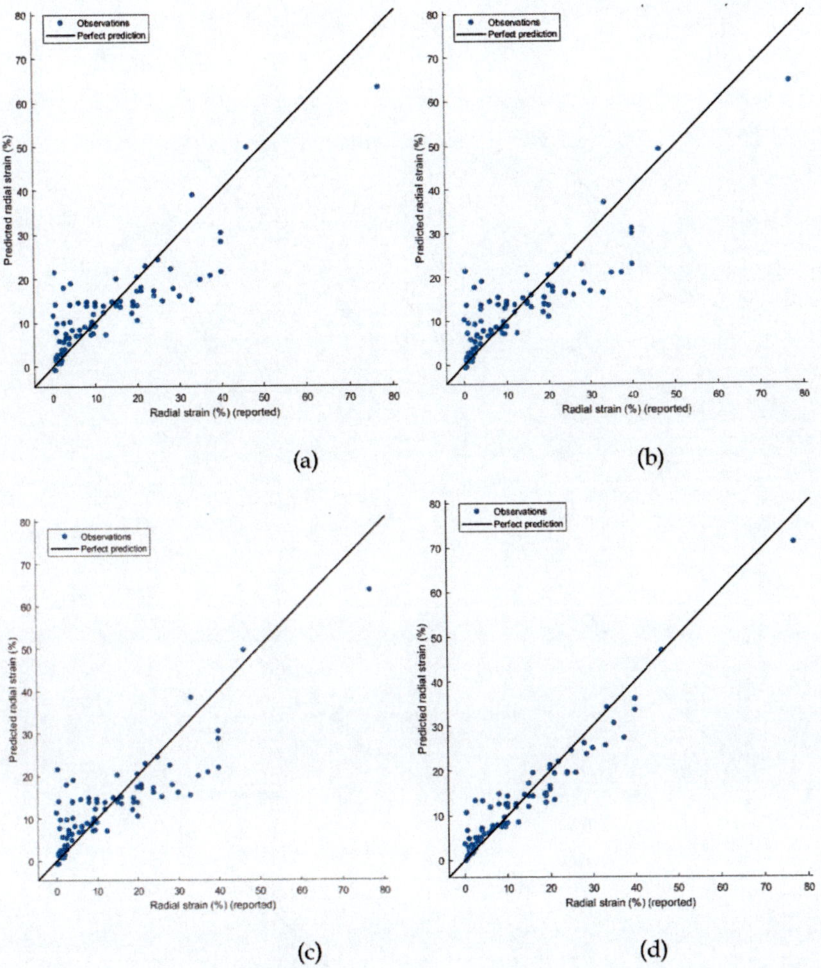

Fig. 4. Predicted vs Reported radial strain as obtained from **(a)** Squared exponential **(b)** Matern 5/2 **(c)** Rotational quadratic **(d)** Exponential GPR model

columns focussing on the failure of columns due to lateral deformation need to be conducted for establishing database for further improvement in ANN model development. It would also enable a reliable sensitivity analysis to understand the principal factors that affect the target parameters. Due to a limited number of datasets sensitivity analysis was not performed in the present study and it is planned to be carried out in further studies.

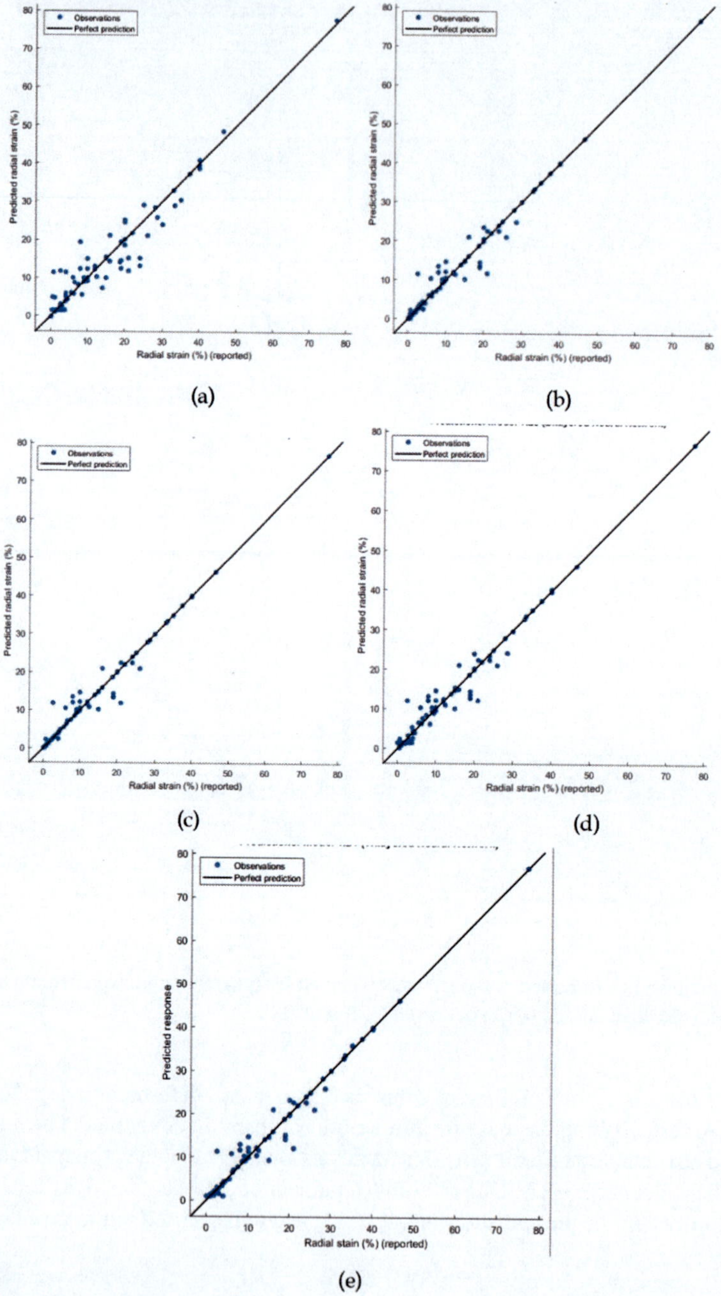

Fig. 5. Predicted vs Reported radial strain as obtained from **(a)** Narrow **(b)** Medium **(c)** Wide **(d)** Bilayered **(e)** Trilayered ANN model

Table 2. Values of RMSE, R-square, MSE and MAE of different models

Model type	RMSE	R-squared	MSE	MAE
Linear regression	9.6588	0.48	93.235	7.1823
Interactions linear regression	5.7451	0.82	33.006	4.2865
Robust linear regression	10.311	0.41	106.31	6.5791
Stepwise linear regression	7.8935	0.65	62.307	5.9342
Linear SVM	10.326	0.41	106.63	6.5033
Quadratic SVM	6.679	0.75	44.609	4.0724
Cubic SVM	4.1814	0.9	17.484	2.6865
Fine Gaussian SVM	8.0808	0.64	65.3	4.2765
Medium Gaussian SVM	8.7979	0.57	77.403	5.4252
Coarse Gaussian SVM	11.917	0.21	142.01	7.7625
Squared exponential GPR	7.1356	0.72	50.916	5.0399
Matern 5/2 GPR	6.785	0.74	46.036	4.7668
Exponential GPR	3.7181	0.92	13.824	2.6707
Rotational quadratic GPR	7.0558	0.72	49.785	4.9707
Narrow neural network	3.6821	0.92	13.558	2.3546
Medium neural network	2.2659	0.97	5.1343	1.0773
Wide neural network	2.2062	0.97	4.8672	0.9548
Bilayered neural network	2.3825	0.97	5.6763	1.3729
Trilayered neural network	2.1939	0.97	4.8133	1.1229

6 Conclusions

The radial strain in OSC and ESC were predicted with the help of Linear Regression, SVM, GPR and ANN models. The variance in column characteristics like the column dimensions, geometry and arrangement, area of load application, including the effect of the presence of a stiff stratum below the columns i.e., end-bearing and floating column, were taken into account by the inclusion of 7 input parameters considered for prediction of target values. The datasets were taken from published literature. Such ANN models are valuable, especially in geotechnical engineering applications, since most problems are dependent on a wide range of factors and determining the physical relationship between such factors involves considerable time. The ANN models are quick to provide an overall estimate of the target parameters in investigation and can work to offer a preliminary idea of the problem and its feasibility/applicability as a solution. However, incorporating basic ideas or prior knowledge of the physical laws that govern the relationship between input parameters that influence the target parameter significantly and the target data would help to ignore noisy data and therefore enhance the efficiency and applicability of such models. Such models will help to predict the radial strain even in larger size columns adopted in field conditions based on such critical input parameters. However, such AI techniques can be adopted as an additional tool and are not recommended as a substitute to field studies/tests. The conclusions derived from the present study are presented below:

1. As compared to Linear Regression, SVM and GPR models, ANN models exhibited higher efficiency of prediction of radial strain in OSC and ESC columns.
2. ANN models with multiple layered networks like medium, wide, bilayered, trilayered neural networks yielded predictions with higher accuracy as compared to narrow ANN models.
3. Exponential GPR and Cubic SVM models were also found to give close predictions with a considerable R-squared value of 0.90.

References

1. Greenwood, D.A.: Mechanical improvement of soils below ground surface. In: Ground Engineering Proceedings Conference, Institution of Civil Engineers, London, pp. 11–22 (1970)
2. Barksdale, R.D., Bachus, R.C.: Design and Construction of Stone Columns. Report No. FHWA/RD-83/026, Office of Engineering and Highway Operations Research and Development, Federal Highway Administration, Washington, DC (1983)
3. Murugesan, S., Rajagopal, K.: Geosynthetic-encased stone columns: numerical evaluation. Geotext. Geomembr. **24**(6), 349–358 (2006)
4. Ali, K., Shahu, J.T., Sharma, K.G.: Model tests on geosynthetic-reinforced stone columns: a comparative study. Geosynth. Int. **19**(4), 292–305 (2012)
5. Hasan, M., Samadhiya, N.K.: Experimental and numerical analysis of geosynthetic-reinforced floating granular piles in soft clays. Int. J. Geosynth. Ground Eng. **2**(3), 1–13 (2016)
6. Wood, D.W., Hu, W., Nash, D.F.: Group effects in stone column foundations: model tests. Geotechnique **50**(6), 689–698 (2000)
7. Murugesan, S., Rajagopal, K.: Studies on the behavior of single and group of geosynthetic encased stone columns. J. Geotech. Geoenviron. Eng. **136**(1), 129–139 (2010)
8. Dash, S.K., Bora, M.C.: Influence of geosynthetic encasement on the performance of stone columns floating in soft clay. Can. Geotech. J. **50**(7), 754–765 (2013)
9. Miranda, M., Fernández-Ruiz, J., Castro, J.: Critical length of encased stone columns. Geotext. Geomembr. **45**(2), 68–80 (2021). https://doi.org/10.1016/j.geotexmem.2021.05.003
10. Balaam, N.P., Booker, J.R.: Analysis of rigid rafts supported by granular piles. Int. J. Numer. Anal. Meth. Geomech. **5**(4), 379–403 (1981)
11. Priebe, H.J.: The design of vibro replacement. Ground Eng. **28**(10), 31 (1995)
12. IS 15284: Design and Construction for Ground Improvement: Guidelines Part 1: Stone Columns. Bureau of Indian Standards, New Delhi, India (2003)
13. Hughes, J.M.O., Withers, N.J., Greenwood, D.A.: A field trial of the reinforcing effect of a stone column in soil. Geotechnique **25**(1), 31–44 (1975)
14. Rathod, D., Abid, M.S., Vanapalli, S.K.: Performance of polypropylene textile encased stone columns. Geotext. Geomembr. **49**(1), 222–242 (2021)
15. Malarvizhi, S.N., Ilamparuthi, K.: Comparative study on the behaviour of encased stone column and conventional stone column. Soils Found. **47**(5), 873–885 (2007)

16. Mazumder, T., Ayothiraman, R.: Numerical study on behaviour of encased stone columns with partial content of shredded tyre chips in soft clay bed. Int. J. Geosynth. Ground Eng. **7**(2), 1–14 (2021). https://doi.org/10.1007/s40891-021-00280-z
17. Mazumder, T., Rolaniya, A.K., Ayothiraman, R.: Experimental study on behaviour of encased stone column with tyre chips as aggregates. Geosynth. Int. **25**(3), 259–270 (2018)
18. Almeida, M.S., Hosseinpour, I., Lima, B.: Field studies of stone columns and geosynthetic-encased columns. In: From Research to Applied Geotechnics, pp. 166–180. IOS Press (2019)
19. Shivashankar, R., Babu, M.D., Nayak, S., Rajathkumar, V.: Experimental studies on behaviour of stone columns in layered soils. Geotech. Geol. Eng. **29**(5), 749 (2011)
20. Shahin, M.A.: State-of-the-art review of some artificial intelligence applications in pile foundations. Geosci. Front. **7**(1), 33–44 (2016)
21. Goh, A.T.: Nonlinear modelling in geotechnical engineering using neural networks. Trans. Inst. Eng. Aust. Civ. Eng. **36**(4), 293–297 (1994)
22. Goh, A.T., Kulhawy, F.H., Chua, C.G.: Bayesian neural network analysis of undrained side resistance of drilled shafts. J. Geotech. Geoenviron. Eng. **131**(1), 84–93 (2005)
23. Das, M., Dey, A.K.: Prediction of bearing capacity of stone columns placed in soft clay using ANN model. Geotech. Geol. Eng. **36**(3), 1845–1861 (2018)
24. Chik, Z., Aljanabi, Q.A.: Intelligent prediction of settlement ratio for soft clay with stone columns using embankment improvement techniques. Neural Comput. Appl. **25**(1), 73–82 (2013). https://doi.org/10.1007/s00521-013-1449-0
25. Chik, Z., Aljanabi, Q.A., Kasa, A., Taha, M.R.: Tenfold cross validation artificial neural network modeling of the settlement behavior of a stone column under a highway embankment. Arab. J. Geosci. **7**(11), 4877–4887 (2013). https://doi.org/10.1007/s12517-013-1128-6
26. Dey, A.K., Debnath, P.: Empirical approach for bearing capacity prediction of geogrid-reinforced sand over vertically encased stone columns floating in soft clay using support vector regression. Neural Comput. Appl. **32**(10), 6055–6074 (2019). https://doi.org/10.1007/s00521-019-04092-1
27. Deb, K., Samadhiya, N.K., Namdeo, J.B.: Laboratory model studies on unreinforced and geogrid-reinforced sand bed over stone column-improved soft clay. Geotext. Geomembr. **29**(2), 190–196 (2011)
28. Elsawy, M.B.D.: Behaviour of soft ground improved by conventional and geogrid-encased stone columns, based on FEM study. Geosynth. Int. **20**(4), 276–285 (2013)
29. Gnandji, R.A., Kalumba, D.: Experimental and numerical analysis of the behaviour of rammed stone columns installed in a South African soft soil. Int. J. Eng. Sci. Innov. Technol. **3**(6), 477–499 (2014)
30. Indraratna, B., Ngo, N.T., Rujikiatkamjorn, C., Sloan, S.W.: Coupled discrete element–finite difference method for analysing the load-deformation behaviour of a single stone column in soft soil. Comput. Geotech. **63**, 267–278 (2015)
31. Gu, M., Zhao, M., Zhang, L., Han, J.: Effects of geogrid encasement on lateral and vertical deformations of stone columns in model tests. Geosynth. Int. **23**(2), 100–112 (2016)
32. Demir, A., Sarici, T.: Bearing capacity of footing supported by geogrid encased stone columns on soft soil. Geomech. Eng. **12**(3), 417–439 (2017)
33. Hong, Y.S., Wu, C.S., Kou, C.M., Chang, C.H.: A numerical analysis of a fully penetrated encased granular column. Geotext. Geomembr. **45**(5), 391–405 (2017)
34. Rajesh, S., Jain, P.: Influence of permeability of soft clay on the efficiency of stone columns and geosynthetic-encased stone columns–a numerical study. Int. J. Geotech. Eng. **9**(5), 483–493 (2015)
35. Pandey, B.K., Rajesh, S., Chandra, S.: 3-D finite element study of embankment resting on soft soil reinforced with encased stone column. In: Latha Gali, M., Raghuveer Rao, P. (eds.) Problematic Soils and Geoenvironmental Concerns. LNCE, vol. 88, pp. 451–465. Springer, Singapore (2021). https://doi.org/10.1007/978-981-15-6237-2_38

36. Xu, Z., Zhang, L., Zhou, S.: Influence of encasement length and geosynthetic stiffness on the performance of stone column: 3D DEM-FDM coupled numerical investigation. Comput. Geotech. **132**, 103993 (2021)
37. Matlab R2021a User Interface (2021). https://in.mathworks.com/help/stats/regression-and-anova.html?s_tid=CRUX_lftnav

Research Status and Prospect of Anti-slide Piles for Slope Stabilization

Hong-Hu Zhu[1], B. P. Naveen[2], Jing Wang[1], Gang Cheng[1,3](✉), and Yuxin Gao[1]

[1] School of Earth Sciences and Engineering, Nanjing University, Nanjing, China
chenggang@ncist.edu.cn

[2] Department of Civil Engineering, Amity University Haryana, Gurgaon, India

[3] School of Computer Science, North China Institute of Science and Technology, Langfang, China

Abstract. The research status of anti-slide piles for slope stabilization is summarized in this paper. The calculation method in pile design is comprehensively reviewed, and special attention is paid to the selection of design parameters according to the actual engineering conditions. The key points in the design and application of anti-slide piles are presented in detail, especially the interaction and deformation coordination between anti-slide piles and soil. Finally, some important issues, such as the soil arching effect and internal force calculation, are discussed.

Keywords: Anti-slide pile · Landslide · Slope stability · Soil arching · Pile-Soil interaction · Pile-Soil interface

1 Introduction

Landslides are one of the most widespread and serious geohazards in the Belt and Road countries, which usually lead to a great loss of human life and property [1, 2]. Consequently, the enormous challenges to maintaining the stability of landslides inspired researchers to develop various stabilizing structures [3, 4]. The anti-slide pile, as known as the stabilizing pile, is designed to be installed through the unstable soil layer and embedded below the potential sliding surface in the stable soil/rock strata (Fig. 1). The slope can be effectively stabilized by these piles, which can transfer part of the thrust force from the sliding soil to the stable strata. Anti-slide piles are widely used in various slope retaining projects because of their large anti-slide ability, good retaining effect, flexible pile position arrangement, and various combination forms that can be adapted according to the actual situation.

The application of anti-slide piles to slope stabilization began in the 1930s. With the optimization of anti-slide pile design and application, anti-slide piles have gradually become the main way of landslide treatment, and many different types of anti-slide pile structures have been developed [6, 7]. After the 1970s, the development of anti-slide piles entered a new era. The developed countries, such as Japan, adopted anti-slide piles with different sizes and shapes as anti-slide structures to treat large landslides and found the active role of prestressed anchor cables in these circumstances [8]. In China, anti-slide

H.-H. Zhu et al. (Eds.): BRWSG 2021, LNCE 230, pp. 223–236, 2022.
https://doi.org/10.1007/978-981-16-9963-4_18

piles were first used in 1954 to treat the Shijiaba landslide in the Baocheng Railway. Since the 1980s, the combination use of anchor cables and anti-slide piles has been widely used, transforming anti-slide piles from the passive to the active stress state and effectively reducing the cross-sectional area and burial depth of piles. In engineering practice, due to the differences in landslide types, geological conditions and landform, different types of anti-slide piles are employed to achieve the optimum reinforcing effect. The efficient use of anti-slide piles can significantly save construction costs.

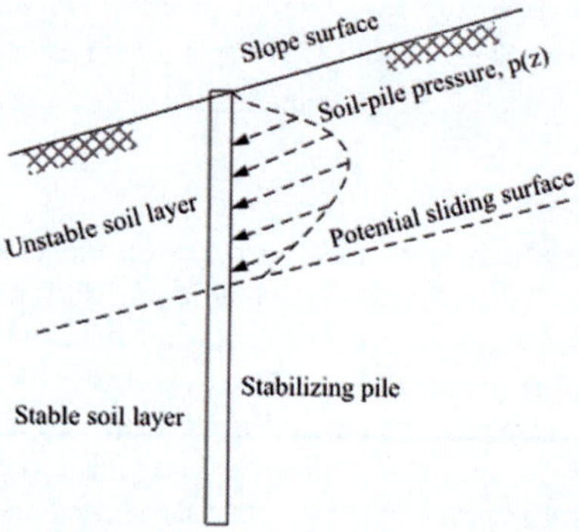

Fig. 1. Stabilizing pile embedded into a semi-infinite slope [5].

Many landslides and pile foundation improvements urge engineers to pay more attention to the design and effectiveness of stabilizing piles [9]. The stability effect of a stable pile is affected by many factors; The driving force acting on the pile and the position of the pile are two important factors. The driving force of pile landslides has an important influence on determining the optimal position of the pile. On the premise of meeting structural safety and stability, it is necessary to fully understand the selection of key parameters and influencing factors in the design of anti-slide piles to reduce the project cost. The design elements of anti-slide piles include the plane position, section size and shape and anchorage depth. Scholars worldwide have researched the design elements and calculation of anti-slide piles in slopes. Among them, the calculation of internal force of anti-slide piles is mainly divided into two methods: displacement method and pressure method, which have their characteristics and scope of application [10–12]. Based on different calculation methods, selecting optimized design parameters, especially pile spacing, landslide stability, and pile plane layout, can improve the stability coefficient of landslide mass and reduce the project cost [13].

The working state of anti-slide piles can be evaluated through field monitoring. At present, the conventional monitoring methods of anti-slide piles mainly include inclinometer, reinforcement gauge, earth pressure gauge, total station and distributed optical fiber technology [14]. Inclinometer, earth pressure gauge and reinforcement gauge belong to point monitoring, and their monitoring results can be directly substituted into the existing design and calculation method of anti-slide piles to evaluate the working state of anti-slide piles. Compared with conventional monitoring methods, distributed optical fiber sensing technology has significant advantages, such as distributed, long monitoring distance, little interference, a large amount of data acquisition, high data acquisition efficiency, real-time monitoring and long-term stability, which can meet the requirements of safety monitoring of slope reinforcement engineering and early warning of landslide [15].

This paper summarizes and evaluates the research progress and some problems existing in the research from the aspects of calculation method of anti-slide piles, pile-soil interaction, optimal layout and field monitoring technology, which can provide a useful reference for future study.

2 Anti-slide Pile-Soil Interaction

The interaction between anti-slide piles and landslide mass is mainly concentrated between anti-slide piles and landslide mass. Under the action of self-weight or upper load of soil between piles, the soil between piles produces uneven displacement, resulting in soil arch between piles. The research history of the soil arching effect can be traced back to the granary effect discovered by Roberst in 1884. In 1895, German engineer H. A. Janssen quantitatively explained this phenomenon with the continuous medium model. The soil arching effect was first proposed by Terzaghi and verified by the movable door test, and then the classical Terzaghi relaxation earth pressure calculation method [16] was derived. With the continuous development of soil arching theory, the arching effect of anti-slide pile and soil around pile has gradually attracted the attention of engineers and technicians, and more and more related research problems are involved.

2.1 Theoretical Analyses

The interaction process between anti-slide pile and soil is very complex. Anti-slide piles are subject to the load transmitted by soil, and the deformation and displacement of soil are also subject to the reaction of anti-slide piles. The interaction process of the two is mutual, which makes it more difficult to analyze the soil arch effect theoretically. At present, some scholars have studied the relationship between longitudinal soil arching and active earth pressure on the back of rigid retaining wall [17], while the soil on the back of the pile sheet wall is more arched in the horizontal direction. Bransby and Shelke respectively studied the deformation and soil arching effect of soil around pile sheet wall and laterally loaded pile group foundation and obtained some regular understanding [18, 19]. In view of the more extensive application of column plate structures such as pile sheet wall in China, and the soil arching effect between piles directly affects the load acting on the retaining structure between piles, some scholars have studied the relationship between the soil arching effect and the earth pressure acting on the baffle.

2.2 Model Tests

Physical model tests can provide the deformation characteristics of the landslides stabilized by anti-slide piles [20, 21]. A series of centrifuge model tests were conducted to investigate the failure mechanism of pile-reinforced slopes under self-weight loading and vertical loading conditions [22]. Hu took the Majiagou landslide and its pile system as a real prototype, and using a model testbed with 57-cm-long test piles made of reinforced concrete and polyesteramide to simulate rigid and flexible piles (Fig. 2), the displacements of two physical models were monitored during progressive loading to simulate the landslide-stabilizing pile system [23].

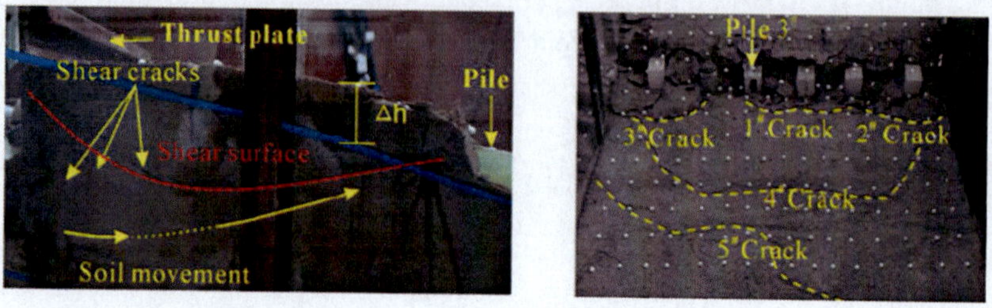

(a) (b)

Fig. 2. Deformation characteristics of model piles after failure. (a) Side view of the model between the thrust plate (left) and the piles (right). (b) Oblique front view, looking uphill, showing crack development of the arcuate shear cracks downslope of the piles (after [23]).

3 Designs and Calculation of Anti-slide Piles

After years of development, the calculation theory of ordinary anti-slide pile (laterally loaded pile) has formed a relatively complete system (Fig. 3) [24]. According to different states of foundation, it is mainly divided into elastic theory, foundation reaction, and numerical simulation methods. According to the different assumptions of soil around the pile, the foundation reaction method is divided into limit foundation reaction method, elastic foundation reaction method and composite foundation reaction method.

3.1 Calculation Methods of Internal Forces in Anti-slide Piles

General Calculation Methods

The calculation methods of internal forces of anti-slide piles widely used in many practices include the cantilever pile method, foundation coefficient method, finite element method and matrix analysis method. These methods only consider the stress of an anti-slide pile when the soil around the pile is in the linear elastic stage.

The elastic foundation coefficient method is generally used abroad, and the internal force calculation of anti-slide piles is divided into two parts above and below the sliding surface. The static equilibrium structure method is used to calculate the internal force of anti-slide piles on the sliding surface, and the finite difference method is used to calculate the internal force below the sliding surface.

The traditional cantilever pile method and foundation coefficient method are widely used in China. The cantilever pile method is characterized by the pile section above the sliding surface as a cantilever beam. The pile section below the sliding surface is calculated according to the Winkler elastic foundation beam model. However, the simplification of the actual stress state of the pile by this method is safe, so the calculation

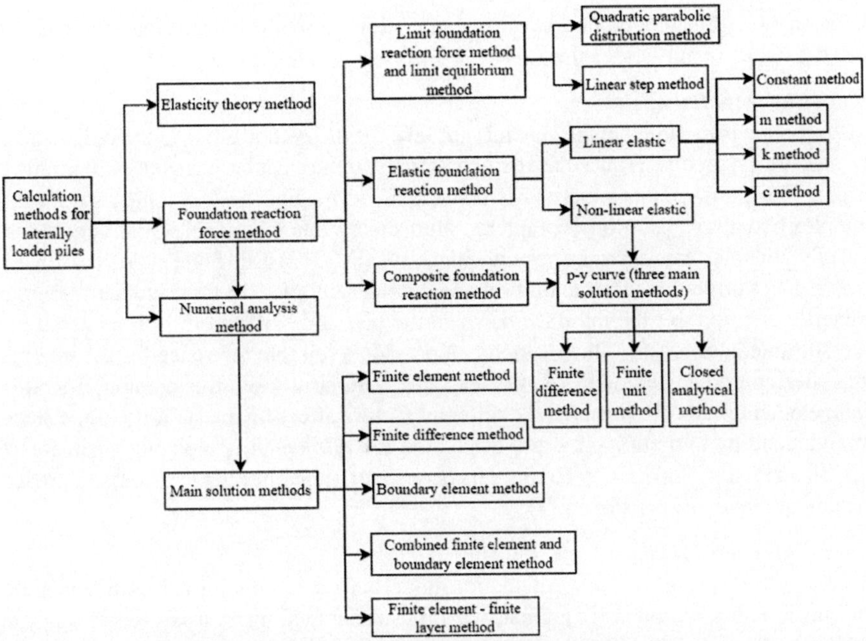

Fig. 3. Current calculation theories of anti-slide pile

results are conservative. The foundation coefficient method regards the whole pile as an elastic foundation beam, which is relatively close to the actual stress state of anti-slide piles.

Recent Developments in Calculation Methods

Although the existing design method of anti-slide piles can meet practical engineering requirements, it does not mean that it meets the actual stress state of anti-slide piles. Therefore, relevant researchers have made many efforts to find a more reasonable calculation mode and method for anti-slide piles.

Improved Cantilever Pile Method

Dai and Shen (2003) assumed that the part of anti-slide piles above the sliding surface is an elastic directional hinged cantilever beam and the part below the sliding surface is a Winkler elastic foundation beam [25]. The improved foundation coefficient "m" method is used to calculate the internal force of the pile. The displacement of the pile at the sliding surface depends on the calculation results below the sliding surface. Therefore, the displacement of anti-slide piles above the sliding surface is superimposed by the calculation results of the upper cantilever beam and the calculation results at the sliding slide to establish a unified coordinate system and make the calculation results more in line with the actual stress and deformation. Compared with the traditional cantilever beam method, this method improves the calculation accuracy and analysis efficiency of internal forces in anti-slide piles.

Finite Element Method
For anti-slide piles, whether the general cantilever beam method or the improved method, the calculation process is also quite cumbersome, and the calculation results will produce errors. The finite element method is convenient in dealing with complex structures, complex boundary conditions, complex stratum conditions and complex load conditions, and pile-soil interactions can be considered. Dai et al. (2012) used finite element software to establish a three-dimensional finite element model, adopted the finite element strength reduction method, set the interface between the pile and soil, and fully considered the pile-soil interaction [26]. This method can provide a reliable reference for the internal force design and calculation of anti-slide piles. Zhu et al. used the strength reduction finite element method to analyze the different factors affecting the stability of the slope strengthened by anti-slide piles and calculated the pile bending moment, shear force, displacement and slope safety factor of the slope anti-slide piles in the process of graded loading at the top of the slope [27].

Finite Difference Method
Dai et al. (2003) proposed a new calculation model and corresponding calculation method for elastic anti-slide piles [25]. First, the finite difference method is used to analyze the internal force of the whole pile, the corresponding calculation method is deduced, and the program for calculation and graphic processing is compiled. The calculation results of this method are consistent with the laws obtained from theoretical analysis and field tests. At the same time, compared with the existing common calculation methods, the maximum bending moment is significantly reduced, which releases the application potential of anti-slide piles. This makes anti-slide piles design and calculation more economical, reasonable and effective.

Finite Difference Method
Cundall proposed the particle discrete element method for rock mechanics in 1971, proposed the particle discrete element method for soil mechanics in 1979, and introduced the soft particle model [28, 29]. The particle discrete element method can overcome the problems that the finite element method cannot predict fracture development and large deformation.

Calculation of the Lateral Force Acting on the Pile
The evaluation of earth pressure under the action of stable piles is of great significance to the study of slope stability. The lateral force acting on the pile depends largely on the relative movement between the pile and soil (Fig. 4) [5]. In the design and calculation of anti-slide piles, the first step is to determine the failure mode, and the failure mode is the basis for establishing the calculation method [30]. There are two types of failure modes of slope strengthened with anti-slide piles: pile structure failures, such as pile dumping or damage; The other is the failure of slope soil, which slides away from the pile. Based on the assumptions of different failure modes and pile-soil interaction methods, a series of calculation methods for the lateral force acting on the pile and corresponding

design methods of anti-slide piles have been developed. The common methods are the load-structure method, Viggiani method, Ito method, and Poulos method.

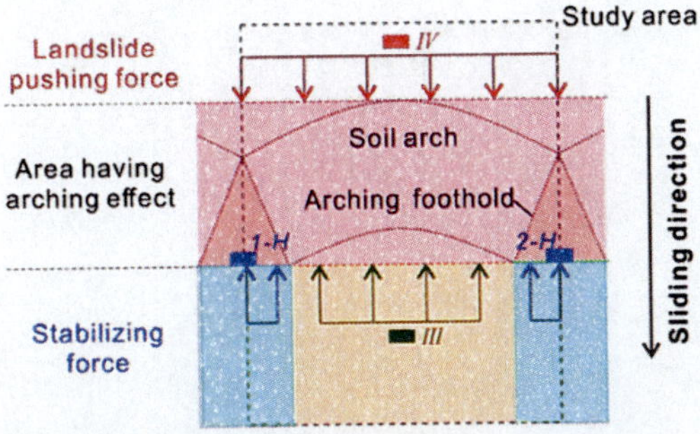

Fig. 4. Calculation of the lateral force acting on the pile

Load-Structure Method

The load-structure method is currently commonly used in anti-slide piles in China. The forces acting on piles are calculated in the two parts below and above the sliding surface, respectively. Then, the sliding force behind the pile and the anti-slide force before the pile were calculated according to the limit equilibrium method of the rigid body. The recursive method is widely used to calculate the sliding force and anti-sliding force in China.

Viggiani Method

The Viggiani method assumes that 1) the pile is rigid; 2) the soil around the pile and the sliding bed is saturated clay; 3) the cohesion of the saturated undrained shear is c_1 and c_2, respectively; 4) the lateral ultimate bearing capacity between piles and soil is $P_y = k{\cdot}c{\cdot}d$ (k is the lateral bearing capacity coefficient, and d is the pile diameter) [31]. Viggiani later proposed a calculation model of pile-soil failure and improved the Viggani method [32]. However, the modified Viggiani method still takes the yield value of pile-soil and materials as the constraint condition of static equilibrium and does not consider the pile-soil interaction.

Ito Method

Ito and Matsui (1975) established an equation for calculating the ultimate lateral pressure of sliding soil based on plastic deformation theory and the force model of a single row of piles [33]. The model is suitable for the infinite rigid pile. It is assumed that the soil around the pile is in a plastic state and meets the Mohr-Coulomb yield criterion. The method considered the effects of pile diameter, pile spacing and parameters of the surrounding soil. The Ito Tomio method considers the rigidity of piles and the bearing capacity of the soil and does not consider the pile-soil interaction. Kumar and Hall (2006) studied the lateral landslide thrust behind the pile in the process of landslide instability, and failure strengthened by anti-slide pile and pointed out that when considering the horizontal soil arching effect behind the pile, the lateral landslide thrust behind the pile was calculated according to the local plastic deformation theory (Fig. 5) is significantly greater than the measured value [34].

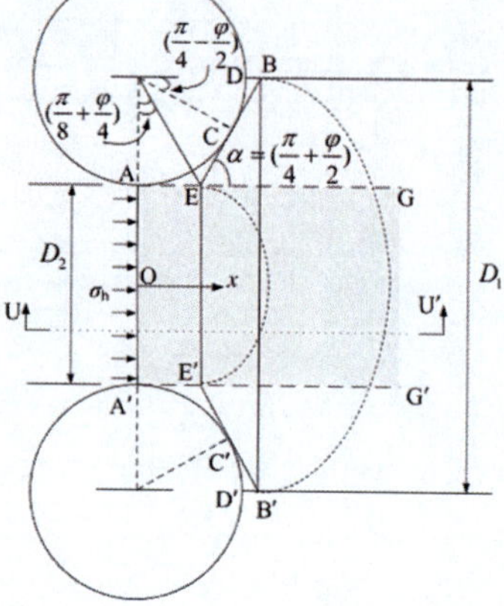

Fig. 5. Soil arch between neighboring piles [33].

Poulos Method
The Poulos method takes the displacement between piles and soil as the basic unknown quantity to analyze the whole pile above and below the sliding surface. In this method, the continuity and behavior of soil with depth are added into the analysis, and various diagrams for calculating the stress of a single pile are obtained. At the same time, the soil properties, boundary conditions, and stiffness of piles are considered. Poulos used an analysis method in which a simplified form of the boundary element method was employed to study the response of a row of passive piles incorporated in limit equilibrium solutions of slope stability [35]. This method revealed the existence of three modes of failure: (1) "flow mode", (2) "short-pile mode", and (3) "intermediate mode". This finding contributed to the practical design of stabilizing piles. Poulos highlighted that the flow mode created the least damage effect of soil movement on piles; if the piles required protection, efforts should be made to promote this mode of behavior.

Limit Equilibrium Method (LEM)
However, these methods are limited to homogeneous soil slopes, where the ground or sliding surface geometry is relatively regular and has no special characteristics (e.g., uneven ground or sliding surface). In addition, these studies do not consider the effects of pore water pressure and shear strength in the slip zone on the lateral force and its distribution. In actual landslide engineering, the most widely used method to evaluate the slope stability and the lateral force acting on the pile is still the limit equilibrium method (LEM) combined with the slice method (Fig. 6) [36, 39]. This technology can adapt to complex slope geometry with different soil properties, pore water pressure conditions, different sliding surface shapes and the effects of external boundary loads [37].

The finite element model combined with the shear strength reduction method (FE-SRM) is widely used in comprehensive slope stability analysis. One of the advantages of FE-SRM is that there is no need to assume the shape and location of the critical failure surface. The location, shape and size of the plastic deformation area can be used to quantify the sliding surface and safety factor [38].

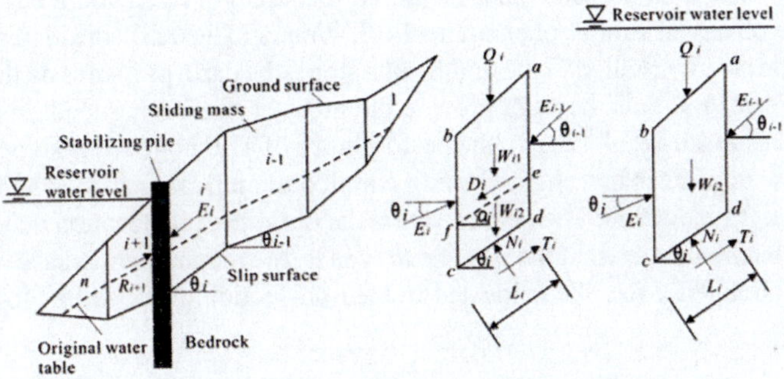

Fig. 6. Schematic representation of the LEM method. [39]

4 Optimal Designs of Anti-slide Piles

4.1 Optimum Pile Spacing

The formation of soil arch between piles is mainly due to the uneven displacement of soil between piles under the action of self-weight or upper load of soil between piles. The soil arching effect is realized through the exertion of the shear strength of the soil. At present, the works on the strength design of anti-slide piles have been relatively mature, but the arching effect is not fully considered in the design and calculation of anti-slide piles. The calculation of the critical spacing of anti-slide piles is a key problem in the design of anti-slide piles, which involves the safety and economy of anti-slide piles. Making full use of the soil arching effect between anti-slide piles can achieve the purpose of economy and efficiency. In recent years, studies on pile spacing under the soil arching effect have been hot, and many results have been obtained.

Wang et al. (2001) deduced the calculation method of pile spacing according to the fact that the sum of total friction resistance between anti-slide piles is beyond the thrust force between piles [40]. Zhou et al. (2004) analyzed the soil arching effect of anti-slide piles using theoretical analyses. They concluded that the static balance condition, the strength condition of the across-section at the arch crown and the arch toe are critical factors that must be considered when determining the pile spacing [41]. Zhao et al. (2010) assumed that the soil arch has the shape of a parabola, and the static balance and strength conditions are met in the horizontal and vertical planes. Then, they deduced the pile spacing calculation equation and obtained the optimal pile spacing [42]. He et al. (2014) inferred the maximum anti-slide pile spacing according to the restriction that the soil strength of the arch toe and the soil shear stress at any point on the pile-soil contact surface should be less than the maximum anti-slide force of the piles [43].

4.2 Optimal Pile Position

The position of anti-slide piles significantly influences the effect of anti-slide piles in stabilizing slopes. Zhu et al. claimed that the factor of safety of the slope is the largest when the pile is located between the middle of the slope and the slope top, and the safety margin of the pile is large (Fig. 7). In this case, the efficiency of anti-slide piles is the highest [27]. Based on the analyzing results of the strength reduction method, Cai and Ugai (2000) observed similar phenomena [44]. Won et al. (2005) considered that when anti-slide piles were installed in the middle of a slope, the earth pressures on the pile sides were the largest, and thus they can provide the greatest reinforcing loading [45]. In the study of Shooshpasha et al. (2020), the performance of a homogeneous slope reinforced with one row of piles is investigated using coupled numerical analysis and the method of shear strength reduction. They claimed that the optimum pile location depends on the pile length, but no matter what the piles head type is, the maximum increase in the factor of safety is obtained if the pile is located in the mid-section of the slope [46].

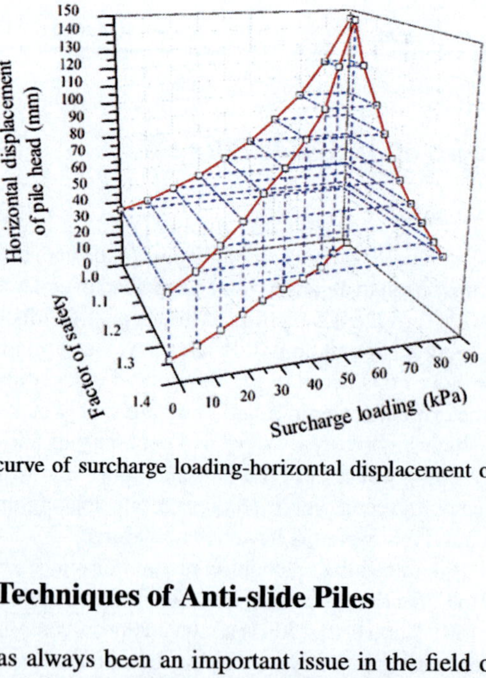

Fig. 7. Relationship curve of surcharge loading-horizontal displacement of pile head-factor of safety [27].

5 Monitoring Techniques of Anti-slide Piles

Slope monitoring has always been an important issue in the field of modern geology and rock engineering. It is an important means to understand and master the evolution process of the slope, timely capture the information of landslide disaster, and provide reliable data and scientific basis for the correct analysis, evaluation, prediction, prediction and treatment of landslide. Because the deformation field of the slope, especially the reservoir bank slope, is the result of the comprehensive influence of many factors, the effects of groundwater, rainfall, temperature, geophysics, chemistry and other field factors cannot be ignored. At present, slope monitoring focuses on deformation monitoring and begins to develop towards multi-field and multi-parameter monitoring. In addition, the monitoring technology is also developing from simple observation of ground macro features and ground instrument monitoring to deep slope monitoring. The development of space telemetry and remote sensing technology provides a new means for mastering slope stability and timely evaluating landslides [47].

The working state of anti-slide piles can be evaluated through field monitoring. At present, the conventional monitoring methods of anti-slide piles mainly include inclinometers, reinforcement gauges, earth pressure cells, total stations and distributed fiber optic sensing technologies. Inclinometer, earth pressure cells and reinforcement gauges are based on point measurement, and their monitoring results can be directly substituted into the existing design and calculation method of anti-slide piles to evaluate the working state of anti-slide piles. These monitoring methods are relatively widely used and mature in engineering with many results. Conventional monitoring methods often have problems in the monitoring process, such as large error, low data acquisition efficiency, improper number of monitoring points, easy-to-produce structural effects [32]. At present, the deep displacement of an anti-slide pile is mainly obtained by an inclinometer. However, the inclinometer sampling interval and data acquisition efficiency are low, and the comparison error of the manual inclinometer at different time points is relatively large.

With the development of DFOS technology, it has been widely used in geotechnical field tests. At present, fiber optic sensing technology has been used to study on-site monitoring of anti-slide piles [48, 49]. Compared with conventional monitoring methods, distributed fiber optic sensing technology has significant advantages, such as full distribution, long monitoring distance, little interference, a large amount of data acquisition, high data acquisition efficiency, real-time monitoring and long-term stability, which can meet the requirements of safety monitoring of slope reinforcement engineering and early warning of landslides [50, 51]. Zhu et al. (2009) investigate the application of fiber Bragg grating technology to monitoring the stability of a reinforced highway slope in Hong Kong [52]. The DFOS technologies are becoming more refined through many trial tests and have shown great potential in capturing anti-slide pile behavior in different working conditions.

6 Conclusions

Anti-slide piles have been widely used in various slope stabilization. Based on the summary and discussion of the current results, this paper analyzes the research status of anti-slide piles and proposes some problems to be solved in the design and field test of anti-slide piles.

(1) The design of anti-slide piles needs to consider the pile-soil interaction and make full use of the soil arching effect between anti-slide piles to achieve the purpose of economy and efficiency.
(2) The calculation methods of internal force of anti-slide piles widely used in engineering practices mainly include the cantilever pile method, foundation coefficient method, finite element method, matrix analysis method, etc. Among them, the discrete element method have great potential in simulating the failure mechanism of

anti-slide piles, which can effectively solve the shortcomings of traditional methods in modeling large deformations.

(3) The working state of anti-slide piles and slope stability can be evaluated through field monitoring. The capability of the DFOS technique in measuring strain profiles is highly advantageous in understanding and detecting any abnormalities in anti-slide piles behavior.

Acknowledgments. This research was funded by the National Key Research and Development Program of China (Grant No. 2018YFC1505104) and the National Natural Science Foundation of China (Grant No. 42077235).

References

1. Wang, F., et al.: Displacement monitoring and physical exploration on the Shuping landslide reactivated by impoundment of the three gorges reservoir, China. In: Sassa, K., Fukuoka, H., Wang, F., Wang, G. (eds.) Landslides. Springer, Heidelberg (2005). https://doi.org/10.1007/3-540-28680-2_40
2. Huang, R.Q.: Some catastrophic landslides since the twentieth century in the southwest of China. Landslides **6**(1), 69–81 (2009)
3. Sun, H.Y., Wong, L.N.Y., Shang, Y.Q., Shen, Y.J., Qing, L.: Evaluation of drainage tunnel effectiveness in landslide control. Landslides **7**(4), 445–454 (2012)
4. Pulko, B., Majes, B., Mikoš, M.: Reinforced concrete shafts for the structural mitigation of large deep-seated landslides: an experience from the Macesnik and the Slano blato landslides (Slovenia). Landslides **11**(1), 81–91 (2012). https://doi.org/10.1007/s10346-012-0372-2
5. Ashour, M., Ardalan, H.: Analysis of pile stabilized slopes based on soil–pile interaction. Comput. Geotech. **39**, 85–97 (2012)
6. Wang, L.P., Zhang, G.: Centrifuge model test study on pile reinforcement behavior of cohesive soil slopes under earthquake conditions. Landslides **11**(2), 213–223 (2013). https://doi.org/10.1007/s10346-013-0388-2
7. Sun, X., Lü, H.: Centrifuge modeling of the dynamic behavior of pile-reinforced slopes during earthquakes. J. Cent. South Univ. Technol. **17**(5), 179–183 (2010)
8. Song, Y.S., Hong, W.P., Woo, K.S.: Behavior and analysis of stabilizing piles installed in a cut slope during heavy rainfall. Eng. Geol. **130**(3), 234–238 (2012)
9. Li, C.D., Tang, H.M., Ge, Y.F., Hu, X.L., Wang, L.Q.: Application of back-propagation neural network on bank destruction forecasting for accumulative landslides in the three gorges reservoir region. China. Stoch. Env. Res. Risk A **28**(6), 1465–1477 (2014)
10. Byrne, P.M., Anderson, D.L., Janzen, W.: Response of piles and casings to horizontal free-field soil displacements. Can. Geotech. J. **21**(4), 720–725 (1984)
11. Jeong, S., Kim, B., Won, J.: Uncoupled analysis of stabilizing piles in weathered slopes. Comput. Geotech. **30**(8), 671–682 (2003)
12. Randolph, M.F., Houlsby, G.T.: The limiting pressure on a circular pile loaded laterally incohesive soil. Geotechnique **34**(4), 613–623 (1984)
13. Li, C., Wang, X., Tang, H., Lei, G., Yan, J., Zhang, Y.: A preliminary study on the location of the stabilizing piles for colluvial landslides with interbedding hard and soft bedrocks. Eng. Geol. **224**, 15–28 (2017)
14. Liu, Y.L., Sun, H.Y.: Deformation monitoring and displacement determination of anti-slide piles. Chin. J. Rock Mech. Eng. **32**(10), 2147–2153 (2013)

15. Sun, Y.J., Zhang, D., Shi, B.: Distributed acquisition, characterization and process analysis of multi-field information in slopes. Eng. Geol. **182**, 49–62 (2014)
16. Terzaghi, K.: Theoretical Soil Mechanics. John Wiley & Sons, New York (1943)
17. Shubhra, G., Patra, N.R.: Effect of arching on active earth pressure for rigid retaining walls considering translation mode. Int. J. Geomech. **8**(2), 123–133 (2008)
18. Bransby, P.L., Milligan, G.W.E.: Soil deformations near cantilever sheet pile walls. Geotechnique **25**(2), 175–195 (1975)
19. Shelke, A., Patra, N.R.: Effect of arching on uplift capacity of pile groups in sand. Int. J. Geomech. **8**(6), 347–354 (2008)
20. Tang, H.M., Hu, X.L., Xu, C., Li, C.D., Yong, R., Wang, L.Q.: A novel approach for determining landslide pushing force based on landslide-pile interactions. Eng. Geol. **182**(19), 15–24 (2014)
21. Zhu, H.H., Shi, B., Yan, J.F., Zhang, J., Wang, J.: Investigation of the evolutionary process of a reinforced model slope using a fiber-optic monitoring network. Eng. Geol. **186**, 34–43 (2015)
22. Zhang, G., Wang, L.: Integrated analysis of a coupled mechanism for the failure processes of pile-reinforced slopes. Acta Geotech. **11**(4), 941–952 (2015). https://doi.org/10.1007/s11440-015-0410-z
23. Hu, X., Zhou, C., Xu, C., Liu, D.: Model tests of the response of landslide stabilizing piles to piles with different stiffness. Landslides **16**, 2187–2200 (2009)
24. Zeng, J.X.: Study on the mechanism and calculation theory of slope reinforcement by plate connected beam reinforced micro anti-slide piles. Ph.D. thesis, Southwest Jiaotong University (2019). (in Chinese)
25. Dai, Z.H., Xu, X.: Three-dimensional finite element method for design and calculation of slope anti-slide piles. Chin. J. Rock Mech. Eng. **31**(12), 2572–2578 (2012). (in Chinese)
26. Dai, Z,H., Shen, P.S., Peng, Z.B.: New model for calculating internal forces of elastic anti-slide piles and its finite difference decomposition method. China Civ. Eng. J. **36**(4), 99–104 (2003). (in Chinese)
27. Zhu, Y., Zhu, H.H., Zhang, W., Shi, B.: Parameter analysis of influencing factors of slope stability strengthened by anti-slide piles. J. Eng. Geol. **25**(03), 833–840 (2017). (in Chinese)
28. Cundall, P.A.: A computer model for simulating progressive large scale movements in blocky system. In: Proceedings of Symposium of International Society of Rock Mechanics, pp. 8–12 (1971)
29. Cundall, P.A., Strack, O.D.L.: The distinct element method as a tool for research in granular media: Report to The National Science Foundation Concerning. Department of Civil and Mineral Engineering, Institute of Technology, University of Minnesota (1979)
30. Li, T.L., Li, P.: Research status of design and calculation of anti-slide piles. J. Eng. Geol. **15**(S), 191–197 (2007). (in Chinese)
31. Viggiani, C.: Ultimate lateral loads on piles used to stabilized landslides. In: Proceedings of 10th International Conference on Soil Mechanics and Foundation Engineering, Stockholm, pp. 555–560 (1981)
32. Lirer, S.: Landslide stabilizing piles: experimental evidences and numerical interpretation. Eng. Geol. **149–150**, 70–77 (2012)
33. Ito, T., Matsui, T.: Method to estimate lateral force acting on stabilizing piles. Soils Found. **15**(4), 43–59 (1975)
34. Kumar, S., Hall, M.L.: An approximate method to determine lateral force on piles or piers installed to support a structure through sliding soil mass. Geotech. Geol. Eng. **24**, 551–564 (2006)
35. Poulos, H.G., Lee, C.Y.: Pile response due to excavation–induced lateral soil movement. J. Geotechnical Geoenviron. Eng. **123**(2), 94–99 (1997)

36. Zhou, X.P., Cheng, H.: Analysis of stability of three-dimensional slopes using the rigorous limit equilibrium method. Eng. Geol. **160**, 21–33 (2013)
37. Yamin, M., Liang, R.Y.: Limiting equilibrium method for slope/drilled shaft system. Int. J. Numer. Anal. Methods Geomech. **34**, 1063–1075 (2010)
38. Cheng, Y.M., Lansivaara, T., Wei, W.B.: Two-dimensional slope stability analysis by limit equilibrium and strength reduction methods. Comput. Geotech. **34**, 137–150 (2007)
39. Zhou, C., Shao, W., van Westen, C.J.: Comparing two methods to estimate lateral force acting on stabilizing piles for a landslide in the three gorges reservoir, China. Eng. Geol. **173**, 41–53 (2014)
40. Wang, C.H., Chen, Y.B., Lin, L.X.: Analysis of soil arch characteristics and maximum pile spacing between anti-slide piles. Mt. Res. **19**(6), 556–559 (2001). (in Chinese)
41. Zhou, D.P., Xiao, S.G., Xia, X.: Discussion on reasonable pile spacing of anti-slide piles in slope engineering. Chin. J. Geotech. Eng. **26**(1), 132–135 (2004). (in Chinese)
42. Zhao, M.H., Liao, B.B., Liu, S.S.: Calculation of pile spacing of anti-slide piles in slopes based on arch effect. Rock Soil Mech. **31**(4), 1211–1216 (2010). (in Chinese)
43. He, L.D., He, Y.F., Liu, J.L.: Soil arch analysis and reasonable pile spacing between anti-slide piles in clayey soil. Sci. Technol. Eng. **14**(4), 90–93 (2014). (in Chinese)
44. Cai, F., Ugai, K.: Numerical analysis of the stability of a slope reinforced with piles. Soils Found. **40**(1), 73–84 (2000)
45. Won, J., Kwangho, Y.: Coupled effects in stability analysis of pile-slope systems. Comput. Geotech. **32**, 304–315 (2005)
46. Shooshpasha, I., Amirdehi, H.A.: Factors influencing the critical pile length in reinforced slope. J. Phys. Conf. Ser. Int. Conf. Eng. Syst. **1687**, 14–16 (2020)
47. Sun, Y.J., Shi, B., Chen, S.E., et al.: Feasibility study on corrosion monitoring of a concrete column with central rebar using BOTDR. Smart Struct. Syst. **13**(1), 41–53 (2014)
48. Zhu, H.H., Ho, A.N.L., Yin, J.-H., Sun, H.W., Pei, H.F., Hong, C.Y.: An optical fibre monitoring system for evaluating the performance of a soil nailed slope. Smart Struct. Syst. **9**(5), 393–410 (2012)
49. Zhu, H.H., Shi, B., Yan, J.F., Zhang, J., Zhang, C.C., Wang, B.J.: Fiber Bragg grating-based performance monitoring of a slope model subjected to seepage. Smart Mater. Struct. **23**(9), 095027 (2014)
50. Liu, Y.L., Sun, H.Y.: BOTDR monitoring analysis of the internal force of anti-slide piles. J. Zhejiang Univ. (Eng. Sci.) **46**(2), 243–249 (2012). (in Chinese)
51. Liu, Y.L., Sun, H.Y.: Determination of landslide thrust on anti-slide piles based on BOTDR monitoring technology. J. Zhejiang Univ. (Eng. Sci.) **46**(5), 798–803 (2012). (in Chinese)
52. Zhu, H.H., Yin, J.H., Hong, C.Y.: Slope engineering monitoring technology based on fiber optic sensing. Geotech. Investig. Surv. **3**, 6–14 (2010). (in Chinese)

Study on the Dewatering Effect of In-Situ Dredged Mud Under the Combined Action of Flocculant-Absorbent

Yanjie Wang[1], Zhongjian Yang[1], Fuqiang Zhu[2], Yue Ma[2], and Jincheng Ren[3](✉)

[1] Transportation Bureau Port, Aviation, Railway and Airport Service Center, Tai'an 271000, China
[2] China State Construction Port Engineering Group Corp., Ltd., Beijing 266011, China
[3] Key Laboratory of Ministry of Education for Geomechanics and Embankment Engineering, Hohai University, Nanjing 210098, China

Abstract. In order to overcome the problems of high construction costs and soil alkalinization associated with chemical curing methods for treating soils with high moisture content, a combined flocculant-absorbent treatment method with super absorbent polymer (SAP) was proposed. The suitability of the optimum flocculant cationic polyacrylamide (CPAM) (0.4‰) for dredged mud in its original state was verified and a comparative study of the drying effect of the absorbent for SAP was carried out. The test results show: after 24 h of standing, the percentage of clear liquor precipitated in the CPAM-0.4, APAM-0.4 and NPAM-0.8 groups was 36%, 32% and 31% of the slurry volume respectively, with the CPAM-0.4 group precipitating the most and the pH of the as-built dredged slurry already approaching neutral pH. Therefore CPAM-0.4 is the optimum flocculant and its corresponding optimum dose. In the second step of the model test, the three absorption methods of SAP need to be reasonably arranged in conjunction with the actual project. Finally, the construction process guidance of the combined flocculant-absorbent treatment method for very high moisture content mud is proposed, which does not cause alkalinization of the soil and has the advantages of short construction period, low difficulty and low cost. This method does not alkalize the soil and has the advantages of short construction period, low difficulty and low cost.

Keywords: Dredged spoil · Flocculants · Polymeric absorbent resins · Recycling · Process guidance

1 Introduction

With the development of infrastructure projects in China, the disposal of dredged mud generated from lakes and rivers has become an inevitable problem. Dredged mud has problems such as high moisture content, low strength and is contaminated to varying degrees [1]. On the other hand, dredged mud has considerable economic value, and as long as it is treated and reused in a reasonable way, it will be turned into a treasure, generating certain economic benefits and embarking on a path of sustainable development. At present, the main treatment methods for dredged mud are marine disposal,

H.-H. Zhu et al. (Eds.): BRWSG 2021, LNCE 230, pp. 237–246, 2022.
https://doi.org/10.1007/978-981-16-9963-4_19

land blowing, natural drying, mechanical dewatering, geotechnical pipe belt dewatering, incineration and heat treatment, chemical curing method, solar sludge drying technology, etc. [2–4]. The chemical curing method is a widely used method for the disposal of large quantities of dredged mud with high moisture content in inland areas at home and abroad. Chemical curing treatment is carried out by mixing the appropriate amount of cement, fly ash, waste gypsum, blast furnace cinder, slag, steel slag and other curing agents, and then fully mixing the original mud and additives to make the water and clay minerals in the mud and the curing agent carry out a series of physical and chemical reactions to reduce the content of free water in the mud [5], increase the strength and improve the engineering mechanical properties of the mud. The treated soil can be used as a construction material for roads, embankment fills, building foundations and other projects [6]. However, the curing process is expensive and the alkaline curing material used can change the soil properties, making it unsuitable for greening and contaminating the surrounding healthy soil. Physical treatment methods such as dewatering of geotechnical pipe strips have the problem of long treatment cycles. To address these problems, this paper investigates a combined flocculation-absorbent treatment method by adding a suitable amount of flocculant and extracting the upper clear layer after the flocculation has stabilized. The slurry is then further dried and treated with SAP absorbent to obtain a slurry with a water content below the liquid limit. This method does not alkalize the soil and has the advantages of short construction time, low difficulty and low cost.

2 Test Materials and Programmes

According to the literature from Su [7] the optimum flocculant type and admixture quantity for this experimental slurry and the optimum pressure for the water absorption test have been clarified. In this paper, dredged mud samples in their original state are studied to analyze their dewatering effect under the combined flocculation-absorption action. The optimum flocculant type (cationic polyacrylamide) and dose (0.4‰) for self-mixed very high moisture content bentonite slurries obtained in the SuFan test are verified for in-situ dredged slurries by simple flocculation tests. Model tests were carried out on the in-situ dredged mud to analyze the final dewatering effect of the mud under different absorption methods and to determine the most suitable absorption method among them. Finally, a construction process guide for the combined flocculation-absorption treatment of very high moisture content t slurries is presented.

2.1 Laboratory Test Materials and Programmes

The in-situ dredged mud used in this paper was from the Jiaxing South Lake Ecological Restoration Project (Phase I) in Zhejiang Province, with an initial moisture content of 316.0%, a liquid limit of 53.0%, a plastic limit of 29.0% and a pH value of 7.2.

For comparison, the three flocculants and their respective optimum dosing in the SuFan test were selected for control, and the protocols for the specific flocculation tests are shown in Table 1.

The test was carried out at room temperature of 20 °C. 250.00 g of raw dredged slurry was added to each beaker, then different types and doses of flocculants were added to

Table 1. In-situ dredged slurry flocculation test programmes

Group	Types of flocculants	Flocculant admixture (‰, mass fraction relative to slurry)
CPAM-0.4 (in-situ)	Cationic polyacrylamide	0.4
APAM-0.4 (in-situ)	Anionic polyacrylamide	0.4
NPAM-0.8 (in-situ)	Non-ionic polyacrylamide	0.8

each beaker and the change in liquid level after 24 h was recorded by rapid stirring with a glass rod for 5 min.

2.2 Model Test Programmes and Methodology

2.2.1 Test Programme

After the flocculation test results were obtained, the model tests could be carried out. In the model tests, three different water absorption methods were set up, the specific test protocols are shown in Table 2. The model tank used for the tests was a 500L rectangular thickened PE plastic tank, the specific dimensions of which are shown in Fig. 1.

Table 2. Test programmes for water absorption methods

Group	Filter bag arrangement	Initial slurry depth (m)	Total mass of SAP used (kg)	Total number of filter bags used (pcs)	Loading pressure (kPa)
Group 1	Upper surface	0.6	3	15	3
Group 2	Upper surface + vertical				
Group 3	Vertical				

For the first group of tests, take 15 filter bags with SAP inside, in 5 layers of 3 each, and lay them flat on the upper surface of the substrate after flocculation and filtration (there is a layer of geotextile between the upper surface of the substrate and the filter bags), see Fig. 2 for the arrangement.

For group 2 tests, 9 filter bags with SAP inside, 3 in 3 layers, 3 in each layer, were inserted vertically inside the flocculated and filtered clear substrate, and 6 filter bags in 3 layers, 2 in each layer, were laid flat on the upper surface of the substrate (with a layer of geotextile between the upper surface of the substrate and the filter bags), see Fig. 3 for the arrangement.

For the third group of tests, 15 filter bags with SAP inside, in 5 layers of 3 each, were inserted vertically inside the substrate after flocculation and filtration of the clear liquid, see Fig. 4 for the arrangement.

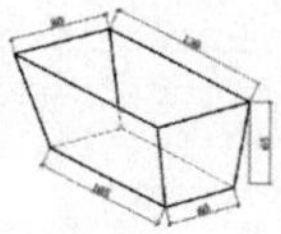

Fig. 1. Dimensions of the model box

Fig. 2. Schematic diagram of the first group of test

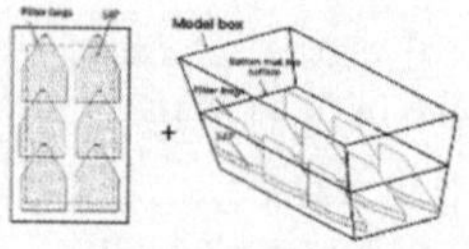

Fig. 3. Schematic diagram of the filter bag arrangement for group 2 test

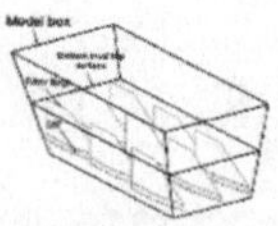

Fig. 4. Schematic diagram of the filter bag arrangement for group 3 test

2.2.2 Test Methods

The steps of the water absorption mode test are as follows.

(1) Add in-situ dredged mud to the model box until the depth of the mud in the box reaches 0.6 m. Calculate the mud mass and the mass of cationic polyacrylamide (CPAM) required based on the density and volume of the mud. After adding the flocculant to the slurry, the tool is used to mix for approximately 5 min. After mixing, the slurry is left to stand for 24 h and wait for the slurry to flocculate.
(2) The supernatant precipitated from the flocculated slurry was filtered out using a tool and the water content of the flocculated substrate was tested.
(3) For test group 1, a layer of geotextile is laid on the surface of the substrate and the filter bag with SAP inside is laid on top of the geotextile in a predetermined arrangement. For group 2 tests, after inserting the filter bags vertically into the substrate, a layer of geotextile was laid on the surface of the substrate and the remaining filter bags were laid on top of the geotextile in a predetermined arrangement. For group 3 tests, after inserting the filter bags vertically inside the substrate, a layer of geotextile is laid on the surface of the substrate.
(4) A PE sheet is placed on top of the filter bag (or geotextile for test group 3) so that the substrate is evenly stressed. A bucket of water is placed on top of the PE plastic sheet for loading.
(5) After loading, soil samples are taken at different depths of the subsoil at certain times to test the moisture content. For Group 1 samples, the horizontal position of the sample was taken at the centre of the subsoil, for Group 2 and 3 tests, the sample was taken in the middle of the two adjacent filter bags in row 2. After the sampling is completed, the filter bags and geotextile are put back in place and the water absorption is continued.
(6) When the water content of the substrate is relatively close to that measured for 2 consecutive times, end the test and recover the instruments and tools.

3 Study on the Effect of Combined Flocculation-Absorption Treatment Method

3.1 Analysis of Flocculation Effect

After the addition of flocculants to the in-situ dredged mud, it can be seen from the graphs that after one day and one night under the action of three different types of flocculants, a more obvious flocculation phenomenon occurred in all groups of mud. Figure 5 shows the comparison of the mud-water separation level scale before and after flocculation of each group of in-situ dredged mud. which. From Fig. 6, the percentage of clear liquid precipitation in the CPAM-0.4, APAM-0.4 and NPAM-0.8 groups to the volume of mud was 36%, 32% and 31% respectively, still the CPAM-0.4 group had the most precipitation, which is consistent with the results obtained from SuFan's flocculation test.

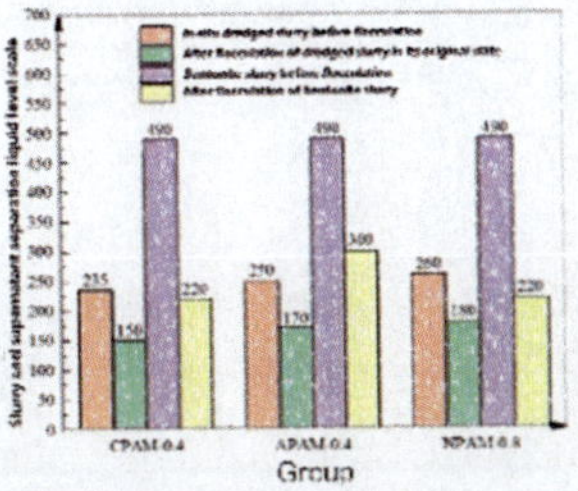

Fig. 5. Comparison of mud-water separation level scales before and after flocculation of each group of in-situ dredg ed mud

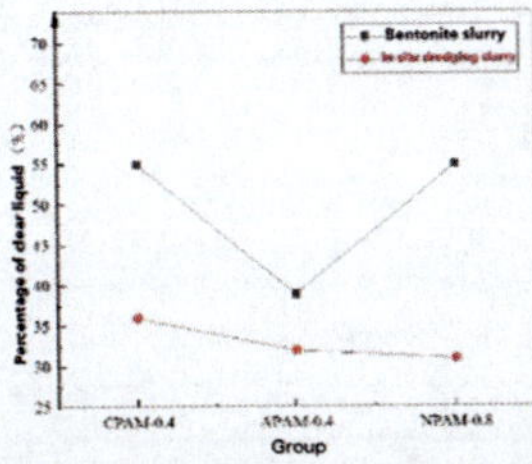

Fig. 6. Comparison of the percentage of clear liquor precipitation after flocculation of each group of bentonite slurry with that of the as-dredged slurry

3.2 Analysis of the Dewatering Effect with Different Water Absorption Methods

The results of the tests on the water absorption pattern of the in-situ dredged mud are shown in Figs. 7, 8, 9 and 10. It can be seen from the figures that the initial water content of the test substrate was around 120.0% in all groups, and the water content could be reduced to approximately 60.0%–70.0% after one round of water absorption.

In all three sets of tests, the rate of decline in the water content of the substrate showed a general trend of "fast and then slow", which can be roughly divided into the "fast water absorption stage", "continuous uniform water absorption stage "and "slow water absorption stage". After the start of water absorption to the 8th hour is the "rapid water absorption stage", when the rate of water content decreases fastest. From the 8th to the 72nd hour is the "continuous uniform absorption stage", where the rate of decrease in moisture content slows down. In the 72 h to 96 h is the "slow water absorption stage", the water content drop is very small, and its relationship curve with time is close to the level, the water absorption process has nearly completed.

In terms of depth, the pattern of variation of water content with time at different depths in the same group of tests is basically the same. For Group 1 tests, which only

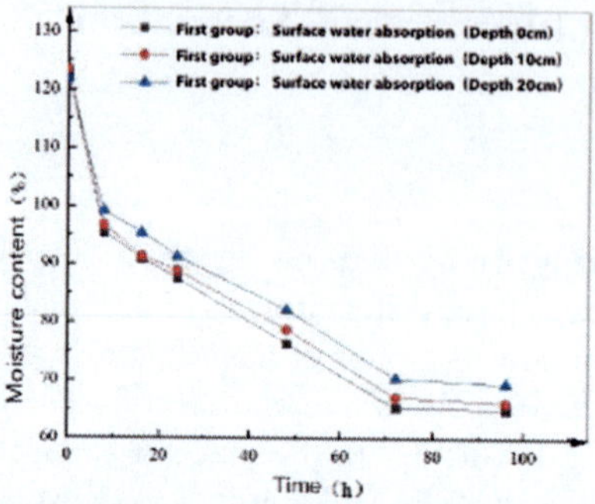

Fig. 7. Plot of water content as a function of time at different depths for group 1 tests

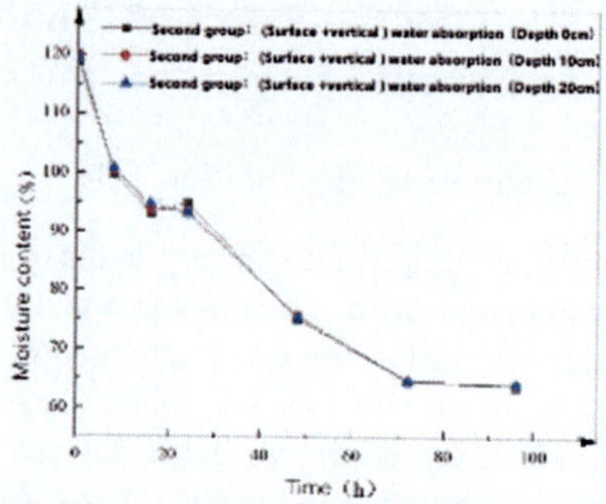

Fig. 8. Plot of water content as a function of time at different depths for group 2 tests

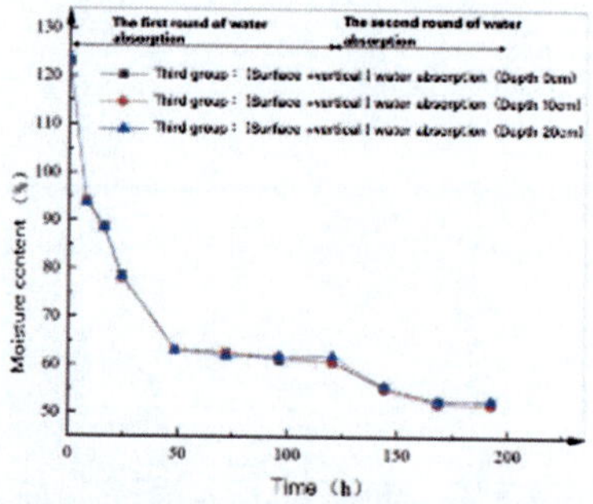

Fig. 9. Plot of water content as a function of time at different depths for group 3 tests

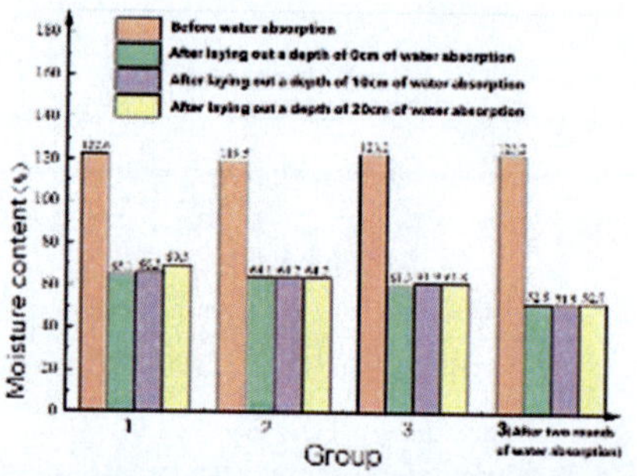

Fig. 10. Comparison of the water content of the substrate before and after each group of tests was absorbed

used surface absorption, only the upper part of the substrate could have sufficient contact with the SAP after the absorption started, and the absorption effect was good, but at the same time, as the depth increased, the water content increased slightly, indicating that the distance of the substrate from the SAP had a certain influence on the change of the water content, the further the distance, the smaller the change of the water content. For Group 2 and Group 3 tests, because both surface and vertical absorption were used, the substrate was in full contact with the SAP at all depths, so the effect of depth on the water content of the substrate was minimal and the water content was almost the same at all depths.

It can be seen from Fig. 10 that the bottom sediment of group 3 has the lowest water content after one round of absorption, so for group 3 bottom sediment after 96 h of absorption, the SAP and filter bags are replaced again and the absorption is still arranged in the original way. As can be seen from Fig. 8, after the second round of absorption, the water content of the substrate started to decrease again, but the rate of decrease was not as fast as in the "rapid absorption stage" and "continuous uniform absorption stage". It was not until about 168 h later that the water content stopped falling. This indicates that after the first round of water absorption, there is still a small amount of water left in the substrate that is difficult to be absorbed. For the SAP that has already absorbed water in the first round and swelled in volume by several tens of times, its water absorption capacity is not as good as the newly added SAP particles, which cannot absorb this residual water.

At 192 h, the replaced SAP only absorbed about a dozen times more water and still had a strong water absorption capacity, but the water content of the substrate no longer decreased, so it can be judged that the water inside the small and medium pores in the substrate had basically been absorbed completely and the water absorption process had been completed.

As can be seen from Fig. 10, after the second round of water absorption, the water content of the Group 3 test substrate is approximately 52.0%, which is already slightly below the liquid limit of the as-built dredged mud, which is consistent with the conclusion obtained in the water absorption test of the bentonite mud, that is, after the water absorption effect of SAP, the water content of the mud can be reduced to at most below the liquid limit close to the liquid limit.

In summary, all three absorption methods are effective in reducing the water content of the dredged bottom mud in its original state after flocculation and filtration of the clear liquor. The vertical absorption method was the most effective, the vertical + surface absorption method was the second most effective and the surface absorption method was slightly less effective, but the overall difference between the three was not significant. In terms of the uniformity of the treatment effect, both the vertical and vertical + surface absorption methods can achieve a more uniform drying effect at different depths, while the water content of the substrate treated by the surface absorption method increases slightly as its depth increases.

In practice, the surface absorption method is the simplest process, although the reinforcement effect decreases with depth, while the vertical absorption method is less affected by depth, but is more difficult to arrange filter bags and requires more manpower and machinery. Therefore, in practice, the absorption method needs to be chosen flexibly according to the site conditions. For mud ponds with little depth, it is advisable to choose surface absorption or surface + vertical absorption treatment, while for mud ponds with greater depth, it is advisable to choose surface + vertical absorption or vertical absorption. For sites where this is possible or where there are stringent requirements for the moisture content of the substrate, it is recommended that a second round of absorption be added after the first round.

4 Combined Flocculation-Absorption Treatment Process for Very High Moisture Content Slurries

In this paper, three mud ponds with bottom dimensions of about 4 m × 5 m and depth of about 1.5 m were arranged by means of mechanical excavation at the mud drying site near South Lake, Jiaxing, Zhejiang, China. After that, through the grouting hose slurry South Lake dredging mud into the pool, in the injection process has been added to the pipeline mud through the tee joint flocculant CPAM. 24 h after the pool mud flocculation and sedimentation, the use of pumps to extract the upper layer of clear liquid. Finally, it is planned to dry the mud in the pond by using polymer absorbent resin according to the three absorption methods in Sect. 2.2.2: upper surface, upper surface + vertical and vertical.

In summing up the experience of the field test, this paper proposes the following field test construction process with a view to providing guidance for the subsequent field practice of this method.

(1) Preparation of slurry tank

For sites with conditions, some abandoned large ponds or other ponds with impermeable sidewalls can be selected nearby to be used directly as mud ponds. Otherwise, excavators and other equipment can be used to excavate the mud pond on site. After the mud pond excavation is completed, use mechanical equipment for the sidewalls and bottom of the pond to level and remove the sharp gravel, masonry, etc. It is recommended to leave the excavated mud pond for one day to determine the leakage of the mud pond. For mud ponds that do not leak or have minimal water seepage, one or more layers of impermeable membranes are laid according to the actual situation.

(2) Grouting

After the preparation of the mud pool is completed, the mud is injected into the pool through the grouting pipe. For the mud pond with a very small amount of seepage at the bottom or side wall, it is advisable to extend the mouth of the grouting pipe to the bottom of the pond and keep the mouth as horizontal as possible to prevent the mud from being injected too fast and impacting the impermeable membrane, resulting in a gap between the impermeable membrane and the bottom or side wall of the pond due to friction. During the grouting, the pre-configured flocculant solution is mixed with the mud in the grouting pipe according to the predetermined ratio using a tee joint and injected into the mud pool together.

(3) Flocculation resting

After the grouting is completed, the mud in the pool is flocculated and left to settle. The resting time is generally more than one day, depending on the site conditions. Measure the suspended matter index in the supernatant every 12 h, when the change of index is small in two consecutive measurements, it can be considered that the mud-water separation has been completed. Use the pump to pump the supernatant, and measure the suspended matter index at the outlet of the pump pipe at any time. When the indicator value rises rapidly or can clearly see the mud-water separation surface in the pool, it is appropriate to stop pumping.

(4) Water absorbent arrangement

After filtering off the supernatant, the water absorbent can be arranged to further dry the bottom mud. The amount of absorbent should be determined according to the water content, density and target dryness of the substrate, generally 1.5–4.5 kg per square of mud. For the first time, as the substrate is still wet after extracting the supernatant, it is advisable to lay a layer of geotextile on the surface of the substrate first, and then use machinery such as cranes to lay the filter bag wrapped with water absorbent on the geotextile, so that the water absorbent can absorb water on the surface of the substrate. The first round of water absorption time is generally short, and can be completed within one day and night. After completion, the crane and other machinery will be used to transfer the filter bags wrapped with absorbent to the outside of the mud pond for subsequent treatment.

In the second arrangement, the geotextile laid in the first arrangement is firstly cleaned up by manual or mechanical means, and then the absorbent is arranged. According to the actual situation on site, the absorbent can choose surface arrangement, longitudinal and three-dimensional arrangement. In the longitudinal or three-dimensional arrangement, the filter bag can be tied to the reinforcement and inserted into the bottom mud for water absorption, and the bottom impermeable membrane should be prevented from being punctured during the insertion. The time of the second round of water absorption is generally 1–5 days, which is decided by the site soil condition.

(5) Loading

In the second round of water absorption, the loading can be carried out manually or mechanically to promote the drying of the substrate. The optimal loading pressure is 3 kPa, and considering the effect of depth on the pressurization effect, the pressure can be increased appropriately in practice, but should not be too large.

(6) Closing work

When the water content of the substrate no longer decreases the absorbent no longer absorbs water and swells, it can be considered that the second round of water absorption has been completed. Use cranes and other machinery to transfer the steel bars and filter bags wrapped with water absorbent to the outside of the mud pond. For the bottom mud after water absorption, use the dredger and other machinery to dig it out and transport it out by tanker truck.

Since the combined flocculation-absorption treatment method is highly susceptible to external recharge, it is necessary to pay extra attention to the weather conditions in use and avoid construction in rainy days as far as possible; otherwise, it is necessary to take protective measures to prevent rainwater or surface water from entering the slurry pond undergoing drying.

5 Conclusions

In the course of this paper, the following conclusions were reached.For all three flocculants, there is an optimum dose, which is 0.4‰ for CPAM, 0.4‰ or 0.6‰ for APAM and 0.8‰ for NPAM, which is consistent with the conclusions from Su. The optimum flocculant types and dosing levels obtained in the bentonite mud tests are also applicable to as-built flocculated muds. Regardless of the type of slurry, the water content obtained after water absorption is slightly below the liquid limit. In practice, the completion of the water absorption process can be judged by whether the water content of the substrate is below the liquid limit or whether it has been reduced to 1.8–1.9 times the optimum water content.

Surface, surface vertical and vertical absorption methods all have a good drying effect on the in-situ dredged substrate but are affected differently by the depth and therefore need to be used flexibly in practical applications. For mud ponds with a small depth, surface absorption or surface + vertical absorption is preferred, while for mud ponds with a larger depth, surface + vertical absorption of vertical absorption is preferred. For sites where this is possible or where the requirements for the moisture content of the substrate are more stringent, it is recommended that the second round of absorption be added after the first round of absorption.

The cost of the material required to treat one side of in-situ dredged mud by this method is approximately $28.1, which is approximately 1.3 times that of the geotechnical pipe bag method. The flocculation-absorption cycle is approximately 5–9 days, which is a short cycle time. In future applications, costs can be reduced by contacting cooperating businesses for the mass supply of flocculants and absorbents or by choosing cheaper filter bags.

Acknowledgements. The authors would like to acknowledge the financial support provided by the Fundamental Research Funds for the central Universities (Grants Nos. B210202034 and B200204036).

References

1. Ji, B., Xiao, X.M., Li, Z.: Solidification technology and resource utilization of dredged mud. Saf. Environ. Eng. **17**(02), 54–56 (2010)
2. Song, C., Li, B., Qiu, Y.: River sludge treatment and resourcing application. China Water Resour. **23**, 35–37 (2018)
3. Ma, J.: Research on medium and small river silt disposal technology and resource utilization. Water Construction and Management **36**(09), 64–67 (2016)
4. Gao, Y., Sun, K., Tan, Y., Wang, X.: Typical application and analysis of various dredged silt dehydration technology. JiangSu Water Resour. **09**, 51–54 (2020)
5. Zhang, C., Guan, F., Li, L., Huang, Y.: The progress in the reutillzation treatment and disposal of dredged sediments in china. Solid Waste Treatment and Disposal
6. Yun, W.R., Cui, J., Chen, Y.R.: The current situation and prospect of research on the resourceization of river and lake dredging sludge. Jiangsu Water Resour. **03**, 15–17 (2015)
7. Su, F.: Study on the combined treatment effect and dehydration mechanism of flocculation and water absorption for extremely high water content slurry. Hohai University, April 2021

Research on Real-Time Monitoring System of Power Transmission and Transformation Slope Based on Hybrid Data Transmission Network

Zeqin Chen(✉), Fan Wu, and Yatao Lin

Electric Power Research Institute of State Grid Fujian Electric Power Co. Ltd., 350007 Fuzhou, China

Abstract. In geotechnical engineering test technology, there is a certain distance between the test site and storage device, and remote data transmission needs to be considered. Hence, this study uses a hybrid data transmission network to transmit measurement data from the lower computer to the upper computer to establish a real-time monitoring system. In addition, this paper selects typical power transmission and transformation slopes to establish a real-time safety monitoring system. The system adopts wireless sensor network to build on site, and its basic composition mainly includes three parts: sensor system, data acquisition and transmission system, and data management system. According to the current monitoring results, the relationship between the settlement of the slope, the horizontal displacement of the deep soil, the vertical earth pressure, the lateral earth pressure, and the pore water pressure with time is obtained. The results show that the impact of rainfall on the slope is limited, and the slope has excellent drainage and good stability.

Keywords: Power transmission and transformation slope · Real-time safety monitoring system · Displacement · Earth pressure

1 Introduction

In geotechnical testing technology, there is a certain distance between the test site and storage and other equipment, and remote data transmission needs to be considered. The process in which different independent devices use lines to exchange data with each other is called signal transmission (or data transmission), and the line that constitutes the entire transmission is called a network [1, 2]. At the physical level, a simple data transmission network includes an upper computer, a lower computer, a data conversion interface and a data transmission carrier, as shown in Fig. 1. In the geotechnical engineering test, the lower computer refers to the special equipment that is directly connected to the sensing device and has processed the analog signal into the digital signal. In order to meet the requirements of complex test conditions on site, a single-chip microcomputer is generally used. The upper computer refers to a computer that issues measurement commands and can receive measurement data, and it can be a personal computer, a workstation, or a server. Due to the different device types of the upper computer and the lower

H.-H. Zhu et al. (Eds.): BRWSG 2021, LNCE 230, pp. 247–257, 2022.
https://doi.org/10.1007/978-981-16-9963-4_20

computer, the supported data transmission carrier may be different. With the development of communication technology, there are various data transmission carriers between the upper computer and the lower computer, such as cables (serial data transmission network, twisted pair network), optical fiber (optical fiber network) and wireless network (WIFI network, Cellular mobile network) etc. Different data transmission carriers require the upper computer or lower computer to convert the data into a signal form recognized by the carrier. This conversion device is called a data conversion interface. Diverse transmission carriers and interfaces bring certain difficulties to the construction of data transmission networks.

Fig. 1. Schematic diagram of the physical composition of the data transmission network

2 Application of Real-Time Safety Monitoring System for Slopes of Power Transmission and Transformation Facilities

2.1 Indoor Construction of Hybrid Data Transmission System

In geotechnical engineering testing technology, the measuring equipment is often far away from the upper computer (such as inter-regional, inter-city, inter-provincial, or even cross-border distance), and the network composed of serial data transmission lines cannot meet the requirements of distance and anti-interference and the cost of self-installation is relatively high [3]. For this reason, access to existing mature networks (see Fig. 2), such as wireless local area network, Internet, mobile cellular mobile network, etc., to transmit measurement data from the lower computer to the upper computer, not only can extend the data transmission distance, but also It can enhance the anti-interference in the process of data transmission and avoid the cost of wiring and network. It is an economical, efficient and practical networking method.

The hybrid data transmission system constructed in this experiment uses the Zigbee network to replace part of the serial data transmission line, and uses the WIFI wireless network for long-distance connection to the upper computer.

The current Zigbee technology is a short-range wireless communication technology developed based on the IEEE 802.15.4 protocol, a wireless method with low power consumption. The protocol used in Zigbee network communication determines that it can only be used for local networks instead of long-distance transmission, and is used to replace some of the lines in the serial data transmission network. The Zigbee coordinator, router and terminal hardware devices are the same. One end is a serial port to connect to

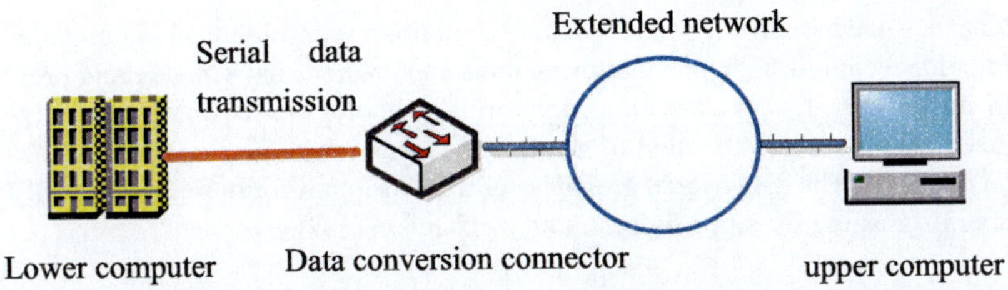

Fig. 2. Extended network for serial data transmission

a serial data transmission network; the other end is an antenna to build or join a Zigbee network. The lower computer is connected to the serial port of the Zigbee router or terminal, the router or terminal is connected to the coordinator through the network, and the serial port of the coordinator is connected to the upper computer. During data transmission, when the router or terminal receives data on the serial data network, it immediately sends the data to the coordinator through the Zigbee network. When the coordinator receives the network data, it is immediately converted into serial data and sent to the upper computer; when the coordinator receives the data on the serial data network, it immediately sends the data to the corresponding router or terminal through the Zigbee network. When the router or terminal receives the data from the Zigbee network, it is immediately converted into serial data and sent to the lower computer. It is necessary to determine whether the device connected to the lower computer adopts a router or a terminal according to the geographical location and low-consumption requirements.

The construction of the hybrid data transmission network involves the mutual conversion and networking of the serial data network and other networks. In order to simplify the steps of on-site construction and improve the efficiency of construction, the necessary wireless coordinator, DTU, and timer are integrated, and the serial can be connected. Data network and wireless network, the data in the two networks can be transformed into each other. In order to reduce power consumption and extend working time, the wireless timing box is equipped with a timing device, such as working once an hour for 2 min each time. The wireless timing box power supply is provided by an external solar panel. The whole box has only one 4-core wire connected to the lower computer, of which 2 cores are excitation (external) power wires, and 2 cores are signal wires. Before starting the measurement work, connect the power cord to the excitation power interface of the lower computer, and connect the signal line to the signal output interface of the lower computer. During the measurement process, the wireless timing box is turned on regularly according to the timing rules. After opening, the box converts the signal from the lower computer through the wireless coordinator and sends it to the upper computer to complete the network transmission of data.

2.2 On-site Construction of a Real-Time Safety Monitoring System for Slopes of Power Transmission and Transformation Facilities

The effective grasp of the slope deformation law comes from the accurate analysis of high-quality monitoring data, and the rich monitoring data relies on a scientific and reasonable slope monitoring system. With the continuous development of computer and information technology, slope monitoring tools and methods have made rapid progress. Multi-information dynamic real-time monitoring technology based on wireless networks has been widely used in the field of slope monitoring, which greatly enriches the evolution of slopes. The data source provides solid data support for more scientifically and accurately grasping the slope deformation mechanism [4–7].

Fig. 3. Test site 110 kV Guanjia Line #58-#59 in Quanzhou City, Fujian Province

In order to realize the field research of wireless real-time monitoring, this paper selects the 110kV Guanjia line #58-#59 in Quanzhou City, Fujian Province as the field test site of the real-time safety monitoring system for the slope of the power transmission and transformation facility, the test site is shown in Fig. 3. 110 kV Guanjia Line# in Quanzhou City, Fujian Province. 58-#59 has complicated geological conditions, and the slope here is a landfill area, which lacks effective reinforcement and protection measures. After reviewing the original design and carrying out the construction according to the original protection design plan, there is a risk of slope instability. In-situ monitoring of it is a very urgent task.

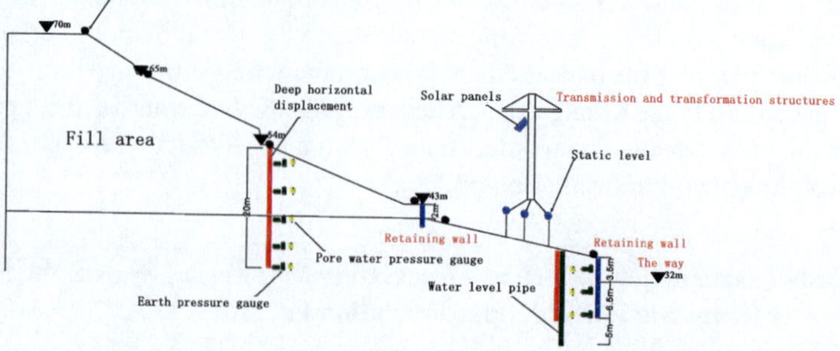

Fig. 4. Sectional view of the layout of the slope safety monitoring sensor system

The slope monitoring system adopts wireless sensor network to build on site, and its basic composition mainly includes three parts: sensor system, data acquisition and transmission system, and data management system.

This experiment uses the slope monitoring system developed for monitoring and data collection for subsequent work. In order to ensure the comprehensiveness of the influencing factors, in addition to the conventional stress and deformation parameters, the sensor layout in the monitoring system is also equipped with small weather stations around it to measure the changes of rainfall intensity, rainfall time, wind speed, wind direction and other parameters in real time. The sensor system layout is shown in Fig. 4.

The data acquisition and transmission system is shown in Fig. 5. The basic process is that the sensor collects physical data and transmits it to the sensor node next to the measuring point through the data line. Each node transmits the data to the site base station through the on-site Zig Bee wireless network [8]. Then the base station uploads the data to the Internet through the 4G mobile data network (CDMA), and finally on the server in the remote campus, you can log in to the client, that is, the data management system, to obtain the data and perform processing analysis, message sending and other follow-up work.

Fig. 5. Physical map of monitoring device

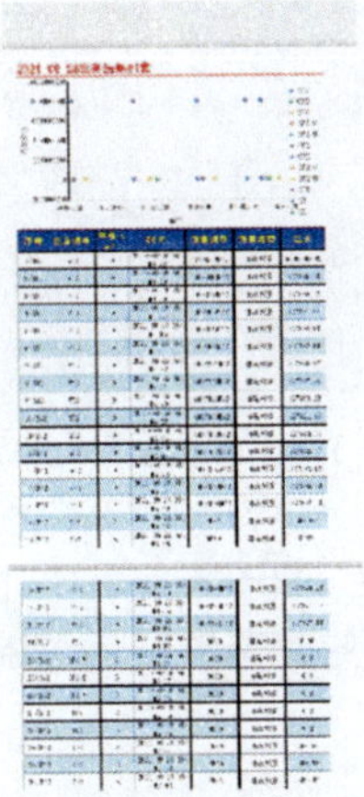

Fig. 6. Data management system

In the data management system shown in Fig. 6, by selecting the query content of the cross-section, axis, measurement type and measurement number where the data is located, the searched monitoring results and profile can be obtained. In addition, data can be viewed remotely via mobile phones, which is faster and more convenient than traditional monitoring methods.

3 Analysis of Monitoring Results

3.1 Rainfall and Slope Surface Settlement

In order to analyze the change law of the settlement of the soil mass, a soil surface settlement observation point is arranged on the slope top, the slope waist and the slope toe respectively. Combined with rainfall, the cumulative settlement-rainfall-time variation curve of each measuring point is drawn as shown in the Fig. 7, the time period is selected as 2021/2/27–2021/7/21. The overall surface settlement of the soil is closely related to the rainfall. The regular characteristics of body subsidence changing with rainfall are prominent, showing a typical stepwise growth pattern, and with intermittent characteristics with rainfall changes, forming two basic change components of short-term rapid deformation and long-term slow deformation. The measurement points of the slope foot and the slope waist change more obviously with the rainfall than the slope top measurement points. By 2021/7/21, the cumulative subsidence of the slope foot is 14.145 mm, the cumulative subsidence of the slope waist is 11.298 mm, and the cumulative subsidence of the slope top is 7.513 mm.

A heavy rainstorm occurred on June 23, 2021, the amount of settlement during rainfall increased sharply in a short period of time, and then stabilized; however, the time of settlement at the top of the slope was earlier than that of the waist and top of the slope.

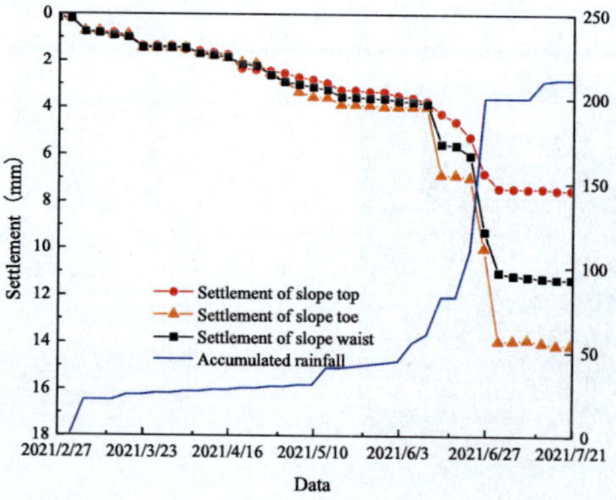

Fig. 7. Comparison of cumulative displacement curves of the surface of each measuring point with time

Observe the settlement of the four foundations (A, B, C, D) of the electrical tower at the foot of the slope. Figure 8 is a graph of the change of the foundation settlement of the electric tower over time. It can be seen from the figure that the foundation settlement of the electric tower increases slowly with time, and the settlement of the foundation does not change much during the rainfall period. The reason is that the foundation of the electric tower itself is deeply buried, and the rock foundation is at a depth of 5 m on the surface, so the foundation of the electric tower is relatively safe.

3.2 Analysis of Results of Horizontal Displacement of Deep Soil

This paper mainly selects the deep horizontal displacement of the slope soil as the research object to analyze the location of the sliding surface of the slope. The two displacement measurement points of the slope monitoring system are respectively arranged

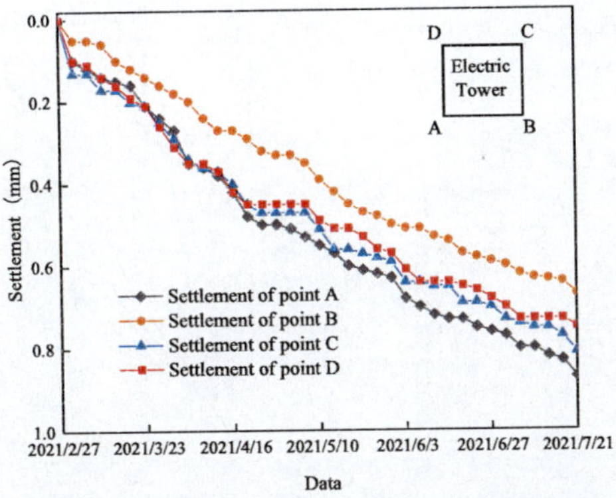

Fig. 8. Curve of electric tower foundation settlement with time

at the slope waist and the slope bottom, and the horizontal displacement to the downslope is specified as a positive value. The time period selection is the same as in the previous section. The test data from 2021/2/27 to 2021/7/21 are used to draw the cumulative displacement-depth-time variation curve inside each measuring hole as shown in Fig. 9. It can be seen from the figure that by July 21, 2021, the maximum horizontal displacement of the slope waist reached 12.2 mm, and the maximum horizontal displacement of the slope foot reached 3.3 mm. Due to the existence of electric towers at the foot of the slope, the 4 ft. of the tower have deep pile foundations, and the slope toe is all rocks below 5 m, and the soil at the foot of the slope is limited to slide, so the horizontal displacement at the foot of the slope is small. In addition, the horizontal displacement-depth curves of each measuring point have obvious inflection points, which correspond to the sliding surface area at their respective positions. The slope waist sliding zone is located in the depth range of about 6 m below the slope surface, and the slope foot sliding zone it is located within a depth range of about 3 m below the slope. The horizontal displacement value of the soil along the depth direction of the slope waist and the slope toe decreases continuously, and the position of the slip surface has a sudden change. The horizontal displacement value of the position above the slip surface is much larger than that of the position below it, and this situation continues to become prominent with the increase of time. After the rain began, the soil at the waist of the slope shifted horizontally toward the bottom of the slope. The horizontal displacement of the soil layer within 6 m depth is much larger than that below 6 m, that is, the depth of influence of rainfall infiltration is mainly within 2 m depth. At the end of the first rainfall, the horizontal displacement was the largest, but the surface displacement did not exceed 5 mm, which can be considered to be within the range of elastic deformation. The soil in the slope did not undergo overall plastic deformation failure after rainfall. The heavy rain did not cause overall instability of the soil.

3.3 Soil Pressure

In order to understand the distribution of soil pressure, two observation sections were set up at the waist and toe of the slope. Set 5 lateral soil pressure observation points and 5

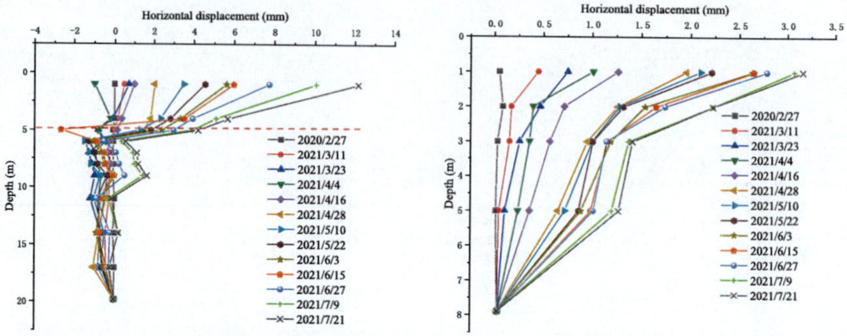

(a) Slope waist **(b)** Slope foot

Fig. 9. Comparison of cumulative displacement curves with time for different depth positions of various measuring points

vertical soil pressure observation points along the depth of 2 m, 5 m, 10 m, 15 m, and 20 m on the slope waist; set 2 lateral soil pressure observation points 2 vertical soil pressure observation points along the 2 m, 5 m depth of the slope foot. The distribution of vertical soil pressure is shown in Fig. 10. Figure a shows the change of vertical soil pressure on the slope waist with time, and figure b shows the change of vertical soil pressure on the foot of the slope with time. It can be seen from the figure that the maximum vertical soil pressure of the measured slope waist is 415.83 kPa, and the maximum vertical soil pressure of the slope toe is 103 kPa. It can also be found from the figure that the vertical soil pressure of the slope increases with the increase of the buried depth. Observing the lines in the figure, it can be found that the greater the buried depth, the greater the measured value of the vertical soil pressure, which is basically the same as the distribution law of the theoretical calculation of the soil pressure. In addition, the extremely heavy rain that occurred on June 23, 2021 affected the soil pressures at 2 m and 5 m. As the rainfall increased the bulk density of the soil, the soil pressure increased, and due to the presence of stagnant water at the foot of the slope, the earth pressure at the foot of the slope increases more than the slope waist. The earth pressure at 2 m after rainfall has a smaller change than that at 5 m, which reflects that the depth of the impact of rainfall on the slope is smaller. Due to the excellent drainage of the slope, the earth pressure quickly recovered and stabilized, so the earth pressure did not detect large fluctuations in the small-scale rainfall that usually occurred.

The distribution of lateral soil pressure is shown in Fig. 11. Figure a shows the change of the lateral soil pressure on the waist of the slope with time, and Figure b shows the change of the lateral soil pressure on the toe of the slope with time. It can be seen from the figure that the measured maximum lateral soil pressure at the waist of the slope is 366.41 kPa, and the maximum lateral soil pressure at the toe of the slope is 89.88 kPa. It can also be found from the figure that the value of the lateral soil pressure of the slope increases with the increase of the buried depth. It is calculated that the average soil lateral pressure coefficient K0 of the slope waist is 0.68, and the average soil lateral pressure coefficient K0 of the slope foot is 0.67. Before rainfall, the measured stress ratios of the

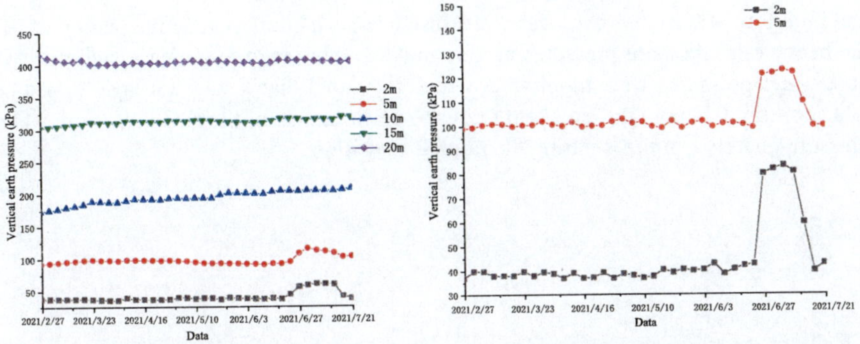

(a) waist of slope

(b) toe of slop

Fig. 10. The distribution of vertical soil pressure on the slope

lateral and vertical soil pressure boxes were both less than 0.7. After rainfall occurs, the lateral earth pressure changes are similar to the vertical earth pressures. They have not been monitored in small-scale rainfall, and they increase after heavy rains. The slope toe increase is larger than the slope waist. The scope of influence is mainly above 5 m.

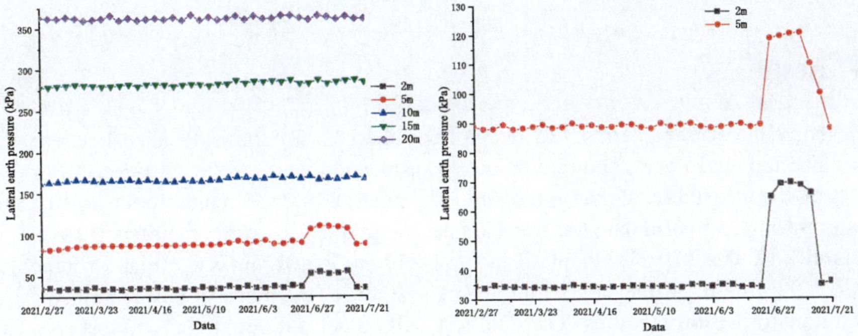

(a) waist of slop

(b) toe of slop

Fig. 11. The distribution of lateral earth pressure on the slope

3.4 Pore Water Pressure

Figure 12 is the curve of slope pore water pressure change. Figure a is the relationship curve of the pore water pressure at the slope waist with time, and Figure b is the relationship curve of the pore water pressure at the foot of the slope with time. It can be seen from the figure that before the start of the first rainfall, the pore water pressure of each depth in the soil layer is low, and the soil is almost dry. After a small area of rainfall, the pore water pressure changes at a depth of 2 m, and because the temperature and humidity of the surrounding environment are greatly affected by the temperature and humidity at 2 m, the pore water pressure has been continuously fluctuating. After the heavy rain, the pore pressures at 2 m and 5 m continued to increase, and positive pore water pressures were monitored, which indicated that a stagnant layer appeared in a short time. Due to the excellent drainage performance of the slope, the pore water pressure quickly stabilizes during the rainfall transition.

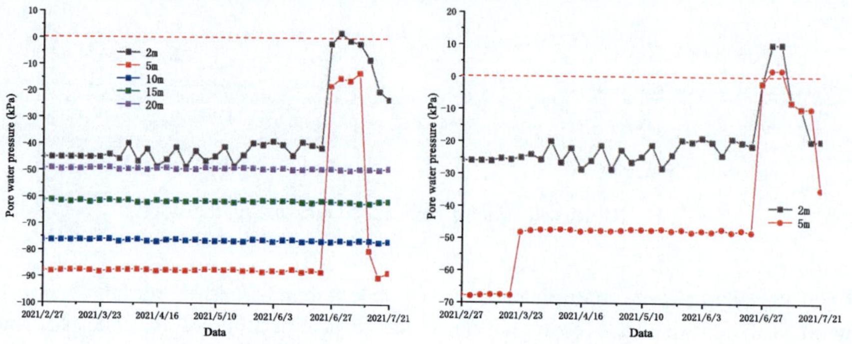

(a) waist of slope **(b)** toe of slope

Fig. 12. The distribution of pore water pressure on the slope

4 Result

In geotechnical engineering test technology, there is a certain distance between the test site and equipment, and remote data transmission needs to be considered. For this reason, the measurement data is transmitted from the lower computer to the upper computer through a hybrid data transmission network, and a real-time monitoring system is established. The 110 kV Guanjia Line #58-#59 in Quanzhou City, Fujian Province, is selected as the real-time safety monitoring system for the slope of power transmission and transformation facilities. The field test will be carried out at the field test site. Get the following conclusions:

(1) Integrate wireless coordinator, DTU, and timer to connect the serial data network and the wireless network, and the data in the two networks can be converted into each other. Simplify the steps of on-site construction and improve the efficiency of construction.
(2) The slope monitoring system is built on-site with a wireless sensor network, and its basic composition mainly includes three parts: a sensor system, data acquisition and transmission system, and a data management system.
(3) The power transmission and transformation slope monitoring system based on a hybrid data transmission network measures the relationship of slope settlement, deep soil horizontal displacement, vertical earth pressure, lateral earth pressure, and pore water pressure over time. The results show that the impact of rainfall on the slope is limited, and the slope has excellent drainage and good stability.

Acknowledgments. The authors would like to acknowledge the financial support provided by the science and technology project of State Grid Fujian Electric Power Co., Ltd. (No. 52130419002H).

References

1. Tang, S.D., Zhang, S.D., Chen, A.S.: Distributed monitoring system based on PC and PLC. Autom. Instrum. **26**(4), 54–55 (2005)
2. Yin, Y., Wang, H., Gao, Y.: Real-time monitoring and early warning of landslides at relocated Wushan town, the three gorges reservoir. China Landslides **7**(3), 339–349 (2010)
3. Xue, C.L., Zhang, D.X.: Research on highway slope monitoring and early warning system based on Internet of Things technology. High. Transp. Sci. Technol. (Appl. Technol. Edn.) **15**(11), 64–67 (2019)
4. Luo, J.H., Mi, D.C., Huang, H.F., Zhang, T., Sun, G.H., Chen, D.Q.: Intelligent monitoring, stability evaluation, and landslide treatment of a carbonaceous mudstone and Shale Slope in Guangxi, China. Int. J. Saf. Secur. Eng. **10**(3), 373–379 (2020)
5. Pei, H.F., Yin, J.H., Zhu, H.H., Hong, C.Y., Jin, W., Xu, D.S.: Monitoring of lateral displacements of a slope using a series of special fibre Bragg grating-based in-place inclinometers. Meas. Sci. Technol. **23**(2), 1–8 (2012)
6. Zhang, S.B., Chen, T., Yan, P., Wang, K., Wang, J.H., Wang, Q.H.: Application of real-time automatic monitoring system in the monitoring of a landslide in the reservoir area. Chin. J. Undergr. Space Eng. **6**(S2), 1714–1719 (2010)
7. Huang, Q.X., Wang, J.L., Xue, X.: Interpreting the influence of rainfall and reservoir infilling on a landslide. Landslides **13**(5), 1139–1149 (2016)
8. Ahmed, M.M., Pothalaiah, S., Rao, D.S.: Real-time monitoring of partially stable slopes for Landslide prediction by using Wireless Sensor Networks. In: 2016 Online International Conference on Green Engineering and Technologies (IC-GET), pp. 1–19. IEEE, India (2016)

Study of Solidification Technology of Marine Sludge by MICP Combined with Portland Cement

Yan-ning Wang[1,2,3](✉), Bi-xia Zhang[1,2], and Bin-song Jiang[3]

[1] Department of Civil and Environmental Engineering, Shantou University, Shantou 515063, Guangdong, China
wangyn@stu.edu.cn

[2] Guangdong Structural Safety and Monitoring Engineering Technology Research Center, Shantou University, Shantou 515063, Guangdong, China

[3] State Key Laboratory for Geomechanics and Deep Underground Engineering, China University of Mining and Technology, Xuzhou 221116, China

Abstract. Granite residual soil is widely distributed in South China, which often brings landslide and other geological disasters because of its obvious characteristics of water disintegration. Microbial Induced Calcite Precipitation (MICP) technology could improve the engineering properties of granite residual cemented-soil in the process of calcite precipitation. This study presents a laboratory tests to verify the process and to explore the influence of MICP technology on the mechanical properties of cement-soil such as strength and stress-strain relationship. Various technological factors such as cement mixing ratio (CMR), calcium ion concentration (CC) and calcium source (CS) were studied for the mechanical property improving of reinforced soil. Unconfined compressive strength (UCS) test was conducted to analyze the mechanical strength of cemented-soil strengthened by MICP. Based on these works, we have the following conclusions: (1) MICP technology can significantly enhance and improve the engineering properties such as strength, stiffness and toughness of cemented-soil; (2) Compared with the control group, the maximum growth rate of strength in the test group is 87.5%, and the most economical cement mixing ratio is about 15% (3) Calcium chloride and calcium acetate can improve the toughness of the sample, but the effect of calcium acetate is better, compared with the control group, the maximum growth rate of the toughness of the test group is up to 69.67%, when the calcium ion concentration is 0.5 mol/L.

Keywords: Geotechnical engineering · Bacillus pasteurii · Microbial Induced Calcite Precipitation (MICP) · Cement-soil · Mechanical properties

1 Introduction

Granite residual soil formed on the seashore often brings difficulties to engineering construction because it has high water content and large compressibility. In the past, the problem was solved by replacing the cushion, stacking, dynamic consolidation, vacuum

H.-H. Zhu et al. (Eds.): BRWSG 2021, LNCE 230, pp. 258–270, 2022.
https://doi.org/10.1007/978-981-16-9963-4_21

preloading and other physical methods. At present, there are few chemical or biological reinforcement methods.

MICP is a new microbiological technology for strengthening soft soil, with environmental protection and high efficiency. It uses the metabolic activities of microorganisms in cement soil to produce cemented calcium carbonate, thereby solidifying materials. MICP was first used to repair cracks in stone by Adolphe [1]. At present, MICP is also used in soil treatment. Whffin [25] discovered that MICP could be used for soil cementation, which can improve the strength, stiffness and other macro mechanical properties of sand. In addition, Zamani Atefeh [28] used the centrifugal model to study the reinforcement of loose sand by MICP. These results indicate that MICP can improve soil strength and porosity. Therefore, MICP has important application value in geotechnical engineering.

In addition, adding Portland cement into the soil to improve soil properties is also a common method in engineering. Gao [9] proved that adding Portland cement into the soil can improve its compressive strength and reduce its permeability in a certain range. At present, a cement-soil mixing pile is often used to treat foundation in engineering [11].

So far, MICP has not been combined with soil cement to treat granite residual soil. The paper will use MICP combined with Portland cement to solidify marine silt. By measuring the shear strength, shear strength index, moisture content, and dry density, to show the improvement and studying the effects of curing days, nutrient solution concentration, and bacteria on the reinforcement effect. It is hoped that this technology can improve the efficiency of silt solidification and reduce the amount of cement consuming so as to achieve a green and sustainable geotechnical treatment effect.

2 Materials and Methods

2.1 Soil Sample

The soil is granite residual soil 30cm below the surface of Sangpu Mountain in Shantou City, Guangdong Province. They are dried at 105 °C for 24 h, after crushing, sieved with a round hole sieve with a diameter of 2 mm. Refer to Standard for Geotechnical Test Methods (GB/T50123-1999). Measure its physical and mechanical indexes, as shown in Table 1.

The cement is P.O 42.5, and the water is deionized water.

2.2 Strains and Culture

The bacteria is Bacillus pasteurii (purchased from the China Common Microbial Culture Collection and Management Center, number CGMCC 1.3687). The medium was prepared according to Table 2, and the pH value of the medium was adjusted to 8.0 by adding sodium hydroxide solution, and then the culture medium was sterilized in a high-pressure steam sterilizer for 8 h, after that placed in an aseptic operating table to cool. Bacillus pasteurii was inoculated into a 250 mL Erlenmeyer flask, removed 200mL culture medium from it, then placed the culture solution in a constant temperature shaking incubator (30 °C, 150 r/min) after sealing. Finally, bacterial suspension was obtained for 36 h, and its OD600 was measured to be 1.0.

Table 1. Basic physical properties of granite residual soil

Physical index	Value
Moisture content (%)	28.5
Natural density/g.cm^{-3}	1.93
Void ratio	0.98
Liquid limit W_L(%)	34.2
Plasticity Index I_P	11.5

Table 2. 0907 Sporosarcina pasteurii mediums

Number		0907	
Ingredients	Peptone	5.0	g
	Meat slurry	3.0	g
	Urea	20.0	g
	DI water	1.0	L

Note: Add 15 g/L agar

2.3 Treatment Methodology

The solution is a mixture of bacteria suspension: urea: calcium source with a volume ratio of 2:1:1, which includes 2 mol/L of urea and the required concentration of calcium source. The bacteria solution is mixed with the urea-calcium source solution and added to the treated sample immediately. After curing for 7 days, 14 days and 28 days, three parallel samples were prepared for each concentration, respectively.

2.4 Prepare Sample

The specimen was made in a cylindrical mold with a height of 80 mm and a diameter of 39.1 mm. Firstly, mix the sieved soil and cement according to the ratio required for the test, and then add the bacterial solution and the urea-calcium source solution. After that, stir until the cement soil and the solution are completely mixed. Secondly, install the mold, pad with permeable stone and filter paper. Then put the cement soil into the mold 4 times (1/4 each time) to ensure that the amount of soil obtained is almost the same every time and the hammer falls at the same height. After dropping 5 times, pound the soil in the mold before adding soil, and scrape the cement soil that is higher than the mold to ensure the top is flat. Then smooth it until there are no gaps at the edges, put filter paper and permeable stone on it. Thirdly, remove the mold after 6 h (Figs. 1 and 2). Finally, put the specimens in the curing box, and prepare 3 parallel samples for each concentration for 7 days, 14 days and 28 days (Fig. 3).

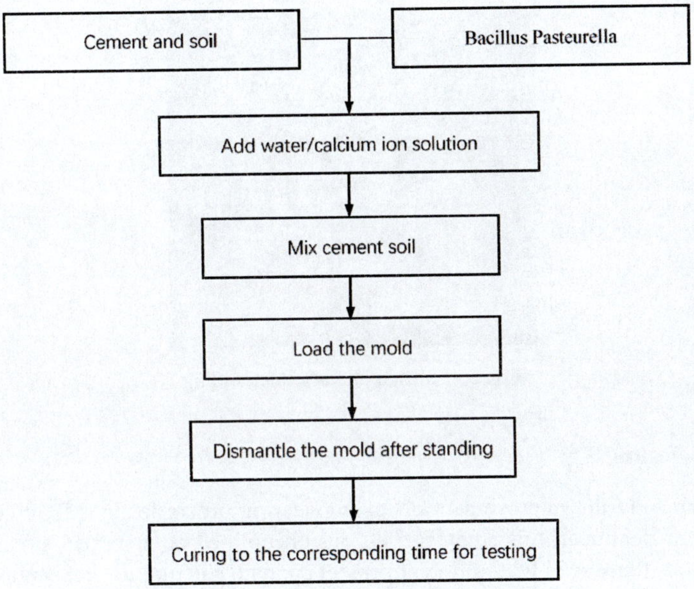

Fig. 1. Test block preparation flow chart

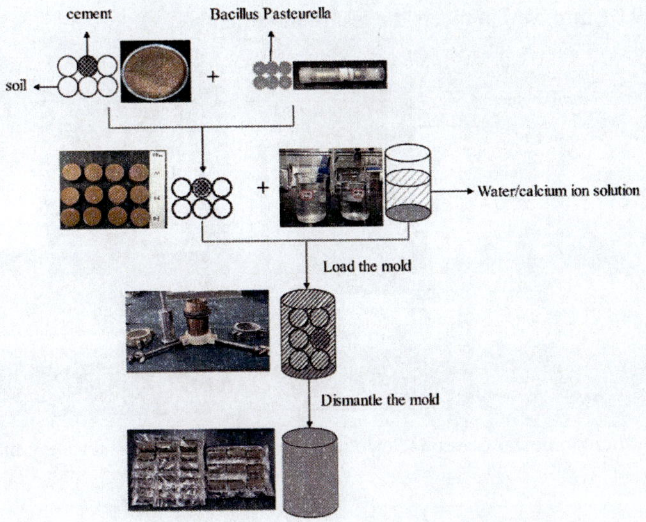

Fig. 2. Test block preparation flow schematic diagram

Fig. 3. Soil samples after demolding

2.5 UCS Testing

In order to study the improvement of its engineering properties by MICP combined with cement treatment, this study carried out unconfined compressive strength tests on specimens reinforced with different cement content, calcium ion concentration and calcium source types. The test equipment is a Byes 3100 microcomputer-controlled electro-hydraulic servo universal testing machine (as shown in Fig. 4). The maximum test force is 1000kN, the test force resolution is 0.01 kN, and the deformation measurement accuracy is 0.01 mm, which meets the test requirements.

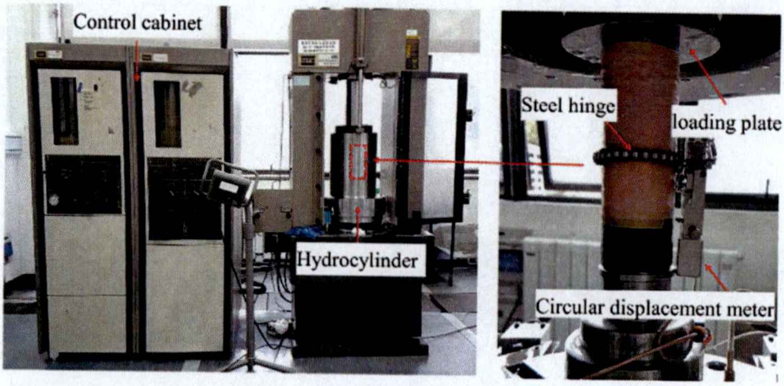

Fig. 4. Microcomputer controlled electro-hydraulic servo universal testing machine

3 Results and Discussion

3.1 Cement Content

Huang Xiaoman [11] showed that the strength of cement-soil mixing piles on soft soil foundations increases with the increase of cement content. Actually, it is reasonable to control the cement content within 8–20%. The study by Ji Xiaolei [13] showed that the cement-soil test block made of silt clay had the largest increase in strength when the cement content was 15%. As a consequence, experiments on microbial modified soil-cement with different contents 10%, 15%, 20% are designed in this section. Take the sterile cement soil as the blank control group, the test method is as mentioned above.

Figures 5 and 6 show the typical stress-strain curves of specimens with different cement contents under the curing age of 28 days. It can be seen that the stress-strain curves of the control groups and the test groups have gone through the pore compaction stage, the elastic stage, the plastic stage in the initial stage, peak value, and the specimens began to fail. In the test, the average value of three parallel specimens is taken as the strength value. When the amount is increased, the peak stress of the sample has been improved, indicating that the amount of cement can significantly enhance the peak stress of the sample.

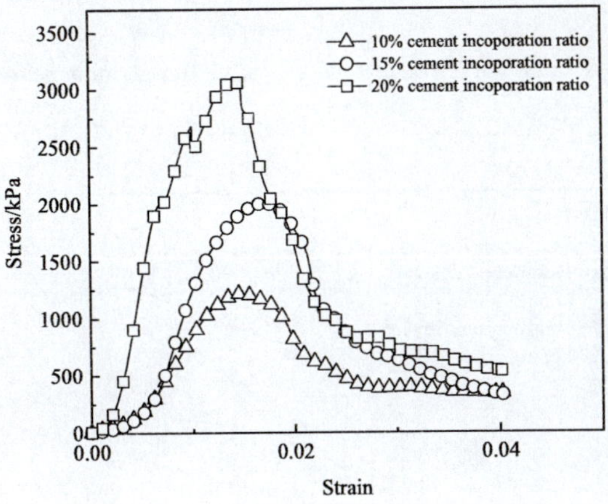

Fig. 5. Typical stress-strain curves of test pieces with different amounts of cement in the control group

Figure 7 shows the variation curve of the UCS of the specimen with the curing time. It can be found that (1) Under the same age, the UCS value of microbial-reinforced cement soil is larger than that of ordinary cement soil. And the growth rates are 51.85%, 87.5%, 47.37% when the curing age is 7 days, and that are respectively 19.77%, 50.3%, 23.26% at 14 days, 22.81%, 64.46%, 41% at 28 days. (2) When the curing age is 7 days, the growth rate is the largest, indicating that the metabolic activity of microorganisms is the most vigorous. (3) Compared with other text groups, when the cement content is 15%, the growth rate of 7 days, 14 days and 28 days is the highest, indicating that the cement content of 15% is a more economical reinforcement amount.

Figure 8 shows the variation curve of the UCS of the specimen with the cement mixing ratio. It can be seen from the figure that (1) Under the same cement mixing ratio, the UCS values of the cement-soil of the two groups increase with the increase of the curing age. Of the two groups, the growth rate of the test group is greater. When the cement content is 10%, the difference between the two experiments is not large. However, the difference gradually increases with the increase of the cement content. (2) When the UCS of the control group was increased from 10% to 15% with cement content, the growth rate is

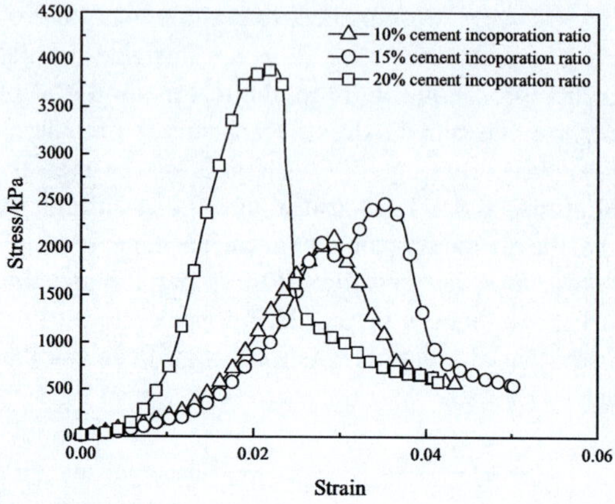

Fig. 6. Typical stress-strain curves of test pieces with different amounts of cement in the test group

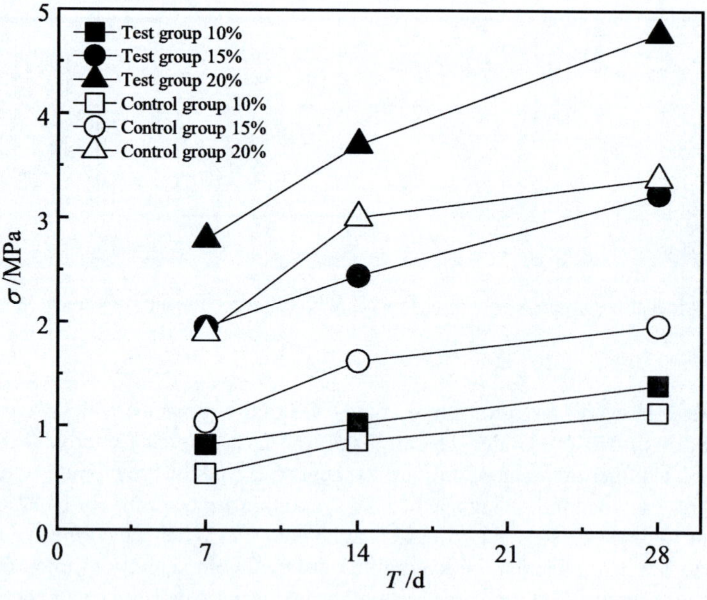

Fig. 7. UCS changes with maintenance age(numbers in the legend are cement incorporation ratios)

significantly lower than that of the test group. This phenomenon is basically consistent with Gao Songhe[9], Xiao Taoli [26], and the difference in soil quality leads to slight differences in the UCS.

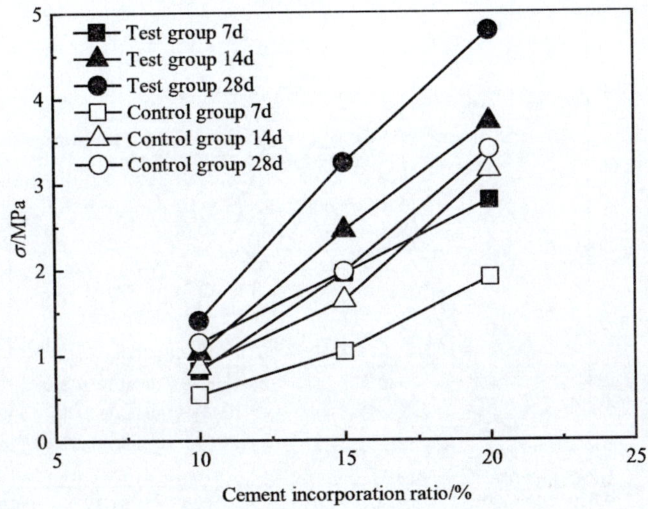

Fig. 8. Change line of test block UCS with cement incorporation(numbers in the legend are maintenance ages)

3.2 Calcium Ion Concentration

The concentration of calcium ions will affect the activity of Bacillus pasteurella, which will affect the amount of urease produced by Bacillus Pasteurella. Besides, as an important raw material in the MICP process, the amount of urease will affect the amount of $CaCO_3$ produced. In turn it will affect the amount of urease produced. In this section, calcium chloride was used as the calcium source to design the composite experiment of MICP and cement under different calcium ion concentrations (0.5 mol/L, 1.0 mol/L and 1.5 mol/L). The sterile cement soil was used as the blank control group, the urea concentration was 1 mol/L, the cement mixing amount was 15%.The test method was as described above. Figure 9 shows the typical stress-strain curves of the specimens under different calcium ion concentrations with a curing age of 28 days. It can be seen that the samples of the typical stress-strain curve of cement-soil in the two previous experiments have gone through the compaction stage, linear elastic stage, plastic stage, peak value, and sample failure stage.

The following conclusions can be drawn from Fig. 9: (1) As the calcium ion concentration is 0.5 mol/L, 1.0 mol/L, 1.5 mol/L, the peak stress of the sample is 4393.92 kPa, 5267.65 kPa, 5106.92kPa and 4473.13 kPa, indicating that low calcium ion concentration is more conducive to improving the strength of cement soil. (2) The strain growth rate of the test group was 17.21%, 48.36%, 43.44%, respectively, indicating that as the calcium ion concentration increases, the strain corresponding to the peak stress of the sample also increases. When the calcium ion concentration reaches a certain concentration, the peak The strain under stress no longer increases. The result shows that the toughness of the sample is improved after adding calcium chloride solution. Furthermore, the calcium ion concentration that is most suitable for improving the toughness of cement soil should be in the range of 1 mol/L–1.5 mol/L.

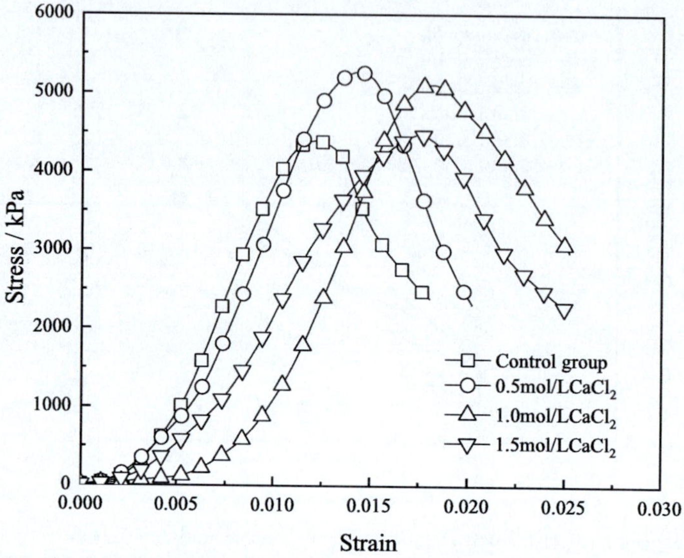

Fig. 9. Typical stress-strain diagrams of samples with different calcium ion concentrations

3.3 Source of Calcium Ions

Due to the different calcium sources, the ions contained in the sample with calcium solution will also be different. The different ion composition will affect the process of Bacillus pasteurianus inducing $CaCO_3$, thus the amount of $CaCO_3$ generated will be distinguishing. Further, there will be a difference in the UCS of the sample. In this section, three groups of experiments with calcium ion concentrations of 0.5 mol/L, 1.0 mol/L and 1.5 mol/L were designed, with the calcium source ($CaCl_2$, $Ca(CH_3COO)_2$) as the independent variable. Urea concentration was 2 mol/L. The sterile cement soil was set as the blank control group. The test method was as described above. Figure 10 is a typical stress-strain diagram with a curing age of 28 days. It can be seen from the figure that the sample has gone through the compaction stage, linear elastic stage, plastic stage, peak value, and sample failure stage as described above.

Figure 10 shows that: (1) The peak stress variation of the test group with $Ca(CH_3COO)_2$ as the calcium source has no obvious rule. While the peak stress decreased with the increase of the calcium ion concentration in the test group with $CaCl_2$ as the calcium source. The experimental group whose source is calcium acetate is large. (2) When the calcium ion concentration is 0.5 mol/L, 1.0 mol/L, and 1.5 mol/L, the strain growth rates of the samples in the $CaCl_2$ test samples under the peak stress are 17.21%, 48.36%, and 43.44%, respectively. While the strain growth rates of $Ca(CH_3COO)_2$ test samples are 26.63%, 54.92%, 69.67% respectively. By comparing the strain growth rate of peak stress, it can be concluded that $CaCl_2$ and $Ca(CH_3COO)_2$ as calcium Sources can improve the toughness of the sample. Among them, $Ca(CH_3COO)_2$ improves the toughness of the sample better.

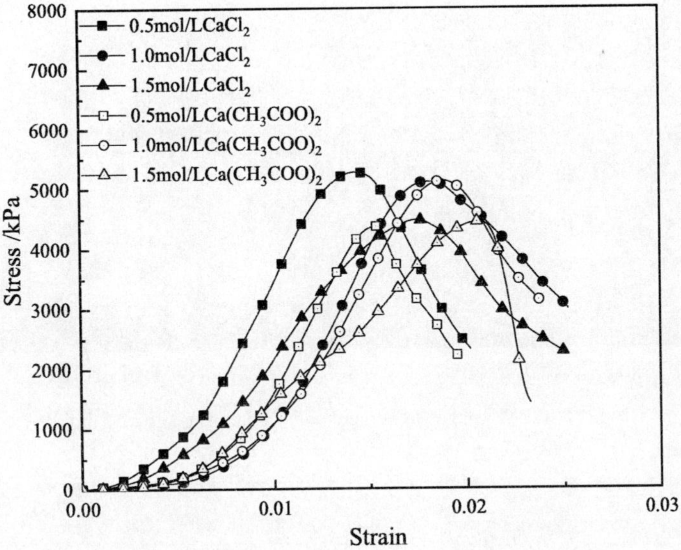

Fig. 10. Typical stress-strain diagram of samples under different calcium sources

3.4 Reinforcement Mechanism

First, cement hydration produces hydration products with cementation, which can fill the intergranular pores. The decrease of porosity can improve the strength of soil-cement. Secondly, MICP technology uses microorganisms to attach to the surface of soil particles, it continuously deposits $CaCO_3$ using itself as a nucleation site, so that the voids of the soil particles are filled and the porosity is further reduced. $CaCO_3$ deposited by microorganisms has a cementing effect to further connect soil particles, thereby increasing the strength of cement soil. Both of them can reduce the porosity of soil-cement and improve the compressive strength of soil-cement. This mechanism can be confirmed by comparing the SEM results of the control group and the experimental group.

Figure 11(a) and (b) is a comparison of the scanning electron microscope (SEM) results of the test group and the control group when the cement content is 15% and the curing age is 7 days. It can be seen that compared with the control group, the pores on the cement surface in the test group are reduced. In addition, the surface density is higher. (a) Cementitious material has been produced between the soil particles of the medium cement soil. Because there are still some soil particles that are not cemented, the cement soil is relatively loose as a whole; (b) There are more cements between the medium soil particles, thus the cement soil presents a relatively dense form as a whole.

Figure 12(a) and (b) is a comparison of the SEM results on 28 days. (a) A large amount of cementitious material has been generated between the soil particles of the medium cement soil, but a small part of the soil particles are still not cemented. The cement soil is relatively dense as a whole, but there are still many small pores; (b) The pores between the soil particles basically disappeared, thus the cement-soil appears dense as a whole. In conclusion, the comparison of test results shows that cement and

(a) Control group (b) Test group

Fig. 11. SEM images of cement soil with curing age of 7 days

(a) Control group (b) Test group

Fig. 12. SEM images of cement soil with curing age of 28 days

MICP have a significant effect on reducing the porosity of soil-cement. This is also the key reason why they can improve the strength of soil-cement.

4 Conclusion

(1) Bacillus Pasteurella modified cement soil can effectively reinforce the UCS of cement soil compared with ordinary cement soil. By comparing the growth rate of the UCS of the sample with different cement content, 15% is the more economical cement content.
(2) When the calcium ion concentration is 0.5 mol/L, the UCS of cement soil increases the most. And when the concentration range is 0.5–1.5 mol/L, the high concentration of calcium ion is beneficial to improve the toughness.
(3) The UCS of the cement-soil samples in the test group where the calcium source is $Ca(CH_3COO)_2$ does not increase significantly, but the toughness of the cement-soil can be improved, and the maximum strain growth rate can reach 69.67%.

The test group with $CaCl_2$ as the calcium source can greatly increase the strength and toughness of the cement soil. The maximum strain growth rate is 48.36%. In addition, $CaCl_2$ is relatively cheap, so $CaCl_2$ is more suitable for engineering applications.

(4) The hydration product of cement has the function of cementation, which can effectively improve the strength of soil cement. In addition, MICP uses the metabolic activities of microorganisms in cement soil to produce cemented $CaCO_3$, which further solidifies the material. Compared with the control group, the combination of MICP and cement can effectively reduce the porosity of soil cement and improve its strength.

References

1. Adolphe, J.P., Loubiere, J.F., Paradas, J., Soleihavoup, F.: Procédé de traitement biologique d'une surface artificielle: EP, EP 0388304 A1(1990)
2. Chen, Z.J.: Engineering characteristics of granite residual soil in Shantou and analysis of slope stability of cuttings. Geology and Mineral Resources of South China (2007)
3. Chen, X.H.: Experimental study on microbial mineralization and fixation of heavy metals and improvement of soil mechanical properties. Tsinghua University, Beijing (2016)
4. Cheng, L., Cord-Ruwisch, R., Shahin, M.A.: Cementation of sand soil by microbially induced calcite precipitation at various degrees of saturation. Can. Geotech. J. **50**(1), 81–90 (2013)
5. Choi, S.G., Wang, K., Chu, J.: Properties of biocemented, fiber reinforced sand. Constr. Build. Mater. **120**, 623–629 (2016)
6. Chu, J., Stabnikov, V., Ivanov, V.: Microbially induced calcium carbonate precipitation on surface or in the bulk of soil. Geomicrobiol J. **29**(6), 544–549 (2012)
7. Dejong, J.T., et al.: Soil engineering in vivo: harnessing natural biogeochemical systems for sustainable, multi-functional engineering solutions. J. R. Soc. Interface **8**(54), 1–15 (2011)
8. Fang, X.G., Shen, C.N., Chu, J., et al.: Experimental study on microbial deposition of calcium carbonate to solidify coral sand. Rock Soil Mech. **36**(10), 2773–2779 (2011)
9. Gao, S.H.: Experimental study on mechanical properties of different cement-soils. Constr. Technol. **43**(S2), 84–89 (2014)
10. Huang, M., Zhang, J.X., Jin, G.X., et al.: Experimental study on NMR characteristics of residual soil MCP grouting stone body freeze-thaw damage. Chin. J. Rock Mech. Eng. **37**(12), 210–219 (2018)
11. Huang, X.M.: Experimental study on the strength characteristics of cement-soil cement mixing piles. West. Transp. Sci. Technol. **5**, 20–23 (2014)
12. Ivanov, V., Chu, J.: Applications of microorganisms to geotechnical engineering for bioclogging and biocementation of soil in situ. Rev. Environ. Sci. Biotechnol. **7**(2), 139–153 (2008)
13. Ji, X.L., Zhu, F.B.: Laboratory study on unconfined compressive strength of cement-soil. Sichuan Build. Sci. Res. **37**(05), 146–148 (2011)
14. Li, M., Li, L., Ogbonnaya, U., Wen, K., Tian, A., Amini, F.: Influence of fiber addition on mechanical properties of MICP-treated sand. J. Mater. Civ. Eng. **28**(4), 04015166 (2016)
15. Li, J.X.: Research on the microscopic characteristics and disintegration mechanism of granite residual soil in Nanyue area. Hunan University of Science and Technology (2014)
16. Li, X.D., Wu, X., Zhang, Z.H.: Research on environmental impact and social willingness to pay based on LCA theory. J. Harbin Inst. Technol. **37**(11), 1507–1510 (2015)

17. Liang, S.H., Zhou, S.Z., Zhang, L., et al.: Statistical analysis of physical and mechanical indexes of granite residual soil in eastern Guangzhou. J. Guangdong Univ. Technol. **1**, 29–33 (2015)
18. Lin, P., Chen, H.M., Wang, Y.Q.: Behavior of the unsaturated granite residual soil and its effects on earth dam project, pp. 68–75. ASCE, Geotechnical Special Publications (GSP), 217. ISSN: 1672-1144 (2011)
19. Liu, S., Chen, Z.B., Chen, W.W., et al.: Experimental study on the improvement of granite residual soil with fly ash. J. Fuzhou Univ. (Natural Science Edition) **46**(05), 115–120 (2018)
20. Liu, S.N.: Environmental impact assessment of cement-concrete system and its application research. Wuhan University of Technology, Wuhan (2003)
21. Stabnikov, V., Chu, J., Ivanov, V., et al.: Halotolerant, alkaliphilic urease-producing bacteria from different climate zones and their application for biocementation of sand. World J. Microbiol. Biotechnol. **29**(8), 1453–1460 (2013)
22. Sun, C.W.: Research on engineering characteristics of granite residual soil and design technology of subway deep foundation pit. China University of Geosciences, Wuhan (2014)
23. Van Paassen, L.A., Ghose, R., van der Linden, T.J.M., van der Star, W.R.L.: Quantifying biomediated ground improvement by ureolysis: a large scale biogrout experiment. J. Geotech. Geoenviron. Eng. **136**(12), 1721–1728 (2010)
24. Whiffin, V.S., Paassen, L.A.V., Harkes, M.P.: Microbial carbonate precipitation as a soil improvement technique. Geomicrobiol. J. **24**(5), 417–423 (2007)
25. Whiffin, V.S.: Microbial CaCO3 precipitation for the production of biocement. Ph.D. thesis, Murdoch University (2004)
26. Xiao, T.L., He, Y.L., Li, Q.F., et al.: Experimental study on unconfined compressive strength of cement-soil. J. Yang. Univ. (Natural Science Edition) **14**(05), 64–66+87 (2017)
27. Yu, J.J., Chen, D. X., Wang, H., et al.: Analysis of the shear strength and slope stability of granite residual soil under dry-wet cycles. J. Xiamen Univ. (Natural Science Edition) **58**(4) (2019)
28. Zamani, A., Xiao, P., Baumer, T., et al.: Mitigation of liquefaction triggering and foundation settlement by MICP treatment. J. Geotech. Geoenviron. Eng. **147**(10) (2021)
29. Zhang, S.: Research on the disintegration characteristics of granite residual soil in Guangzhou area. China University of Geosciences, Wuhan (2009)
30. Zhao, L.J.: Study on the strength and disintegration characteristics of improved granite residual soil. Hunan University of Science and Technology, Changsha (2015)

Hydraulic Properties of Unsaturated Engineering Soils Vegetated with Vetiver Grass in Green Infrastructures

Hao Wang[1,2], Junwen Huang[1,2], Rui Chen[1,2], and Zhaofeng Li[1,2(✉)]

[1] School of Civil and Environmental Engineering, Harbin Institute of Technology, Shenzhen, 518055 Shenzhen, China
lizhaofeng@hit.edu.cn

[2] Shenzhen Key Laboratory of Intelligent Structure System in Civil Engineering, Harbin Institute of Technology, Shenzhen, Shenzhen 518055, China

Abstract. Aiming at green infrastructures which are widely established in countries and regions within the Belt and Road Initiative, both in-situ monitoring and indoor experiments were conducted in the present study. Relationships of hydraulic properties (e.g., soil water characteristic curve and permeability) with root characteristics (e.g., distribution, morphology as well as biomass) and meteorological conditions were investigated. It is found that coarse roots are mainly observed in shallow soils, while fine roots are widely distributed in deep layers. Preferential flow induced by coarse roots and pore clogging due to fine roots are proved in dye-tracer test. The field monitoring is divided into three stages, i.e., T1 (sunny days), T2 (rainy days) and T3 (sunny days after T2). There exist distinct diurnal characteristics of changes of soil suction and water content during the T1 and T3 stages. Moreover, soil suction is negatively relevant with air relative humidity, under absence of rainfalls. Vegetation increases amplitudes of diurnal changes of soil suction and water content. During the T2 stage, diurnal variations of VWC of both vegetated and bare soils are negligible due to rainfalls. Soil air-entry value (AEV) is positively relevant with root biomass. In addition, vegetation improves soil water-holding capability at most vegetation-available suction ranges. As compared to indoor-measured SWCC, field-measured SWCC is more similar to that of ideal coarse-grained soils. Roots improve soil permeability through preferential flows and enhanced water-retention capacity. Results obtained in the present study helps to obtain desired outcomes, i.e., improving feasibility, performance and durability of green infrastructures.

Keywords: Unsaturated soil · Soil water characteristic curve · Permeability · Green infrastructure · Dye-tracer technique

1 Introduction

The Belt and Road Initiative was proposed by China to enhance cooperation with neighbouring countries and improve infrastructures in developing countries. As of January 30, 2021, a totally 205 cooperation agreements about the Belt and Road Initiative were

H.-H. Zhu et al. (Eds.): BRWSG 2021, LNCE 230, pp. 271–287, 2022.
https://doi.org/10.1007/978-981-16-9963-4_22

signed by more than 170 countries. In addition to conventional infrastructures (e.g., ports, expressways and railways, etc.), green infrastructures (e.g., vegetated landfill, vegetated slope and bioretention, etc.) are essential parts of the Belt and Road Initiative. Unlike loose and fertile agricultural soils, engineering soils used in green infrastructures are commonly re-compacted. Besides, effects of changes in climate and weather are more predominant in these infrastructures, under the absence of artificial irrigation. Given these, interactions of soils and roots in green infrastructures might be significantly different from that in agricultural soils.

Soil suction and water content of vegetated soil are two essential hydraulic factors that influence (1) shear strength of unsaturated soil and consequently stability of vegetated soil slope [1], and (2) water-holding capability, gas permeability as well as hydraulic conductivity of vegetated soils used in landfill cover [2, 3]. In a natural environment, matric suction and water content of vegetated soil could be influenced by weather factors such as rainfall [4–6], light intensity [7], air relative humidity [8] and air temperature [9, 10]. Commonly, changes in meteorological factors mentioned show distinct diurnal characteristics. The coupled effects of meteorological factors on matric suction and water content of the vegetated soil could be investigated by conducting field tests under natural weather conditions. Some field studies have been carried out to investigate performances of the vegetated soil layers in Canada, Australia and Italy, etc. [11–14]. Nevertheless,a few field-monitoring on vegetated soil layer was carried out in China, Vietnam and Pakistan, etc. Climatic characteristics, types of native vegetation and engineering requirements in these Asian countries are significantly different from those in North America and Europe. In addition to weather conditions, matric suction and water content of vegetated soil are also influenced by root characteristics. For instance, root density is a critical factor affecting the responses of matric suction and moisture content in vegetated soil under weather changes such as drying and wetting [15].

As far as the authors are aware, the effects of diurnal weather changes on hydraulic properties of vegetated soils in field were rarely reported. Besides, the influences of roots on hydraulic properties of compacted engineering soils were not fully investigated. Thus, the present study intends to (a) comprehensively investigate relationships of hydraulic properties and meteorological conditions (especially diurnal weather changes) in field, and (b) explore effects of root characteristics (e.g., root biomass, root diameter as well as root distribution) on hydraulic properties of soils. To achieve these goals, soil water characteristic curves (measured in both field and laboratory), unsaturated permeability and dye-tracer test of vegetated soils were conducted in this study.

2 Materials, Theory and Methodology

2.1 Field Site and Vegetation Cultivation

The field test site was located at a lawn at University town of Shenzhen, Shenzhen, Guangdong Province, China (longitude 113°58' E and latitude 22°35'N). Basic physical properties of soils used in the present study were summarized in Table 1. As shown in Fig. 1, the gravel, sand, silt and clay contents of tested soil were 31.3%, 28.8%, 26.9% and 13%, respectively. According to the soil classification system by the United States Department of Agriculture (USDA), tested soil was classified as a loam with gravel that was commonly distributed in countries and regions of the Belt and Road Initiative. Soils were crushed and air-dried (refer to Fig. 2b), and then mixed with water to a target water content (i.e., 20%). After that, the slurry was sealed for 1 day using plastic cloth to ensure homogeneity of soil moisture content. Soils were compacted (with 85 ± 1% degree of compaction) in 5 layers (as observed in Fig. 2c and 2d). During compaction, contacting surfaces between different layers were roughened artificially (as shown in Fig. 2e).

Vetiver grass (i.e., *Vetiveria zizanioides*) was vegetated in the test site mentioned in the present study. In southeast China, Thailand as well as Vietnam, etc. (i.e., main regions of the Belt and Road Initiative), vetiver grasses were natively distributed. In addition, vetiver grasses were artificially vegetated nowadays in these regions to achieve financial goals, for instance, papermaking, and productions of animal feeds and commercial flavoring, etc. Before being transferred to the test site, vetiver grasses were pre-cultivated in fertile soil for about 6 month.

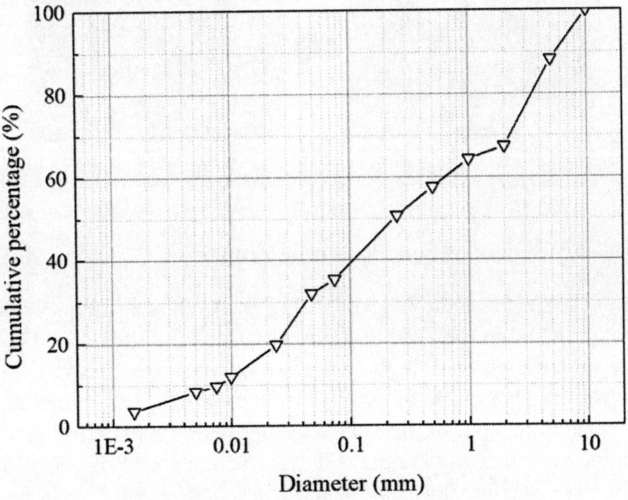

Fig. 1. Pore size distribution of test soil.

Table 1. Basic physical properties of tested soil

Gravel content (%)	Sand content (%)	Silt content (%)	Clay content (%)	Optimum moisture content (%)	Maximum dry density (g/cm^3)
31.3	28.8	26.9	13.0	18.5	1.64

Fig. 2. Preparation of test site.

2.2 In-Situ Measurement of Soil Hydraulic Properties

Arrangement of the field test site is given in Fig. 3. As shown in Figs. 3a and 3b, three soil plots were prepared in the field test site, i.e., G1, G2 and B. Plot G1 and plot G2 represented vegetated plots with high and less root biomasses, respectively. To obtain relatively low root biomass, fine roots of grasses used in G2 were moderately pruned before being transferred from fertile soil to testing field. Plot B represented bare soils, i.e. root biomass was nil. Plots G1 and G2 had the same size of 0.9 m in length and 0.6 m in width. The size of plot B was 0.6 m in length and 0.3 m in width. Prior to the field monitoring, all vetiver grasses were irrigated daily to ensure acceptable growth condition.

A self-designed high capacity tensiometer (HCT) was used to measure soil matric suction in the present study. The HCT was able to measure matric suction from 0 kPa to 500 kPa, with an accuracy of ±2 kPa. Temperature sensitivity of the HCT at 0 kPa was tested by recording the readings of HCT within temperature range from 23 °C to 45 °C. The results showed that the change of the HCT readings within the above temperature range was limited (less than 5 kPa). More details about the HCT could be found in previous study [16]. Meanwhile, soil volumetric water content (VWC) was measured by EC-5 sensor (Decagon Device, Inc.), with an accuracy of ± 0.02 cm^3/cm^3. A high frequency data acquisition system was used to capture the diurnal changes of matric suction and soil water content. The HCT and EC-5 sensors were pre-calibrated, and installed at soil depth of 100 mm with a horizontal spacing of 100 mm (as shown in Fig. 3b). Based on the results about root influence zone reported by previous study [4], it could be regarded that both HCT and EC-5 of plots G1 and G2 in this study were located within the influence zone of grass roots.

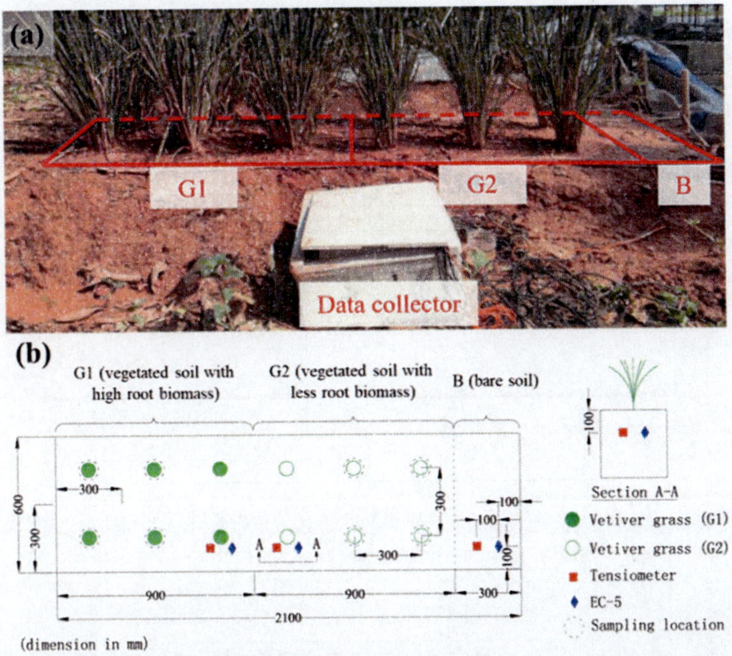

Fig. 3. Arrangement of the field test site: (a) reality image and (b) schematic diagram.

2.3 Meteorological Monitoring During Field Test

The two-week field monitoring test was conducted from 12:00 on September 17, 2017 to 12:00 on September 30, 2017. During the field monitoring, hourly meteorological parameters (e.g., precipitation and air relative humidity, etc.) were recorded by a meteorological station located at the University town of Shenzhen. The measuring height of precipitation and air relative humidity (i.e., RH) was 1.5 m above the ground. Before the two-week field monitoring, three rainfall events occurred during the period from September 12, 2017 to September 14, 2017, followed by two days without rainfall. During the two-week field monitoring, recorded hourly rainfall, air temperature and RH were given in Fig. 4a, 4b and 4c, respectively.

2.4 Analyses of Root Vertical Distribution

Quantitative Analysis of Root Characteristics. After the field monitoring, cylindrical soil specimens (200 mm in diameter and 200 mm in height) with roots were sampled from plot G1. Because roots of grasses vegetated in plot G2 were artificially pruned to obtain a less root biomass, specimens from G2 are not able to describe natural distribution of roots. Soil specimens from G1 were first soaked for 2 h using deionized

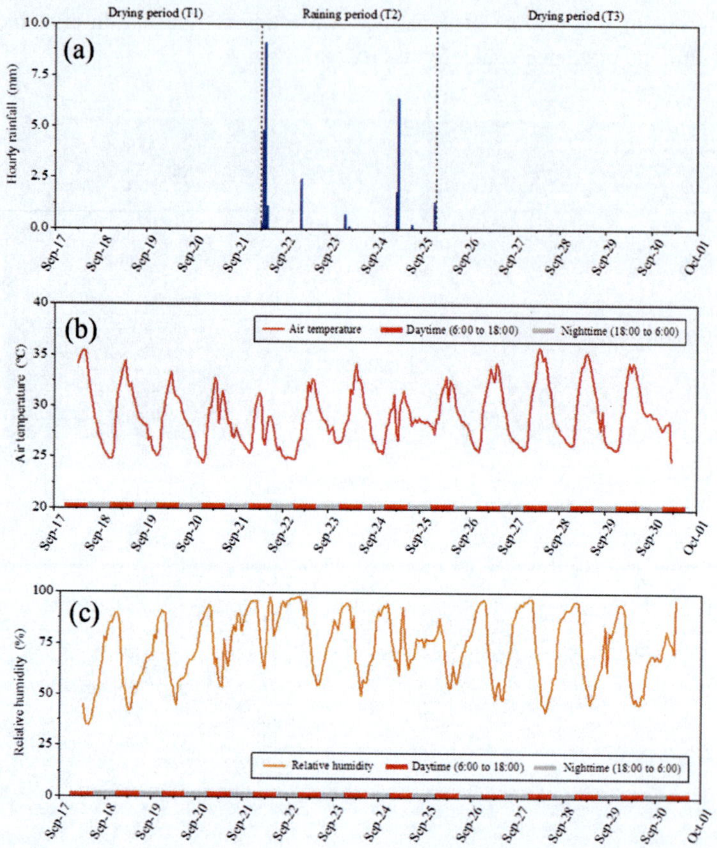

Fig. 4. In-situ meteorological monitoring of (a) rainfall, (b) air temperature and (c) relative humidity.

water, to ensure natural shedding of most soil particles. And then, rhizosphere soils were washed out gently. After that, pictures of roots with high resolution were photographed at controlled environmental conditions (e.g., illumination intensity). Based on captured pictures, root parameters (i.e. root length and root diameter) were determined using WinRhizo Pro software (version 9.0; Regent Instruments, Inc., Quebec, QC, Canada). With assumption that all cross sections of roots were circular, the root volumes were calculated from the measured root length and root diameter.

Visualization of Root Distribution Based on Dye-Tracer Technique. Dye-tracer technique was used in the present study to visually show vertical distribution of roots. In accordance with quantitative analysis of root characteristics mentioned above, soil columns used in dye-tracer method were also from plot G1. In order to minimise boundary effects, following processes were conducted. (1) Vaseline was applied on sides of soil columns (with 15 cm height and 20 cm diameter). (2) Sides of soil columns were sealed using plastic film. (3) Sealed soil columns were placed in metal containers, and kaolinite clay was used to fill gaps between sealed columns and the containers. Dying was conducted using aqueous solution of brilliant blue dye with concentration of 30 g/L. Brilliant blue dye was selected due to (1) non-adherence to soil particles, and (2) obvious colour difference between the dye and tested soils. Three days after the dyeing process, soil specimens were cut into short columns with 30 mm height. Sections were photographed at controlled environmental conditions (e.g., illumination intensity) using digital camera.

2.5 Measurement of Hydraulic Properties Based on Simplified Evaporation Method

Soil water characteristic curve (SWCC) is defined as the relationship of soil moisture content with soil matric suction. In the present study, simplified evaporation method was conducted to measure SWCC and calculate unsaturated permeability. In addition, soil saturated hydraulic conductivity was measured by falling head permeability test.

In simplified evaporation method, soil suction was measured by two sensors at different heights. Suction at the central plane was assumed to be the mean value of suctions measured by upper and lower sensors. Also, mass of soil sample was measured through an electronic scale during evaporation. The change of soil water content at the central plane was calculated according to water mass loss (i.e., mass of water evaporated). If saturated soil water content was known, soil water contents at each moment could be calculated. As shown in Fig. 5, a pressure sensor (ZNBS-P model) was utilized to measure mass of water evaporated. Wireless measurement systems were designed in the present study to minimize negative effects of changes of environment on measurement. The wireless system consisted of analog to digital converter (i.e., ADC; UL10AI, Liankong Intelligent Technology Inc.), wireless signal transmitter (ZigBee DRF2619A, DTK Electronics Inc.) and power supply. Simplified evaporation method was conducted in triplicate for each type of soil.

Moreover, unsaturated permeability of soil was given as Eq. (1).

$$k(\overline{\psi}) = \frac{\delta V}{2A\bar{i}\delta t} \tag{1}$$

where k means hydraulic conductivity during time interval δt, and is related to soil suction at the central plane (i.e., mean suction of the upper and the lower $(\overline{\psi})$). Volume loss of soil water in time interval δt and area of cross section are denoted by δV and A, respectively. During time interval δt, $\bar{i}$ represents mean hydraulic gradient, and is expressed as Eq. (2).

$$\bar{i} = \frac{1}{2}\left(\frac{\psi_{0,up} - \psi_{0,low}}{\delta z} + \frac{\psi_{1,up} - \psi_{1,low}}{\delta z}\right) - 1 \tag{2}$$

where δz indicates vertical distance between two suction sensors. Soil suctions measured by the upper sensor and the lower sensor were represented by $\psi_{time,up}$ and $\psi_{time,low}$, respectively. Soil suctions measured at the initial moment and the end moment of time interval δt were denoted by $\psi_{0,location}$ and $\psi_{1,location}$, respectively.

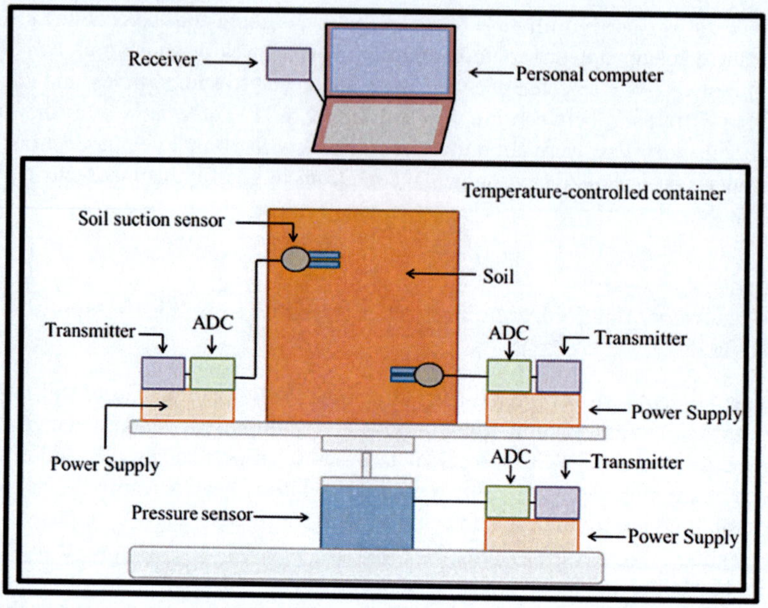

Fig. 5. Wireless measurement of hydraulic properties based on simplified evaporation method.

3 Results and Discussions

3.1 Root Distribution of Vegetated Soils

Soil dyeing conditions at different soil depths are shown in Fig. 6. Dyeing is only observed significantly within 0 cm to 9 cm depth, and it is hardly observed at 12 cm soil section. As shown in Fig. 6a, surface layer of soil is totally dyed. At 3 cm depth, a consecutive dyeing is distributed at relatively concentrated area (i.e., area surrounding root system). Dyeing is hardly observed at rootless area at 3 cm depth where influences of surface cracks are insignificant. At 6 cm depth and 9 cm depth, dyeing distributions become relatively discrete (as observed in Fig. 6c and 6d), as compared to consecutive pattern at 3 cm depth. It could be concluded that with increasing soil depth, dyeing area becomes larger, but dyeing distribution becomes discrete.

Relationships of soil depth with root length density (RLD) and root volume density (RVD) were given in Fig. 7a and 7b, respectively. Based on RLD, percentages of root biomass distributed in 0–5 cm, 5–10 cm, 10–15 cm and 15–20 cm are 52.4%, 30.5%, 14.3% and 2.8%, respectively. Contrarily, biomass percentages calculated by RVD were 71.9%, 17.8%, 7.9% and 2.4%, respectively. Percentage of 0–5 cm calculated by RVD is greater than that calculated by RLD. This implies that coarse roots are mainly distributed in shallow soils (0–5 cm), while fine roots are widely distributed in deep layers (5–20 cm).

In addition, dyeing is hardly observed at 12 cm soil section, but roots are still found at deeper soil depths (as shown in Fig. 7). It is because that roots distributed in deeper layers are mainly fine roots (less than 1mm diameter). Fine roots with less mechanical strength are prone to buckle under stresses from surrounding soil particles [17]. Therefore, fine

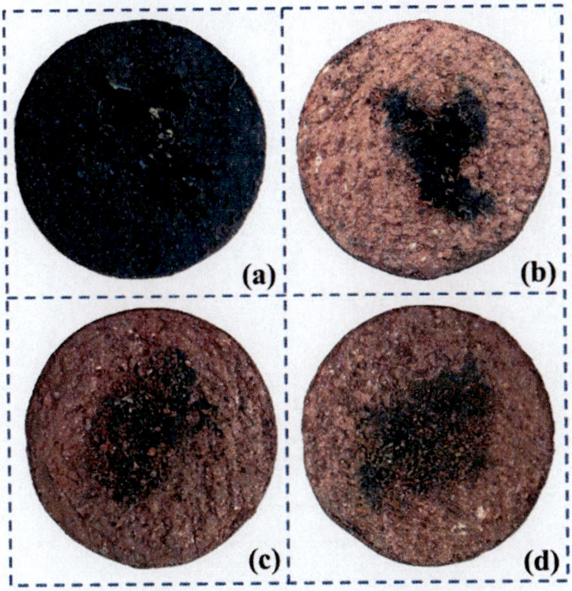

Fig. 6. Dyeing conditions of vegetated soils at (a) surface, (b) 3 cm depth, (c) 6 cm depth and (d) 9 cm depth.

roots tend to grow along tortuous pathways among soil particles, and finally cause blocking of pores [18–20]. According to previous study [21], fine roots cause reduced saturated hydraulic conductivity (K_s) via pore clogging. Pore clogging induced by fine roots likely hinders infiltration of dye solution, which results in disappearance of dye in deeper layers in this study. On the contrary, coarse roots are widely distributed in shallow layer (i.e., 3 cm depth). At 3 cm depth, dyeing is hardly observed at rootless area (i.e., bulk soils). In other words, dye solution infiltrates mainly through soils surrounding coarse roots at 3 cm depth. This proves the preferential flow effect induced by coarse roots.

3.2 Field-Measured Soil Water Characteristic Curves of Bare and Vegetated Soils

Soil water characteristic curves of soils with different root biomass measured in filed are given in Fig. 8. As shown in Fig. 8, air entry value (AEV) of tested soil is positively related to root biomass in soils. This implies that air could entry pores in saturated bare-soils more easily, as compared to saturated vegetated-soils. There are mainly two passages for air to enter the soil from the atmosphere, i.e., surface cracks and connected pores. Roots distributed in shallow soils are mainly coarse ones. Coarse roots are able to constrain desiccation cracks upon soil surface. Moreover, fine roots with less mechanical strength are not able to cause soil structure changes in highly compacted soils. They are prone to buckle under stresses from surrounding soil particles [17], which leads to pore clogging.

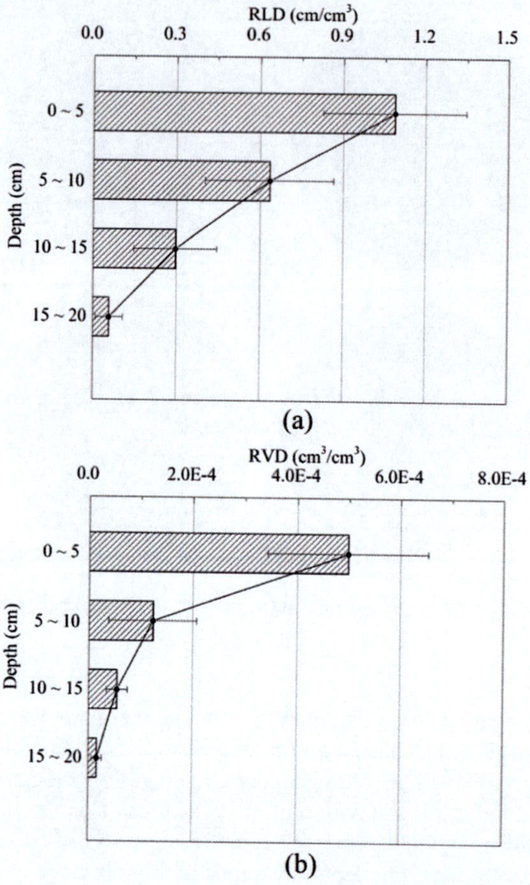

Fig. 7. Root characteristics versus soil depth: (a) root length density and (b) root volume density.

3.3 Field-Measured Soil Suction During Different Stages

Relationship of measured soil matric suction with hourly rainfall and air relative humidity (RH) during in-situ monitoring is shown in Fig. 9. The two-week of field monitoring is divided into three periods of time, i.e., T1, T2 as well as T3. T1 represents the days without rainfall (i.e., from September 17 to September 20), T2 denotes the rainy days from the first rainfall event to the last rainfall event (i.e., from September 21 to September 25), and T3 represents the days without rainfall after T2 (i.e., from September 26 to September 30). Note that data of matric suction were lost from 12:00 to 24:00 on September 25, due to unexpected interruption of power supply of the data logger. Moreover, daytime is defined as the period of time from 6:00 to 18:00 of a day, according to actual sunrises and sunsets in Shenzhen during the period of field monitoring. Correspondingly, nighttime is defined as the period of time from 18:00 of the day to 6:00 of the next day. As shown in Fig. 9, average RH of atmosphere near test site during the two-week field monitoring is approximately 75%. In the daytime, RH decreases dramatically from 6:00 to 12:00 and

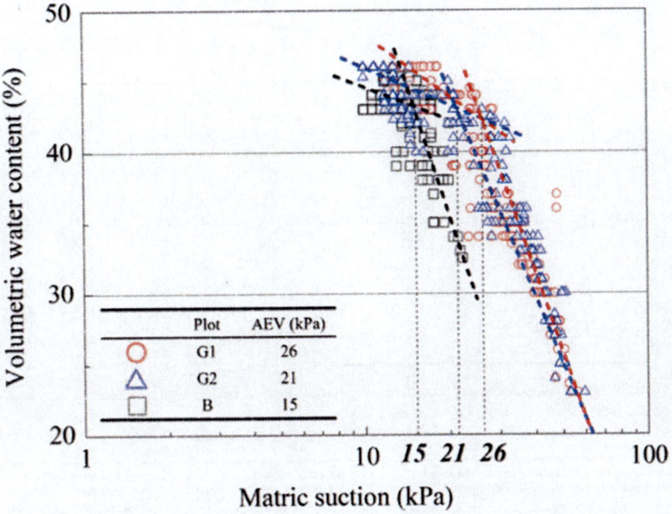

Fig. 8. Field-measured soil water characteristic curves.

reaches a trough value (i.e., 60% or below) at around 12:00. Then RH starts to increase from 12:00 to 18:00. In the nighttime, RH increases significantly and reaches a peak value (i.e., 90% or above) at the end of nighttime (around 6:00). Similar changes of RH were reported in previous study [22], in which the peak value of RH was also observed at 6:00. Besides, it is found that soil suction is negatively related to RH at a given day in T1 and T3 stages.

It is as expected that soil suction tends to increase during drying-stage, while reduce dramatically due to rainfalls. During T1 stage, only matric suction of bare soil increases slightly, while those of vegetated soils fluctuate within suction range of 25 kPa to 60 kPa. Sharp reductions were observed for all types of soils at the beginning of the T2 stage. After that, suctions of all types of soils fluctuate at relatively low suction range (less than 25 kPa). During T3 stage, suctions of all types of soils are prone to increase with fluctuation.

Next, diurnal changes of matric suction of soils during three stages are discussed. During a given day in T1, matric suctions firstly increase from 6:00 to 12: 00, and then reach peak values at around 12:00. After that, matric suctions decrease from 12:00 to 18:00 in the daytime. In the nighttime, matric suctions decrease from 18:00 to 6:00 of the next day, during which the trough values are observed at around 6:00. Amplitude of diurnal change of suction (ADCS) is defined as the difference between the peak value and the trough value of suction. ADCSs are nearly 9 kPa and 22 kPa for bare soil and vegetated soil, respectively. Since the ADCSs are much larger than the magnitude of measurement accuracy of HCT (i.e., ±2 kPa), it could be regarded that diurnal changes of matric suction of both vegetated soil and bare soil were distinct. In addition, it is found that difference of ADCSs of G1 and G2 is minor, i.e., root biomass plays a negligible role in influencing ADCS of vegetated soils. Root biomass seems to be only able to affect suction of vegetated soils, instead of diurnal characteristics of suction changes.

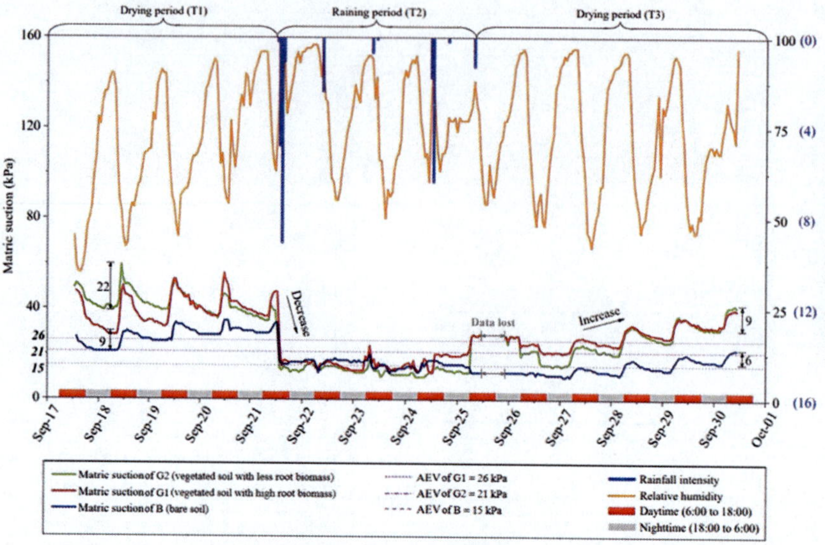

Fig. 9. Field-measured soil suction, rainfall intensity and air relative humidity.

Moreover, diurnal characteristics of vegetated soils (with about 22 kPa ADCS) are found to be more distinct than bare soil (with approximately 9 kPa ADCS). During T2 stage, diurnal changes of soil suction of both vegetated and bare soils are negligible due to rainfalls. Like T1 stage, diurnal characteristics of suction changes are distinct during T3 stage. However, difference between ADCSs of vegetated and bare soils becomes relatively minor during T3 stage, as compared to T1 stage. During T3 stage, ADCS of vegetated and bare soils are 9 kPa and 6 kPa, respectively. This suggests history of weather changes (especially rainfalls) likely affects diurnal changes of soil suction.

3.4 Field-Measured Soil Moisture Content During Different Stages

The evolutions of measured volumetric water content (VWC) and corresponding hourly rainfall intensity and RH during field monitoring are given in Fig. 10. It is observed that VWC is positively relevant with RH during a given day of T1 and T3 stages, particularly for vegetated soils. Like changes of soil suction, distinct diurnal characteristics are also observed in VWC changes. From 6:00 to 12:00 at a given day in T1 stage, VWCs of both vegetated and bare soils firstly decrease. And then, VWCs reach trough values at around 12:00. After that, they increase from 12:00 to 18:00 in the daytime. In the nighttime, VWCs increase from 18:00 to 6:00 of the next day, and peak values are observed at about 6:00 of the next day.

During T1 stage, amplitudes of diurnal change of water contents (ADCW) are nearly 0.12 cm^3/cm^3 and 0.04 cm^3/cm^3 for vegetated soil and bare soil, respectively. As the amplitudes are greater than the magnitude of measurement accuracy of EC-5 sensor (i.e., 0.02 cm^3/cm^3), it could be regarded that diurnal characteristics of VWC changes are distinct (especially for vegetated soils). Besides, difference between ADCW of two types of vegetated-soils is minor. During T2 stage, diurnal changes of VWC of both vegetated and bare soils are negligible because of rainfalls. During T3 stage, diurnal characteristics of VWC changes are also distinct, but difference between ADCWs of vegetated and bare soils becomes relatively limited. This implies that history of meteorological changes (particularly rainfalls) also influence diurnal changes of VWC.

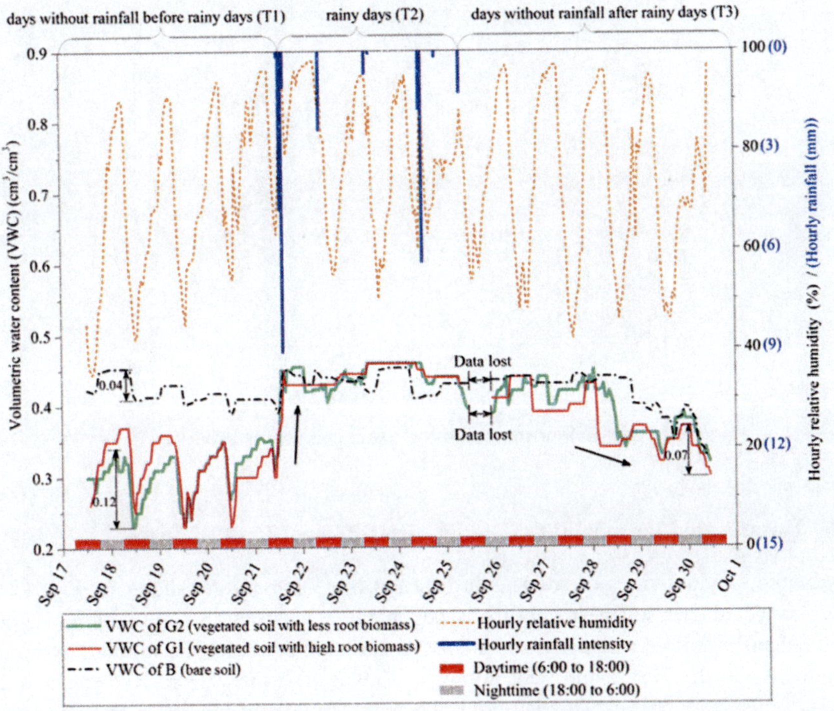

Fig. 10. Field-measured volumetric water content, rainfall intensity and air relative humidity.

3.5 Indoor-Measured Soil Water Characteristic Curves of Bare and Vegetated Soils

Soil water characteristic curves of soils with different root biomass measured in laboratory are shown in Fig. 11. In accordance with in-situ measured SWCCs, indoor-measured AEV of soils are increased due to vegetation. Furthermore, it is found that vegetation improves soil water-holding capability throughout measured suction range. This is because of hydrophilic chemicals and root water potential. According to previous studies [23–25], mucilage, that is exuded by roots, improves water-retention capacity of rhizosphere soils. Driven by water potential difference, water moves from deep soil layer to shallow layer, and then from shallow layer to the atmosphere. Roots retard water movement mentioned through creating a new water potential difference between themselves and surrounding soils [26]. Moreover, desorption rate of indoor-measured SWCC is lower as compared to SWCC measured in the field (refer to Fig. 8). This suggests curve shape of field-measured SWCC is more similar to that of ideal coarse-grained soils (with relatively high desorption rate).

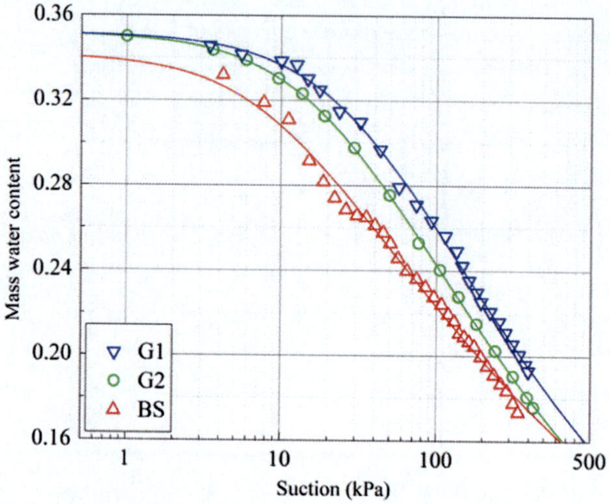

Fig. 11. Indoor-measured soil water characteristic curves.

3.6 Unsaturated Permeability of Soils with Different Root Biomass

Hydraulic conductivities of soils with different root biomass are shown in Fig. 12. It is as expected that soil permeability is negatively relevant with soil suction. Vegetation enhances soil permeability, especially in saturated conditions. Saturated hydraulic conductivities of G1, G2 and BS samples are 3.9×10^{-8} m/s, 7.7×10^{-8} m/s and 2.1×10^{-8} m/s, respectively. This is mainly because of two reasons, i.e., improvement of water-holding capacity and preferential flows induced by roots. Vegetated soils are able to hold more water at given suction (refer to Fig. 8 and Fig. 11), while soil permeability is highly dependent on water content. In addition, roots (especially coarse roots) result in re-orientation of soil particles and cracking of macro-aggregates. Therefore, soils surrounding roots tend to be more fragmented and less cohesive. This means that more channels are provided for water infiltration. Therefore, vegetated soils are able to conduct water more rapidly as compared to bare soils. Besides, macro-pores are formed after decay of coarse roots, which provides a greater permeability [27, 28]. Besides, root-induced changes of permeability are also influenced by root exudates. Mucilage at the interface of root and rhizosphere soil was reported to be able to increase hydraulic conductivity [23].

It is noteworthy that permeability of G2 sample (with less root biomass) is much greater than that of G1 sample (with high root biomass). As mentioned above, fine roots of grasses in G2 are pruned moderately before the field test. Therefore, root-induced changes of soil permeability are mainly caused by coarse roots in G2, while they are caused by both coarse and fine roots in G1. Coarse roots cause re-orientation of soil particles and cracking of macro-aggregates in G2 soil. On the contrary, fine roots are widely distributed in G1 soils (as shown in Fig. 6). As compared to coarse roots, mechanical strength of fine roots is minor. Fine roots lead to clogging of pores and amalgamation of micro-aggregates [21], which definitely reduces soil permeability.

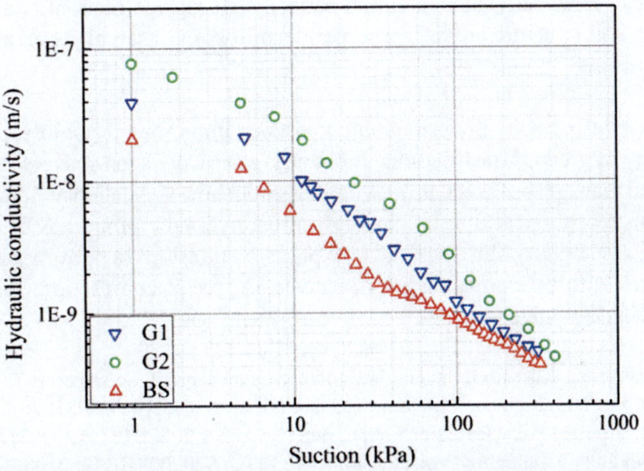

Fig. 12. Indoor-measured relationships of hydraulic conductivity with soil suction.

4 Conclusions

Aiming at green infrastructures which are widely established in countries and regions within the Belt and Road Initiative, both in-situ monitoring and indoor experiments were conducted in the present study. Results obtained in this study helps to understand how plant roots affect hydraulic properties of highly compacted soils commonly used in green infrastructures (e.g., landfill cover and vegetated slopes, etc.).

(1) Maximal depth of root system is approximately 20 cm, but the most of root biomass are in the surface layer (0–5 cm depth). Coarse roots are mainly observed in shallow soils, while fine roots are widely distributed in deep layers. Preferential flow induced by coarse roots and pore clogging due to fine roots are proved in the dye-tracer test. Pore clogging induced by fine roots hinders infiltration of dye solution, thereby dyeing is only observed within 0 cm to 9 cm depth. At 3 cm depth, dyeing is hardly observed at rootless area, i.e., dye solution infiltrates mainly through soils surrounding coarse roots.

(2) There exist distinct diurnal characteristics of changes of soil suction and VWC during T1 and T3 stages of field monitoring. Vegetation increases amplitudes of diurnal changes of soil suction and VWC, but effects of root biomass on diurnal characteristics of vegetated soils are relatively limited. During T2 stage, diurnal changes of soil suction and VWC of both vegetated and bare soils are negligible due to rainfalls.

(3) Soil air-entry value (AEV) is positively relevant with root biomass, irrespective of being measured in the field or the laboratory. In addition, vegetation improves soil water-holding capability at most vegetation-available suction ranges. As compared to indoor-measured SWCC, field-measured SWCC is more similar to that of ideal coarse-grained soils. Moreover, roots improve soil permeability through preferential flows and enhanced water-retention capacity. Because effects of coarse and fine roots on soil structure are different, permeability of G2 sample is greater than that of G1 sample.

This study helps to obtain desired outcomes, i.e., improving feasibility, performance and durability of green infrastructures. It ought to advise decision-makers, geotechnical engineers and managers of green infrastructures about how vegetation and site conditions (especially meteorological diurnal changes) affect hydraulic properties of engineering soils. Future investigations are required to explore relationships of performances of green infrastructures with various factors, for instance, degree of compaction, soil texture as well as layering design, etc.

Acknowledgements. All authors would like to express their gratitude sincerely to the National Key Research and Development Program of China (Grant No. 2019YFC0810702), the Guangdong Basic and Applied Basic Research Foundation (Grant No. 2020B1515120083) and the Shenzhen Key Laboratory Launching Project (Grant No. ZDSYS20200810113601005) for financial supports.

Compliance with Ethical Standards. All authors declare that there is no conflict of interest.

References

1. Chen, R., Ng, C.W.W.: Impact of wetting-drying cycles on hydro-mechanical behavior of an unsaturated compacted clay. Appl. Clay Sci. **86**, 38–46 (2013)
2. Ng, C.W.W., Liu, J., Chen, R.: Numerical investigation on gas emission from three landfill soil covers under dry weather conditions. Vadose Zone J. **14**(8) (2015). https://doi.org/10.2136/vzj2014.12.0180
3. Zhan, L., Qiu, Q., Xu, W., Chen, Y.: Field measurement of gas permeability of compacted loess used as an earthen final cover for a municipal solid waste landfill. J. Zhejiang Univ.-Sci. A **17**(7SI), 541–552 (2016)
4. Ng, C.W.W., Woon, K.X., Leung, A.K., Chu, L.M.: Experimental investigation of induced suction distribution in a grass-covered soil. Ecol. Eng. **52**, 219–223 (2013)
5. Rahardjo, H., Satyanaga, A., Leong, E.C., Santoso, V.A., Ng, Y.S.: Performance of an instrumented slope covered with shrubs and deep-rooted grass. Soils Found. **54**(3), 417–425 (2014)
6. Chen, R., Ge, Y.H., Chen, Z.K., Liu, J., Zhao, Y.R., Li, Z.H.: Analytical solution for one-dimensional contaminant diffusion through unsaturated soils beneath geomembrane. J. Hydrol. **568**, 260–274 (2019)
7. Eymar, E., Oki, L.R., Lieth, J.H.: Continuous measurements of electrical conductivity in growing media using a modified suction probe: Initial calibration and potential usefulness. Plant Soil **230**(1), 67–75 (2001)

8. Zhou, C., Ng, C.W.W.: A new and simple stress-dependent water retention model for unsaturated soil. Comput. Geotech. **62**, 216–222 (2014)
9. Ng, C.W.W., Zhou, C.: Cyclic behaviour of an unsaturated silt at various suctions and temperatures. Geotechnique **64**(9), 709–720 (2014)
10. Zhou, C., Ng, C.W.W.: Simulating the cyclic behaviour of unsaturated soil at various temperatures using a bounding surface model. Geotechnique **66**(4), 344–350 (2016)
11. Marchiol, L., Cesco, S., Pinton, R., Zerbi, G.: Germination and initial root growth of four legumes as affected by landfill biogas atmosphere. Restor. Ecol. **8**(1), 93–98 (2000)
12. Venkatraman, K., Ashwath, N.: Can phytocapping technique reduce methane emission from municipal landfills? Int. J. Environ. Technol. Manage. **10**(1), 4–15 (2009)
13. Garg, A., Coo, J.L., Ng, C.W.W.: Field study on influence of root characteristics on soil suction distribution in slopes vegetated with Cynodon dactylon and Schefflera heptaphylla. Earth Surf. Proc. Land. **40**(12), 1631–1643 (2015)
14. Jalilzadeh, H., Hettiaratchi, J.P.A., Fleming, I., Pokhrel, D.: Effect of soil type and vegetation on the performance of evapotranspirative landfill biocovers: field investigations and water balance modeling. J. Hazard. Toxic Radioact. Waste **24,** (2020). https://doi.org/10.1061/(ASCE)HZ.2153-5515.0000535
15. Ng, C.W.W., Ni, J.J., Leung, A.K., Zhou, C., Wang, Z.J.: Effects of planting density on tree growth and induced soil suction. Geotechnique **66**(9), 711–724 (2016)
16. Chen, R., Liu, J., Li, J.H., Ng, C.W.W.: An integrated high-capacity tensiometer for measuring water retention curves continuously. Soil Sci. Soc. Am. J. **79**(3), 943–947 (2015)
17. Bengough, A.G., McKenzie, B.M., Hallett, P.D., Valentine, T.A.: Root elongation, water stress, and mechanical impedance: a review of limiting stresses and beneficial root tip traits. J. Exp. Bot. **62**(1), 59–68 (2011)
18. Bruand, A., Cousin, I., Nicoullaud, B., Duval, O., Begon, J.C.: Backscattered electron scanning images of soil porosity for analyzing soil compaction around roots. Soil Sci. Soc. Am. J. **60**(3), 895–901 (1996)
19. Maggi, F., Porporato, A.: Coupled moisture and microbial dynamics in unsaturated soils. Water Resour. Res. **43**(7), W07444 (2007). https://doi.org/10.1029/2006WR005367
20. Koebernick, N., Daly, K.R., Keyes, S.D., et al.: High-resolution synchrotron imaging shows that root hairs influence rhizosphere soil structure formation. New Phytol. **216**(1), 124–135 (2017)
21. Lu, J., Zhang, Q., Werner, A.D., Li, Y.L., Jiang, S.Y., Tan, Z.Q., et al.: Root-induced changes of soil hydraulic properties - a review. J. Hydrol. **589**, (2020). https://doi.org/10.1016/j.jhydrol.2020.125203
22. Sriboon, W., Tuntiwaranuruk, U., Sanoamuang, N.: Hourly soil temperature and moisture content variations within a concrete pipe container for planting lime trees in Eastern Thailand. Case Stud. Thermal Eng. **10**, 192–198 (2017)
23. Carminati, A., Schneider, C.L., Moradi, A.B., et al.: How the rhizosphere may favor water availability to roots. Vadose Zone J. **10**(3), 988–998 (2011)
24. Carminati, A., Zarebanadkouki, M., Kroener, E., Ahmed, M.A., Holz, M.: Biophysical rhizosphere processes affecting root water uptake. Ann. Bot. **118**(4SI), 561–571 (2016)
25. Kroener, E., Zarebanadkouki, M., Bittelli, M., Carminati, A.: Simulation of root water uptake under consideration of nonequilibrium dynamics in the rhizosphere. Water Resour. Res. **52**(8), 5755–5770 (2016)
26. Kennedy, D., Swenson, S., Oleson, K.W., et al.: Implementing plant hydraulics in the community land model, Version 5. J. Adv. Model. Earth Syst. **11**(2), 485–513 (2019)
27. Ghestem, M., Sidle, R.C., Stokes, A.: The influence of plant root systems on subsurface flow: implications for slope stability. Bioscience **61**(11), 869–879 (2011)
28. Ni, J.J., Leung, A.K., Ng, C.W.W.: Modelling effects of root growth and decay on soil water retention and permeability. Can. Geotech. J. **56**(7), 1049–1055 (2019)

Study on Disturbance Assessment and Control of Construction Near Existing Cable Tunnel

Wei Yu[1], Rong Pan[2], Gang He[2], Yi Lu[2], and Chengming Cai[2](✉)

[1] State Grid Nanjing Power Supply Company, 1 Aoti Street, Nanjing 210019, China
[2] Nanjing Suyi Industrial Company Limited, 251 Zhongshan Road, Nanjing 210005, China

Abstract. As we all know, the disturbance of adjacent construction will affect the safety of the existing urban power cable tunnel. However, at present, the research on the mechanism of disturbance effect of adjacent construction on existing cable tunnels and the evaluation and control index of tunnel deformation is still blank. In order to solve this problem, numerical analysis is carried out to study the influence of the size of foundation pit and the relative position of foundation pit and tunnel on the displacement and deformation of shallow-buried cast-in-situ reinforced concrete cable tunnel. The results show that the increase of excavation length of foundation pit causes large displacement and bending strain change of cable tunnel. The increase of excavation width of foundation pit has little effect on displacement and strain of tunnel. The change of excavation depth of foundation pit is related to the vertical distance between the bottom of foundation pit and tunnel, and the larger the distance, the smaller the influence of tunnel displacement is with tunnel. The horizontal net distance between road and foundation pit decreases and increases gradually, but the strain changes little. And the sensitivity of the parameters affecting the tunnel is evaluated at last.

Keywords: Power tunnel · Construction disturbance · Influence law · Assessment and control

1 Introduction

In recent years, due to the limitation of urban space, there are more and more construction projects adjacent to or intersecting cable tunnel in urban construction. The close construction conditions are complex and various, and the daily operation and maintenance pressure of cable tunnel is increasing. Many studies show that construction near existing cable tunnel will cause great deformation and displacement of cable tunnel, which will affect the operation safety of power grid. Power cable tunnel belongs to urban lifeline project and shall not be destroyed or interrupted during operation.

In recent years, the research results on the influence of foundation pit excavation and tunnel crossing on adjacent tunnel structures are increasing, but it focuses on the influence of adjacent construction on subway tunnels. It is found that under the same buried depth and working conditions, the displacement and strain values of cable tunnels are obviously larger than those of subway tunnels, while the research on the evaluation

H.-H. Zhu et al. (Eds.): BRWSG 2021, LNCE 230, pp. 288–299, 2022.
https://doi.org/10.1007/978-981-16-9963-4_23

and control indexes of the disturbance influence of adjacent construction on existing cable tunnels is still blank.

Based on the above research, this paper mainly studies the construction disturbance evaluation and control of existing power tunnels by means of numerical simulation. Taking the actual project in Nanjing as the background, the parameter analysis was carried out to study the mechanism and law of the influence of the excavation size (length, width and depth) and the relative position between the tunnel and the foundation pit on the displacement and deformation of the cable tunnel.

2 Numerical Simulation

2.1 Proposed Site

Based on the completed 220 kV cable tunnel project from Mochou–Ninghai Road, Nanjing, this paper establishes a parameter analysis scheme to study the influence of foundation pit excavation in silty clay layer on the displacement and deformation of the shallow cast-in-place reinforced concrete rectangular section cable tunnel located on one side of the foundation pit in the length direction. Meanwhile, we study the influence of different parameters, length, width and depth, of the tunnel on the displacement and deformation of the tunnel, so to show the mechanism and law of cable tunnel response.

The project is located in Gulou District in the center of Nanjing, starting from 220 kV Mochou Substation and ending at 220 kV Ninghai Road Substation. The total length of cable structures in the whole line is about 4.0 km, including pipe jacking section, shield section and excavation section. Among them, the cast-in-place tunnel is about 0.230 km long, with a rectangular cross section and a cross section size of 4.0 m × 3.7 m It goes north along the east side of Shanghai Road to the east side of Ninghai Road and then passes through Shanghai Road to Ninghai Road Substation westward.

2.2 Materials Properties

Considering that the cable tunnel studied in this paper is mainly located in silty clay layer, in order to simplify the calculation, the whole soil will be regarded as silty clay homogeneous foundation in the numerical simulation, and the finite element analysis will be carried out with the cast-in-place tunnel in the excavation section of the 220 kV cable tunnel project from Mochou–Ninghai Road in Nanjing as the background.

Many studies show that soil has basic properties such as nonlinearity, inelastic, stress path dependence and strain path dependence [1–4] and small strain characteristics [5–8], and whether the selected constitutive model can better simulate the properties of materials will directly affect the simulation results. Therefore, the constitutive model adopted in this paper is a clay sub-plastic model which can consider the small strain characteristics and stress path dependence of soil. And the tunnel retaining structure adopts the linear elastic constitutive model.

Some parameters of soil sub-plastic model used in numerical simulation are obtained from geotechnical engineering investigation report, and the other part is obtained through literature data related to the study of sub-plastic parameters in soft soil area [9, 10].

Among them, λ * reflects the softness and hardness of soil, and the larger λ * is, the softer the soil is. λ * and *N* can be obtained by drawing the compression line of ln (1 + *e*)-*lnp*, which are the slope and intercept of the compression line respectively. The value of κ * is generally between 0.01 and λ */4. The value of r ranges from 0.07 to 1.3, and the smaller r is, the greater the shear modulus of soil. κ * and *r* and intergranular strain parameters (m_R, m_T, R, β_r and χ) were obtained by parameter study. Specific parameter values are shown in Table 1.

Table 1. Parameters of soil sub-plastic model.

Parameter	φ'c	λ*	κ*	N	r	mR	mT	R	βr	χ
Value	10	0.206	0.024	3.42	0.95	4.5	4.5	10^{-4}	0.4	5

In this paper, several different depths (*He* = *3. 5, 7, 10. 5, 14* m), lengths (*L* = *14, 21, 28, 35, 42* m) and widths (*B* = *14, 21, 28, 35, 42* m) of foundation pit are selected to provide reference for engineering practice.

2.3 Construction Conditions

In the numerical simulation, the cable tunnel and retaining structure of foundation pit are simulated by "Wished in place" method, which is assumed that the tunnel and retaining structure already exist before foundation pit excavation, and the influence of dewatering before foundation pit excavation and foundation pit support on soil stress field around foundation pit is not considered. The simulation process is as follows:

Firstly, the initial stress field of soil is established, and the acceleration of gravity is g. Then, the foundation pit excavation process is simulated by the unit life-and-death function in ABAQUS, which is divided into two excavations, and the simulated excavation rate is the same as the actual project. The relevant parameters involved in parameter analysis are shown in Table 2, and the dimensions of model, tunnel and retaining structure are shown in Fig. 1.

3 Results of Analysis

3.1 Influence of the Length of the Foundation Pit on the Cable Tunnel

In this paper, the length direction of foundation pit excavation is parallel to the cable tunnel axis direction, and the width direction is perpendicular to the cable tunnel axis direction.

Figure 2 shows the influence of different excavation lengths on the vertical and horizontal displacements along the longitudinal direction of cable tunnel. It can be seen from Fig. 2(a) that the cable tunnel bulges along its longitudinal direction. When the excavation length of the foundation pit is less than 28 m, with the increase of the normalized distance between the cable tunnel and the center of the foundation pit (X/(L/2)), the

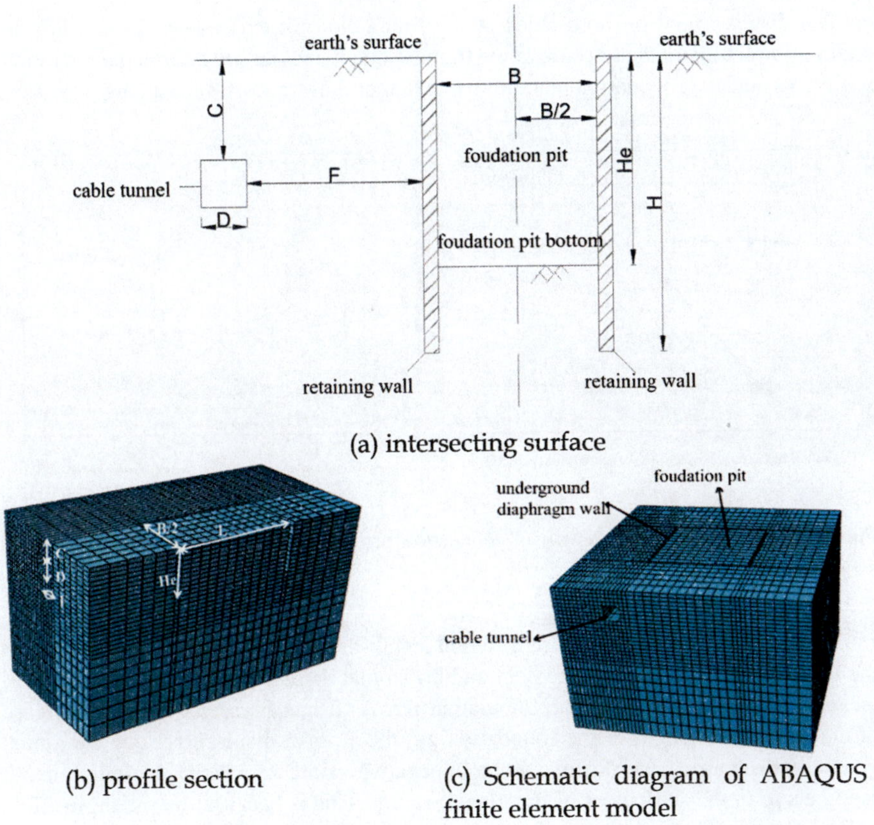

Fig. 1. Definition of geometric parameters involved in finite element analysis.

Table 2. Parametric analysis.

Variable	Length, L (m)	Width, B (m)	Depth, He (m)
L (m)	14, 21, 28, 35, 42	14	7
B (m)	14	14, 21, 28, 35, 42	7
H_e (m)	14	14	3. 5, 7, 10. 5, 14

tunnel uplift value is almost unchanged. However, when the excavation length exceeds 28 m, with the increase of the normalized distance between the cable tunnel and the center of the foundation pit (X/(L/2)), the vertical displacement of the tunnel gradually increases linearly, and the tunnel itself has a large non-uniform vertical displacement. With the increase of excavation length, the uplift value of the tunnel gradually decreases. Excavation will lead to settlement of the soil behind the wall, because the cable tunnel is located on the right side of the excavated foundation pit, resulting in asymmetric vertical displacement on both sides of the foundation pit excavation. In a word, the retaining structure on the left side of the foundation pit and the soil behind the wall show settlement, and the foundation pit bottom and tunnel show upward vertical displacement.

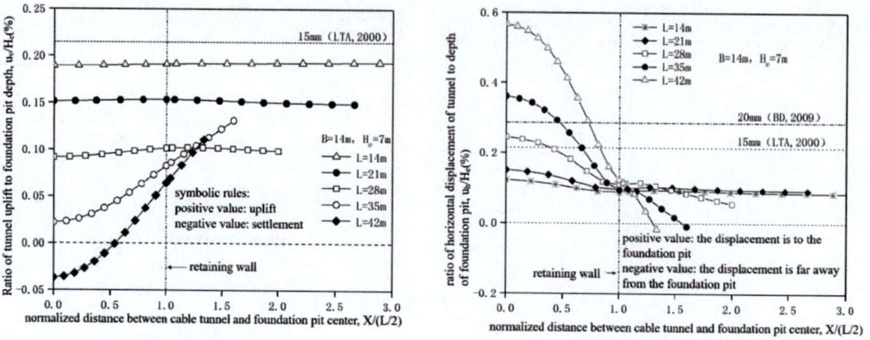

(a) (b)

Fig. 2. The influence of the length of foundation pit on the vertical (a) and horizontal (b) displacement of the tunnel.

Figure 3 shows the influence of different length-depth ratios of the foundation pit on the longitudinal vertical displacement and horizontal displacement of the cable tunnel located at the center side of the foundation pit. As it can be seen, with the increase of the excavation length of the foundation pit, the vertical displacement of the tunnel gradually decreases from positive value to negative value. And when the length-depth ratio gets to 6, the settlement of the tunnel occurs. This is because the pit angle effect of the foundation pit is enhanced with the increase of the length of the foundation pit, and the displacement of the soil on the central axis of the foundation pit is constantly increasing, and the tunnel is constantly moving towards the pit, resulting in settlement.

It can be seen from Fig. 2(b) that the unloading of the foundation pit when excavated leads to the horizontal displacement of the cable tunnel towards the foundation pit. When the length of the foundation pit is less than 21 m, the horizontal displacement decreases slowly with the increase of the normalized distance between the cable tunnel and the center of the foundation pit (X/(L/2)), and that displacement tends to be stable behind the retaining structure. When the length exceeds 21 m, the decrease gets sharply and the tunnel structure itself shows significant uneven horizontal displacement. With the gradual increase of the length of the foundation pit, the horizontal displacement of cable tunnel increases rapidly, and the maximum horizontal displacement of tunnel is located on the central axis of foundation pit. The horizontal displacement of the cable tunnel shows little difference at the retaining structure because the displacement is limited by the structure on its side. It can be seen from Fig. 3(b) that the maximum horizontal displacement of the tunnel increases rapidly with the increase of the length-depth ratio of foundation pit as a result of the increase of the pit angle effect during this process.

The horizontal lines marked in Fig. 3 respectively correspond to the tunnel longitudinal displacement limits specified by the Land Transport Authority of Singapore (LAT, 2000) [11] and the Construction Department of the Hong Kong Special Administrative

Region Government (BD, 2009)[12]. It can be seen that when the length-depth ratio of foundation pit is less than or equal to 6, the vertical displacement value of cable tunnel is within the tunnel uplift limit specified by LAT and BD; When the length-depth ratio is 4, the horizontal displacement of the tunnel exceeds the limit, 15mm, specified by LAT while it is still within the limit, 20 mm, specified by BD. When the ratio is greater than 4, the horizontal displacement of tunnel exceeds the limit specified by LAT and BD.

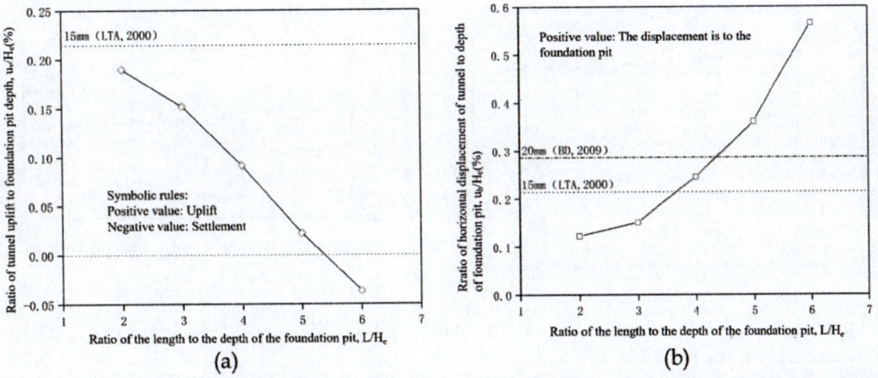

Fig. 3. The influence of the length-depth ratio of foundation pit on the vertical (a) and horizontal (b) displacement of the tunnel.

3.2 Influence of the Width of the Foundation Pit on the Cable Tunnel

Figure 4 shows the influence of different widths of the foundation pit on the longitudinal vertical and horizontal displacement of cable tunnel. Figure 5 shows that influenced by different width-depth ratios of the foundation pit. It can be seen from Fig. 4(a) that the cable tunnel uplifts along its longitudinal direction, and the value is almost the same with the increase of the normalized distance between the cable tunnel and the center of the foundation pit (X/(L/2)). It can be seen from Fig. 5(a) that with the gradual increase of the width of the foundation pit, the vertical displacement of cable tunnel first increases, then decreases, and finally increases slightly. When the width-depth ratio of the foundation pit excavation is 2, the vertical displacement of the tunnel is smaller due to the larger rigidity of the retaining structure system. However, the change of width makes little difference on the vertical displacement of the cable tunnel, which is because the change of width has little influence on the displacement of soil behind retaining structure in the length direction of the foundation pit.

It can be seen from Fig. 4(b) that the unloading of foundation pit when excavated leads to the horizontal displacement of the cable tunnel to the foundation pit. With the increase of the width, the horizontal displacement of the tunnel decreases gradually. At the position of retaining structure, the horizontal displacement of cable tunnel is basically the same, which is because the displacement is limited by the retaining structure. It can be seen from Fig. 5(b) that the maximum horizontal displacement of the tunnel decreases with the increase of the width-depth ratio of foundation pit. This is because the vertical displacement difference between both sides of the foundation pit increases with the increase of the width, and the soil on the side of the tunnel tends to move away from the foundation pit, resulting in the decrease of the horizontal displacement of the tunnel towards the foundation pit.

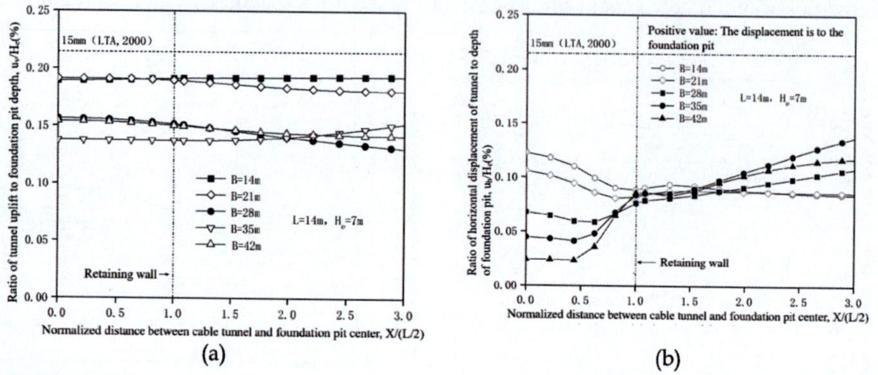

Fig. 4. The influence of the width of foundation pit on the vertical (a) and horizontal (b) displacement of the tunnel.

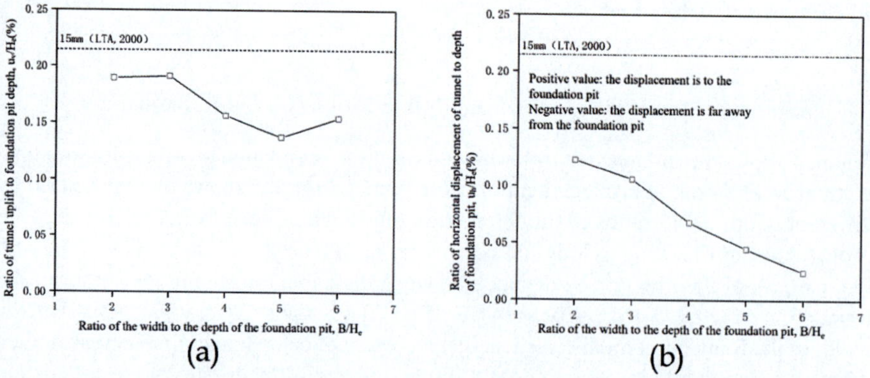

Fig. 5. The influence of the width-depth ratio of foundation pit on the vertical (a) and horizontal (b) displacement of the tunnel.

It can be seen from Fig. 5 and Fig. 6 that when the width-depth ratio of foundation pit is between 2 and 6, the vertical and horizontal displacement of the tunnel and the maximum displacement synthesized by the two vectors are within the limit specified by LAT and BD.

3.3 Influence of the Depth of Foundation Pit on the Cable Tunnel

Figure 7 shows the influence of different depths of the foundation pit on the longitudinal vertical and horizontal displacement of the cable tunnel. The depth of the foundation pit

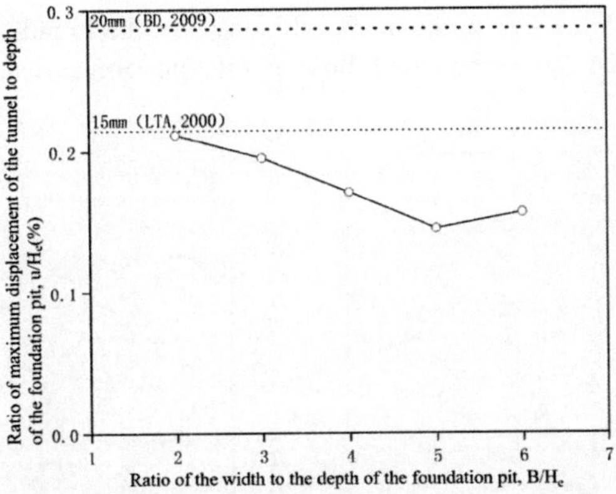

Fig. 6. Influence of width-depth ratio of the foundation pit on maximum displacement of the tunnel.

is 3.5, 7, 10.5 and 14 m in different conditions, and the diameter and buried depth of the tunnel are both 3.5 m. As can be seen, the cable tunnel bulges along its longitudinal direction, and with the increase of the normalized distance between the cable tunnel and the center of the foundation pit (X/(L/2)), the value of the uplift increases slowly. Unloading while excavating leads to horizontal displacement of cable tunnel towards the foundation pit, and the smallest horizontal displacement occurs at the retaining structure.

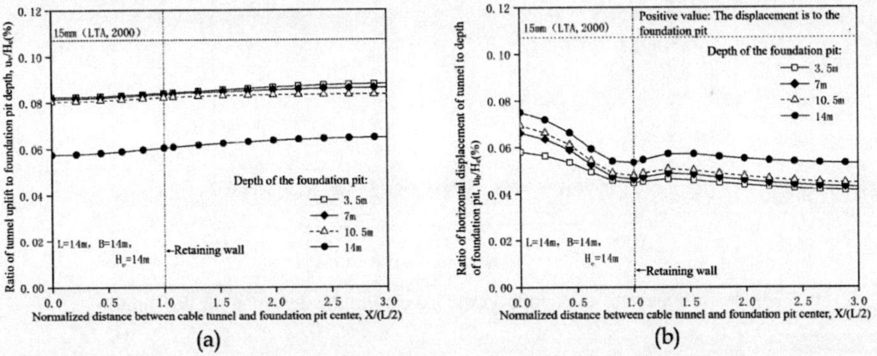

Fig. 7. Influence of depth of the foundation pit on the vertical (a) and horizontal (b) displacement of the tunnel.

Figure 8 shows the relationship between the maximum vertical and horizontal displacement on the central axis of the tunnel lining and the depth of the foundation pit. When the foundation pit is excavated deeper and deeper, the part releasing stress moves down continuously, and the distance from which to the cable tunnel is getting farther and farther, and the soil outside the pit continuously moves into the pit, resulting the influence of the rebound of soil on the "lifting" of the tunnel weakened. In addition, the deformation area of soil outside the pit increases with the increase of the depth.

Therefore, as it is in the figure, the deeper the depth of the foundation pit, the smaller the uplift value of the cable tunnel and the larger the horizontal displacement.

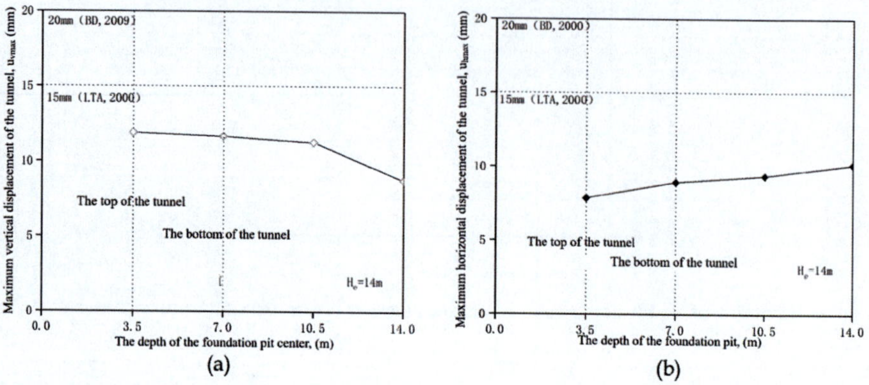

Fig. 8. Influence of depth of the foundation pit on the maximum vertical (a) and horizontal (b) displacement of the tunnel.

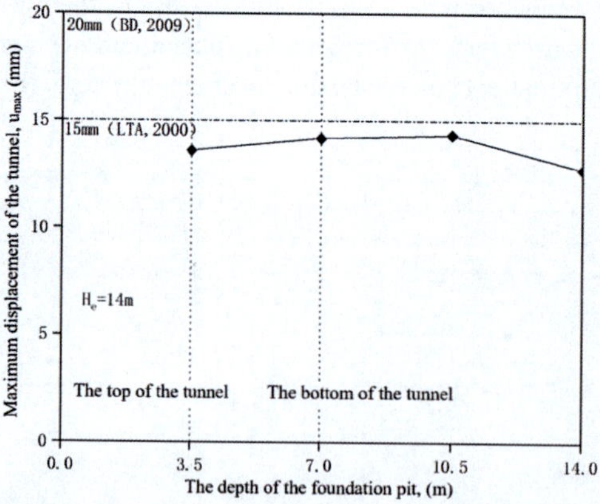

Fig. 9. Influence of the depth on the maximum displacement of the tunnel.

It can be seen from Fig. 8 that with the increase of the depth of the foundation pit, the vertical and horizontal displacement of the cable tunnel do not exceed the limits specified by LAT and BD. Figure 9 shows the curve of the relationship between the maximum displacement of cable tunnel and the depth of foundation pit, and the displacement does not exceed the limit specified by LAT and BD neither. However, it is worth mentioning that when the vertical distance between the bottom of the tunnel and the bottom of the foundation pit gets closer, the displacement of the tunnel gets larger. Therefore, it is worth further research on the case that the bottom of the tunnel and the bottom of the foundation pit are located on the same horizontal plane.

4 Evaluation of the Influence of the Excavation of the Foundation Pit

In this part, we will evaluate the influence of the certain parameters on key indicators quantitatively. If a certain parameter changes slightly and leads to a great change in key indicators, this parameter is a sensitive factor, which needs attention or control. Sensitivity analysis is often used in engineering to evaluate sensitive factors that have an important impact on the safety control indicators of engineering projects, and to analyze and calculate the degree of influence and sensitivity on structural safety.

The most important characteristic parameters of the foundation pit are the size and the depth. As it is said above, it can be seen that the deeper the foundation pit, the larger the area, the greater the influence of foundation pit excavation on the tunnel. In order to reflect the nature of each parameter more clearly and predict the actual project, this section will evaluate the sensitivity of each parameter based on the results of the numerical calculation above. Assuming that the parameters are independent of each other, we change each parameter within the possible range while keeping other parameters unchanged to analyze the influence degree and trend of parameter changes on the displacement and strain of tunnel adjacent, and the sensitivity of parameters is obtained. Detailed analysis results of three parameters, length, width and depth of the foundation pit, are shown in Tables 3, 4 and 5 respectively.

Table 3. Influence of the length of foundation pit on the strain of tunnel.

Index	Percentage increase in length (%)				
	0	50	100	150	200
Vertical disp. (mm)	13.25	10.61	6.40	1.55	−2.54
Growth rate of vertical disp. (%)	/	−19.9	−51.7	−88.3	−119.2
Horizontal disp. (mm)	8.58	10.52	17.20	25.23	39.57
Growth rate of horizontal disp. (%)	/	22.6	100.5	194.1	361.2

Table 4. Influence of the width of foundation pit on the strain of tunnel.

Index	Percentage increase in width (%)				
	0	50	100	150	200
Vertical disp. (mm)	13.25	13.38	10.94	9.63	10.78
Growth rate of vertical disp. (%)	/	0.98	−17.4	−27.3	−18.6
Horizontal disp.(mm)	8.58	7.41	4.74	3.14	1.71
Growth rate of horizontal disp. (%)	/	−13.6	−44.8	−63.4	−80.1

Table 5. Influence of the depth of foundation pit on the strain of tunnel.

Index	Percentage increase in depth (%)				
	−50	0	50	100	/
Vertical disp. (mm)	13.52	13.25	12.81	9.95	/
Growth rate of vertical disp. (%)	2.0	/	−3.3	−24.9	/
Horizontal disp. (mm)	7.53	8.58	8.99	9.72	/
Growth rate of horizontal disp. (%)	−12.2	/	4.8	13.3	/

It can be seen from Tables 3, 4 and 5 that the variation range of the length of foundation pit is 0–200%. When the length is increased by less than 50%, the variation of the displacement of cable tunnel is obviously less than that when the length is increased by more than 50%. Therefore, it should be strictly controlled that the length of foundation pit increases less than 50%. However, when the width and the depth of foundation pit vary, the vertical and horizontal displacement of cable tunnel change little. Therefore, in practical engineering, the length of foundation pit can be set as the main influencing parameter, while the width and depth of foundation pit are the secondary influencing factors, and the change of the main influencing factor need to be paid attention to and controlled during construction.

5 Conclusions

In this paper, the numerical simulation method is applied to study the influence of deformation law, deformation mechanism and evaluation method of tunnel deformation caused by the excavation of foundation pit, and the following conclusions are drawn.

(1) The displacement of the cable tunnel under the influence of the adjacent foundation pit is gradually increased with the increase of the length foundation pit. When the length-depth ratio is greater than 4, the displacement of the tunnel exceeds the limits specified in the code. This is because the pit angle effect of the retaining structure increases with the increase of the length, which leads to the weakening of its limiting effect on the displacement of the soil outside the retaining structure of the foundation pit.

(2) The displacement of the cable tunnel has little response to the change of the width of the foundation pit. With the increase of the width, the tunnel displacement firstly decreases, and then it reaches the minimum value when the width-depth ratio is 5. And after that, the displacement increases. It is because the change of width, whose direction is perpendicular to the tunnel, has little influence on the displacement of the soil behind the retaining structure.

(3) The displacement of the cable tunnel is related to the vertical distance between the tunnel and the foundation pit. With the foundation pit getting deeper, the stress relief position gradually moves down, and the vertical distance between the cable tunnel and the foundation pit constantly changes, which leads to different tunnel

displacements. But the maximum displacement of the cable tunnel changes little with the increase of depth and does not exceed the limit specified in the code.

(4) The length, width and depth of foundation pit are studied as three independent parameters which influence the cable tunnel, and the sensitivity of each parameter affecting the safe operation of existing power tunnel is evaluated. The results show that the length of foundation pit is the main influence parameter.

References

1. Kuwano, R., Jardine, R.J.: A triaxial investigation of kinematic yielding in sand. Géotechnique **57**(7), 563–579 (2007)
2. Gasparre, A.: Advanced laboratory characterisation of London Clay. Ph.D .Thesis. University of London (2005)
3. Richardson. D.: Investiaations of threshold effects in soil deformation. Ph.D. thesis, The City University, London (1988)
4. Atkinson, J.H., Richardson, D., Stallebrass, S.E.: Effect of recent stress history on the stiffness of overconsolidated soil. Géotechnique **40**(4), 531–540 (1990)
5. Burland, J.B.: Small is beautiful-the stiffness of soils at small strains. Can. Geotech. J. **26**, 499–516 (1989)
6. Mair, R.J.: Developments in geotechnical engineering research: application to tunnels and deep excavations. In: Proceedings of the Institution of Civil Engineers. London (1999)
7. Atkinson, J.H., Sallfors, G.: Experimental determination of stress-strain-time characteristics in laboratory and in-situ tests. In: Proceedings of the 10th European Conference on Soil Mechanics and Foundation Engineering. Florence (1991)
8. Romano, M.A.: A continuum theory for granular media with a critical state. Arch. Mech. **26**(20), 1011–1028 (1974)
9. Masin, D.: A hypoplastic constitutive model for clays with metastable structure. Can. Geotech. J. **44**(3), 363–375 (2007)
10. Niemunis, A., Herle, I.: Hypoplastic model for cohesionless soils with elastic strain range. Mech. Cohes.-Frict. Mater. **47**(2), 279–299 (1997)
11. LTA: Code of Practice for Railway Protection. Development & Building Control Department, Land Transport Authority (LTA), Singapore (2000)
12. BD: Practice note for authorized persons APP-24. Technical notes for guidance in assessing the effects of civil engineering construction/building development on railway structures and operations. Building department of the government of HKSAR (BD) (2009)

Urban Transportation System Problems in Context of the Indian Conditions

Akhilesh Kumar[1](✉), Arunava Poddar[2], and Akhilesh Nautiyal[3]

[1] Department of Civil Engineering, Chitkara University Institute of Engineering and Technology, Chitkara University, Patiala, Punjab, India
akhileshsharma54@gmail.com

[2] Department of Civil Engineering, Shoolini University, Solan, Himachal Pradesh, India

[3] Department of Civil Engineering, National Institute of Technology, Hamirpur, Hamirpur, Himachal Pradesh, India

Abstract. In the urban transportation system, towns and cities play an essential role in promoting economic growth and prosperity. In India, where traffic conditions with no lane disciplines generate more complicated scenarios. Extensive hours are getting wasted on the roads due to traffic congestion, which negatively affects the overall economy. In the future, urban areas will play a critical role in sustaining high economic growth rates as India becomes more urbanized. The economic continuity in the growth can be sustained if these towns and cities function efficiently, which ultimately depends upon the effectiveness of its urban transport systems. The existing traffic framework is not meeting the traffic demand due to an exponential increase in population and vehicles. This research aims to access the traffic congestion impacts of the urban transportation network in the Indian scenario of urban transport system issues, challenges, and some mitigation strategies. The article firstly reviews the trends of vehicular growth and transport infrastructure in Indian cities, followed by a discussion on the traffic congestion impacts and nature of urban transport problems such as congestion, road accidents, and health effects. Based on the study, mitigation strategies are proposed to improve the urban transportation system in India.

Keywords: Urban transportation · Traffic · Congestion · Vehicles · Mitigation strategies

1 Introduction

India is urbanizing, and the urban population is growing at an average rate of 3% per year. One specific feature of India's urbanization is the increase in metropolitan cities, which is defined as growth in the number and size of cities with a million-plus population [1]. Rapid urbanization, increased motorization, and a diminishing modal share of non-motorized transport are the main factors impacting the mobility of urban populations in Indian cities. In India, urban areas contribute more than 60% of the national income. In the coming years, more Indian cities become urbanized, urban areas will play a critical role in sustaining high economic growth rates. In India, the importance of urban transportation arises from its role in reducing poverty by enhancing access to labour markets and

H.-H. Zhu et al. (Eds.): BRWSG 2021, LNCE 230, pp. 300–314, 2022.
https://doi.org/10.1007/978-981-16-9963-4_24

therefore raising earnings in deprived areas [2]. Cities and towns' efficiency mainly depends upon the effectiveness of their transport systems. Due to rapid urbanization, most cities in India are experiencing multi-faceted problems. Urban areas congestion is one of the major problems afflicting urban masses in India and has several effects on economies. Urban area congestion is broadly defined as excess demand for travel over its supply. In India, central and state governments are continuously revising their policies for urban mobility due to the growing demand for travel with limited sources of services. Urban congestion restricts the free movement of traffic. With urban transport infrastructure serving as the skeleton of the urban form, mobility flows have become a significant factor in India's fast urbanization process [3].

Various policies and efforts are being implemented to promote urban mobility, with the primary goal of improving and strengthening urban infrastructure. In reality, current levels of urban mobility are already causing a crisis defined by high levels of congestion, pollution, traffic deaths, and unfairness, ultimately leading to an unwelcome accessibility catastrophe [4]. With nearly a quarter of India's metropolitan population living in poverty, poor people's mobility issues are of particular importance [5]. In Asian countries, for restricting private car ownership, various strategies such as congestion pricing are being adopted. However, some of the conventional congestion causes are still rooted in growing cities due to policy overlaps and distorted policy implementation. Despite these strategies, inefficient and insufficient public transportation, low-price parking policies, mixed-use of dedicated roads, lack of connectivity between different modes, lack of transport planning, poor driving behavior, and the absence of intelligent transport systems restrict the implantation ground level. Therefore, the creation of new infrastructure will not meet the urban transportation requirements. This article addresses the factors contributing to congestion in Indian cities. It also concludes the status of congestion in Indian cities.

According to the study, "for every extra one million people in a developing city, an extra 3.5 to 4 million public transport trips per day are generated" [6]. In Kolkata, the average speed during peak hours in the Central Business District area goes down as low as around 10 km/h [7]. The problem of congestion and delays ultimately wastes time, energy, and the scale of opportunities for improvement. A high level of pollution is the second undesirable feature of crowded streets. Statistics indicate that traffic accidents are a primary cause of accidental deaths in Indian cities. Public transport systems, especially in Indian cities, have not been able to keep pace with substantial increases in demand over the past few years. As an outcome, people have twisted to personalized modes such as bikes, scooters, cars, and intermediate public transport modes such as taxis, auto-rickshaws, and tempos.

There is a need to design and adopt various strategies and policies that reduce travel time by personalized modes, public transport, and particularly the bus transport system. The guidelines should be considered to improve the quality and quantity of public transport services and impose restrictions on private vehicles. People should also be encouraged to use non-motorized transport. It must not forget that cities are the major contributors to economic growth, and movement in between cities is crucial for improved quality of life. As of 31st March 2017, the length of urban roads was 5,26,483 km, an 8.93% share of the total roads in India. In 1960–1961 the urban road length was 46361 km. It has increased to 526483 km [8]. The data in Fig. 1 reveals that the total length of urban roads increased from 509730 km on 31st March 2016 to 526483 km on 31st March 2017, increased about 3.29% [8].

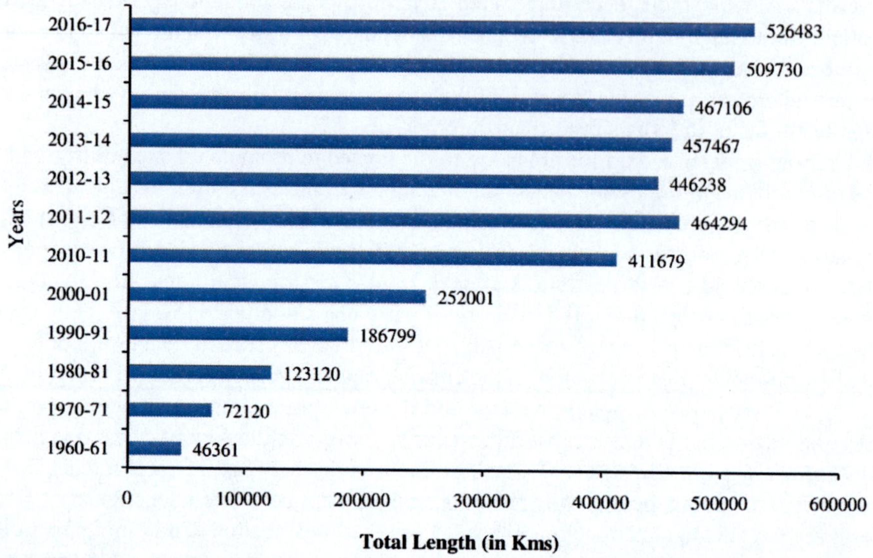

Fig. 1. Total length of urban roads in India 1960–61 to 2016–17. Source: Transport Research Wing of the Ministry of Road Transport & Highways 'Basic Road Statistics of India 2016–17

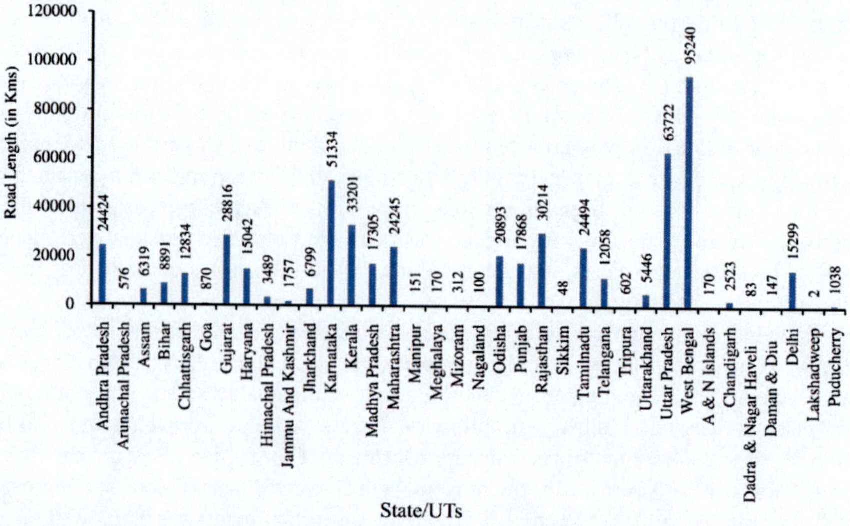

Fig. 2. State/UT-wise length of urban roads (in Kms) [18]. Source: Transport Research Wing of the Ministry of Road Transport & Highways 'Basic Road Statistics of India 2016–17

The State/UT-wise break-up of urban roads Fig. 2 reveals that West Bengal accounted for the largest share (18.09%) in total length of Urban Roads as of 31st March 2017, followed by U.P. (12.10%), Karnataka (9.75%), Kerala (6.31%) and Rajasthan (5.74%) [8]. These five States accounted for about 51.99% of the total length of urban roads. The details of different categories of urban road share have been presented in The facts reveal that within the type of Urban Roads as of 31.03.2017. Municipal roads accounted for the highest share of 93.70% Table 1. Table 2, represents the urban road length by type of surface in India, whereas Table 3, presents the total road length and percentage share of urban roads from 1951 to 2017.

Table 1. Total and surfaced urban road length in India as of 31.03.2017 [8]

	Category I wise road length (% share of total road length)	Category II wise surfaced road length (% age of type I)
Urban roads	526483 (9.27)	406867 (77.3)
Municipal roads	493328	375315 (76.1)
MES roads	18556	18002 (97.0)
Major port roads	781	779 (99.8)
Minor port roads	361	340 (94.2)
Railway roads	13457	12431 (92.4)

Table 2. Urban road length by type of surface in India as of 31.03.2017 [8]

	Total (III+IV)	WMM/WBM (I)	CC/BT (II)	Total (III) (I+II)	Un-surfaced (IV)	Category II wise surfaced road length (% age of type I)
Urban roads	526483	41093	365774	406867	119616	406867 (77.3)
Municipal roads	493328	39227	336088	375315	118013	375315 (76.1)
MES roads	18556	278	17724	18002	554	18002 (97.0)
Major port roads	781	58	720	779	2	779 (99.8)
Minor port roads	361	24	316	340	21	340 (94.2)
Railway roads	13457	1506	10925	12431	1026	12431 (92.4)

Table 3. Total road length and percentage share of urban road during (1951–2017) [8]

Year (as on 31st March)	Urban roads	Year (as on 31st March)	Urban roads	Year (as on 31st March)	Urban roads
1951	0	2003	2,97,259	2010	4,02,448
%	0	%	8.42	%	8.78
1961	46,361	2004	3,01,310	2011	4,11,679
%	8.84	%	8.32	%	8.8
1971	72,120	2005	2,86,707	2012	4,64,294
%	7.88	%	7.53	%	9.54
1981	1,23,120	2006	2,91,991	2013	4,46,238
%	8.29	%	7.52	%	8.53
1991	1,86,799	2007	3,00,580	2014	4,57,467
%	8.03	%	7.48	%	8.47
2001	2,52,001	2008	3,04,327	2015	4,67,1 06
%	7.47	%	7.41	%	8.54
2002	2,50,295	2009	3,73,802	2016	5,09,730
%	7.3	%	8.36	%	9.1
				2017	5,26,483
				%	8.93

Traffic congestion is categorized into two types, (i) recurring type: It takes place due to behavioral and capacity issues (ii) non-recurring type: It takes place due to construction, accidents, or emergencies. These two types are having different causes but have inseparable effects. Traffic jams caused to traffic congestion quickens additional interruption, causing the randomness of travelers to travelers engendering to tension and hazardous transportation circumstances. Scarcity of the workforce, defective traffic signaling structures, and limited roadway width and overtaking may increase traffic jams. Traffic congestion not only affects human lives but also raises pollution. The Mobility and accessibility parameters are influenced by traffic congestion. Fuel costs and travel time is adversely affected by an increase in traffic congestion. Level of service, lane-mile duration index, and roadway congestion index measurements have been used to identify the traffic congestion parameters. A speed performance index is one of the essential parameters for evaluating traffic congestion. Physical and mental stress on drivers due to traffic congestion increase their anger, bringing unfortunate incidents like road crashes. Traffic congestion initiates significant noise, air pollution, and consequently exaggerates the whole atmosphere surrounding it. Traffic congestion in urban transportation can be related to transit costs, fuel expenditure costs, health, and environmental concerns.

2 Literature

Traffic congestion is a significant problem which, the Indian transportation system is suffering. The growth of private vehicles in major cities has increased at a faster rate. The non-uniform roadway geometrics, poor roadway conditions, adverse land use, poor lane discipline, wrong bus stop locations, and uncontrolled on-street parking indicates the cause of congestions. Although the roads are constructed at a fast rate, there have been no serious attempts to quantify the growth of congestion in most cities. The non-availability of funds has seriously constrained the development of the supply side.

A study presented the Intelligent Transportation Systems (ITS) techniques to cater to traffic conditions and congestion [9]. An innovative approach was introduced, which was used to detect the congestion on the urban arterials in India. The Wi-Fi signal emitting device was developed to identify the congestion. It proved a high accuracy of classifying on a congested road [10]. Urban arterial congestion based on metrics measurements such as travel time, speed, and volume was concluded in 2011. Rao et al. 2012 identified the parameters of traffic congestion, travel time reliability measures [11]. Detection techniques for traffic congestion were suggested a Global Positioning System (GPS)-based method is the best one to analyze the result of traffic [12]. A performance index technique was introduced to determine the existing road network conditions of congestion in urban road networks. The study considered the speed performance index as the urban road traffic system [13]. A study was carried out to propose GPS, ITS, and Bluetooth Low Energy (BLE) technologies that have proven to be suitable for use in vehicle positioning in transportation system services [14]. A stereo-vision and GPS for traffic conflict investigation were presented for detecting conflicts between vehicle-pedestrian [15]. A Multinomial logistic regression model was developed to study the path of lateral movement of vehicles on undivided urban roads. The study proved to be influential in developing traffic simulation models for carrying out various traffic control and management measures [16]. The impacts of traffic regimes on accident likelihood and severity are investigated in a study [17]. The study concluded that high occupancy values increase accident probabilities. A methodology for heterogeneous traffic was proposed for traffic micro-simulation calibration. The research suggests that, various sets of optimal driving behavior parameters for different vehicle classes [18]. In literature, multiple studies proposed that GIS can be used as a management system for accident analysis by applying spatial-statistical methods [19–27]. An Emergent intelligence technique was offered, which collects, analyses, shares, and monitors traffic resources [28].

3 Methodology

The authors gathered information on the growth of urban transportation networks in India from the early 1990s through a thorough literature study. National and international research articles, book chapters, working papers, vision documents, planning commission reports, publicly accessible plans such as comprehensive mobility plans and comprehensive development plans, national and state policy briefs, relevant laws, regulations, and notifications, and publicly accessible consultancy reports published by non-governmental organizations were used in the review. During the literature research, a wide range of news, blog postings, and magazine websites, as well as local, national, and worldwide media, were referred to. The important results on common patterns, problems, effects, and difficulties were summarised following the literature study. Following that, semi-structured meetings with specialists in the field of urban transportation were held. To achieve a diverse range of viewpoints on urban transportation in India, efforts were undertaken to obtain enough representation from academic publications across India. The current research focused on three aspects of urban transportation in Indian cities: the evolution of urban transportation in India over the last few decades; major urban transportation deficiencies in Indian cities and their impact on people; and effective strategies for achieving sustainable transportation in the Indian transportation system. Finally, the essay highlighted the issues and offered strategies in the urban transportation system, combining many inputs to form a wide perspective on the current environment, as well as the primary obstacles and strategic intervention areas.

4 Results and Discussions

4.1 Urban Transportation Systems Deficiencies

In urban transport, the various problems were concluded as follows; (i) Congestion: With the increase in population, the intensity of distance and average travel distances. This trend in trip length and frequency is only expected to increase with the increase in income level, a service-oriented economy, and the migrations, the number of trips tends to increase. Subsequently, more people travel together over long distances regularly for employment and education purposes, leading to congestion on the roads. (ii) Parking problems: The shortage of parking spaces in Indian cities increases the period consumed searching for parking spots, ultimately inducing traffic congestion. Top places in India suffer the on-street parking problems. For example, the studies depict that in Delhi, about 14% of total road length is used for on-street parking whereas, in Surat, it is 60%. In Indian cities, commercial development of vacant plots has taken place without following organized planning procedures which induce heavy traffic leading to little congestion and parking problems. There is a need for a systematic planning process implemented through strict enforcement. (iii) Air pollution: Air pollution severity in Indian cities is judged based on CPCB's (Central Pollution Control Board) & SPCB's (State Pollution Control Board) air quality classification. It has been concluded that Air pollution is the fifth chief cause of death in India. (iv) Weakening road safety: In major cities, non-motorized modes like walking and cycling share the same paths as two-wheelers and cars. Lack of adequate pedestrian facilities leads to an increase in pedestrian fatalities. The streets in India are not meeting the requirements designed according to the regulations.

In an urban transportation system, congestion can be measured in many ways. Congestion measurement methodologies adopted should be simple to understand and definite. It should have the ability to predict future changes and describe the existing traffic conditions. With a minimum of data collection, it should have the ability to apply various statistical techniques. The approaches should have pertinence to multiple modes, periods, and facilities.

A few metrics of identifying the congestion measurement and assessing the congestion level are as follows: (i) Travel time and delays: Under free-flow travel conditions, congestion commonly occurs, which delays the travel time and delays. The measurement of time-based congestion offers a diverse perception of congestion. It may provide direction on identifying significant issues, enabling the policymakers to address problems more effectively (ii) Volume: It is one of the parameters through which congestion measurement can be assessed. Generally, congestion narrates an excess of vehicles on a portion of the road at a particular time, resulting in slower vehicles than average speed (iii) Speed: Quality of traffic at a particular section and a particular time is directly affected by the congestion. It involves the severity of road traffic accidents, especially in the urban environment. Congestion is a function of reducing rates, which is the primary reason for the loss of time and leads to increased fuel consumption, vehicle operating costs, and emissions of air pollutants. The usage of the speed range for the study area depicts the lack of consent among urban areas as to the suitable threshold, which reflects local conditions. These days, studies incorporate the efficient vehicle monitoring system, which can be achieved by integrating derived traffic data by GPS into a GIS environment. (iv) Capacity/Demand: The capacity of a particular road network depends upon how freely of quickly vehicles are moving. On a transportation facility, the volume of vehicles when exceeds the capacity/demand of a particular facility ultimately results in a state of congestion (v) Cost: The congestion can reduce accessibility and subsequently increase costs that are directly associated with travel. Travel time and reliability may generally affect the level of access (vi) Level of Service: It is the important parameter for measuring traffic congestion. The level of service can be compared to volume to capacity ratios to measure conditions of congestion. It is considered the representative variable in traffic flow analysis because of ease in collecting data in the field (vii) Delay: It may be defined as the extra time experienced by a road user compared to the acceptable travel time. Delay rate may be used to estimate the difference between system performance and the expectations for those system elements. The measurement of delays could allow transportation specialists to evaluate the improvements which can affect a particular corridor. It calculates the amount of lost travel time due to congestion (viii) Data collection methods: These techniques can be broadly classified into two groups: first is the use of mobile vehicles with data collection equipment, and second is the use of fixed sensors. Fixed sensor-based techniques include magnetic sensors and inductive loop detectors positioned on the road to collect the data. Whereas fixed sensor techniques include dual loop detector-based methods, magnetic sensor-based technologies, image sensor-based techniques, and probe vehicle-based techniques. Because of the many forms of transportation, the numerous sources and destinations, and the volume and variety of traffic, transportation in metropolitan areas is extremely complicated. The focus of urban transportation has always been on passengers since cities are seen as places where people interact. As a result, in addition to commuting, cultural activities and economic transactions have created complex traffic patterns. Inadequate public transportation (road or rail) leads to an increase in the number of individualized automobiles. Transportation networks are inextricably linked to socioeconomic transformations. People and freight mobility, as well as levels of accessibility, are at the center of this interaction. Where transportation infrastructure meets mobility demands and ensures access to markets and resources, economic possibilities are likely to emerge. Different parts of the world have been influenced differently by economic growth, from the industrial revolution in the

19th century through globalization and economic integration processes in the late 20th and early 21st centuries. Transportation networks on a global, regional, and local scale have all become critical components of economic activity. As a result, commerce and distribution are becoming increasingly important. The nation's economic progress will be hampered by an insufficient or non-sustainable transportation system. Congestion, accidents and mobility gaps are all negative effects of insufficient transportation.

Cities, on the other hand, are centers of production, consumption, and distribution, all of which are related to freight transit. As a result, public transportation, private transportation, and freight transportation are essential in any metropolitan center.

4.2 Impact

The impact of congestion on urban transportation systems can be broadly classified into three categories: (i) impact on economy, (ii) impact on the environment, (iii) impact on health. Table 4, shows the percentage of persons who were killed and injured to the total number of accidents. (i) Impact on economy: Due to traffic congestion, the additional economic losses are due to vehicle operating costs, additional transportation costs, additional fuel consumption, and maintenance cost. The recent National Crime Records Bureau (NCRB) data on "Accidental Deaths and Suicides in India" reports that 59.5% and 40.5% of road accidents were reported in rural areas (2,60,379 cases) and urban areas (1,77,017 cases) respectively [29]. The majority of accidents in urban areas were reported near residential areas. In urban areas, 31.5% of the cases (55,742 out of 1,77,017 total) occurred near residential areas. Pedestrian crossings were involved in 6.0% of road accidents in metropolitan areas (10,591 out of 1,77,017 cases). Furthermore, 8.5% of all traffic accidents in the country (37,143 out of 4,37,396 occurrences) occurred near schools, colleges, or other educational institutions. Road accidents near schools, colleges, and other educational establishments in metropolitan areas have claimed the lives of 26.3% of people in Uttar Pradesh and 8.0% of people in Odisha, respectively Table 5. Uttar Pradesh also had the largest number of fatalities related to traffic accidents near residential areas (urban areas), accounting for 19.4% of all fatalities in 2019 [29]. Table 6 shows the growth trend in the number of vehicles and road accidents in India from 2015 to 2019.

Table 4. Percentage of the persons were killed and injured to the total number of accidents in the urban area [29]

Type of area	Total accidents	Person killed	Person injured
Urban roads	1,95723 (42.1% of the total road accidents)	51,334 (34.7% of the total killed in road accidents)	1,83,703 (39.0% of the total injured in road accidents)

Travelers are bound to waste their time due to traffic conditions during the journey. As per the database from the research articles, Fig. 3 shows the time wasted by people due to traffic congestion. The additional time, which they can renovate towards contributing

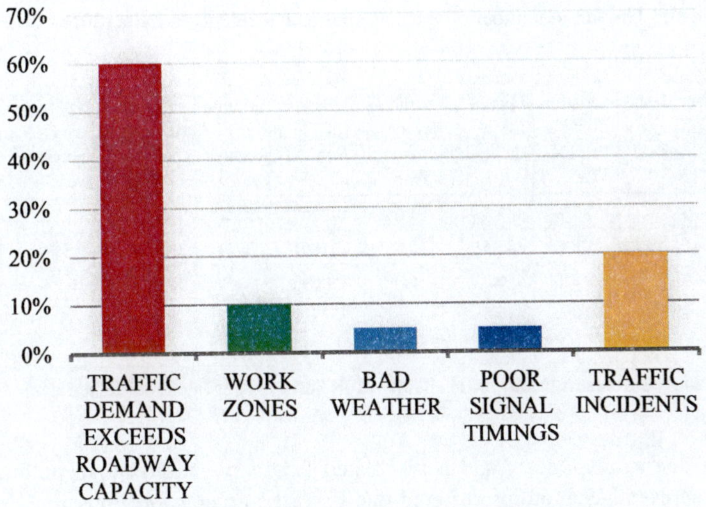

Fig. 3. Wastage of time due to traffic congestion [30]

Table 5. Place of incidence - wise road accident cases, persons injured and died during 2019 (all India) [29]

Sr no.	Cause	No. of cases	Persons injured			Person died			Percentage share (Cases)
			Male	Female	Total	Male	Female	Total	
1	Near school/college/educational institution	18390	14414	3821	18235	5095	1023	6118	4.2
2	Near residential area	55742	41126	10430	51556	14686	2763	17450	12.7
3	Near religious place	10578	8131	2215	10346	2993	556	3549	2.4
4	Near recreation place/cinema hall	11722	8433	2227	10660	2848	495	3343	2.7
5	Near factory/industrial area	14480	10230	2792	13022	3841	668	4509	3.3
6	At pedestrian crossing	10591	6609	1905	8514	2606	514	3121	2.4
7	Others	55514	42820	10171	52991	14724	2558	17282	12.7
8	Total	177017	131763	33561	165324	46793	8577	55372	40.5

to their things and economy of the nation. (ii) Impact on health: Due to congestion, people spend extra time on the road, which directly correlates with health. Impact on the environment: Traffic congestion affects the climate by noise pollution and air pollution. Among the significant contributors to air pollution, the transportation sector plays a vital role. Emissions from vehicles that pollute the air, also affect public health. It is severe that more than 2% of the people aged between 15 to 45 years die prematurely in Delhi

Table 6. Growth in number of vehicles and road accidents in India (2015–2019)

Sr. no.	Year	Road accidents (in thousand)	% variation over Previous Year	Persons injured (in thousand)	% variation over Previous Year	Persons killed (in Nos.)	% variation over previous year	No. of vehicles (in thousand)#	% variation over previous year	Rate of deaths per thousand vehicles
1	2015	464.7	3.1%	482.4	1.0%	148707	5.1%	210023	15.1%	0.71
2	2016	473.0	1.8%	485.5	0.6%	151801	2.1%	230031	9.5%	0.66
3	2017	445.7	−5.8%	456.2	−6.0%	150093	−1.1%	253311	10.1%	0.59
4	2018	445.5	−0.0%	446.5	−2.1%	152780	1.8%	253311	–	0.60
5	2019	437.4	−1.8	439.2	−1.6	154732	1.3	253311	–	0.61

Source: Road Accidents in India - 2019, TRW, MoRT&H, as per latest published data

every year due to respiratory and heart-related disorders caused by air pollution [6]. Figure 4 represents Symptoms suffered due to traffic congestion Emissions from the vehicles depending upon the vehicle's age, kilometers the car is driven, and speed of vehicles. As per a report by CPCB, in Delhi, 76.2% of CO, 96.9% of hydrocarbons, and 48.6% of NOx are caused by emissions from the transport sector only [31].

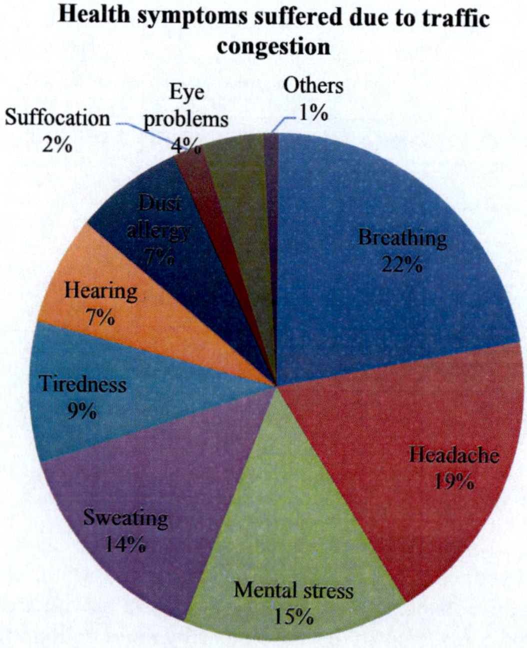

Fig. 4. Symptoms suffered due to traffic congestion

4.3 Suggested Mitigation Measures

Based on various characteristics of the urban transportation system, there are primarily four areas of issues that require attention: engineering organizational, ITS, and policy issues. Administrative issues involve controls in control and implement various aspects of urban traffic and transportation systems. There is a need for multiple policies to be set up with central and state government bodies. There is immediate attention to set up a national-level committee consisting of traffic and transportation personnel, including research institutions and ITS experts in various fields for proper planning and management, for faster and more efficient delivery of the urban transportation network. The National Law and Order management and enforcement issues need to be modernized and strengthened to take care of this efficiently. The requirement is there to promote and implement integrated public transportation systems, particularly metros, monorails with special provisions for non-mortar transport such as bicycles in a planned manner. A policy framework needs to be set up for collecting a database of all motorized vehicles to manage the real-time traffic accident data. A proper online management system is necessary for issuing challans and issue of driving licenses. Efforts should be made towards networking with automation such as automatic toll collection, detection of traffic offenses, and other traffic control systems. Various engineering issues for urban infrastructure assemblies related to traffic shall be designed, planned, and constructed. Engineering issues can be dealt with by engaging professional engineers who may ensure safety, sustainability, aesthetics, and fast construction, keeping an eye on all existing and new geotechnical and geophysical parameters. In addition, to simplify proper planning of infrastructure for urban transportation systems, surveyed data, topographical sheets, and other chief parameters of the traffic and transport be accurately mapped in the digital model to be available to engineering organizations, government officials, and planners.

Further, to achieve sustainability in the urban transport systems, the various strategies may be followed: (i) By identifying and improving specific parameters of public transport system which will influence the private vehicle commuters to shift to the public transport system. It will help in improving the traffic congestion as the number of private vehicles will significantly reduce. (ii) By providing appropriate parking facilities at the public transport terminals which will attract more private vehicle to shift on public transport modes. (iii) By sustainable integrating different public transportation systems in coordination with other modes of transport. (iv) Adopting suitable policy Level sustainable strategies to restrict the usage of private vehicles. However, it may only be implemented if there is a proper and affordable public transport system. (v) Timely evaluating and modifying the existing and new transportation system according to safety, environment, safety, social, and economy (vi) To cater to urban traffic-related problems, metro rails should be developed. In addition to these, a drive on the National level will be conducted to improve safety on the roads as the safety statistics are alarming and getting worse day by day.

5 Conclusions

In Indian cities, sustainable mobility is a critical facilitator of economic progress and the elimination of poverty. To capture synergies, promote interconnectivity, and maximize functioning through multimodal mobility solutions for Indian cities, comprehensive integration of urban transport with land use planning systems is required. Congestion, traffic accidents, pollution, and other current urban transportation concerns cannot be solved by traditional solutions that favor public money and investments in private transportation over public forms of transportation. Non-motorized transportation and high-capacity public transportation infrastructure require more public resources. It's also critical that the urban transportation industry is viewed as a whole in terms of system finance and pricing. The spatial morphology and layout of built forms in cities are influenced by urban transportation networks. Congestion measurement and the impact of congestion were evaluated in India's urban transportation system, and recommendations have been suggested to mitigate traffic congestion. Several factors were taken into account to assess the congestion and their impacts on the overall environment, economy, and health. It has been concluded that congestion cannot be eliminated but can be prevented by different approaches. In urban transportation systems, congestion occurs due to the increase in population and vehicle counts, significantly increasing the count of private vehicles and passenger cars. Among the proposed congestion mitigation measures, it has been concluded that congestion in the urban transportation system cannot only be minimized by applying strict traffic laws or policies. Instead, it may be minimized by controlling roadside activities, adequate parking facilities, and improving geometric designs. With the help of proper coordination between the experts, planners, and engineers involved in the execution part, it may be minimized to some extent. In India, where heterogeneous traffic conditions prevail, congestion is a serious issue and is much complicated. Our transportation system needs to cop up and develop comprehensive and new dimensions to tackle these problems. A sustainable urban transportation system in Indian cities can only be realized if strong, integrated, and participative institutions are established and enabled via clear roles, legislative power, financial independence, and professional competence. Above all, systems for openness, monitoring, and accountability of such institutions to their constituents must be established. All of this is only achievable if there is significant political will and public demand for change.

References

1. Shaban, A., Kourtit, K., Nijkamp, P.: India's urban system: sustainability and imbalanced growth of cities. Sustainability **12**(7), 2941 (2020)
2. Estache, A.: Infrastructure and development: a survey of recent and upcoming issues. In: Annual World Bank Conference on Development Economics, pp. 47–82. World Bank, Washington DC (2007)
3. Ki-moon, B.: Planning and design for sustainable urban mobility global report on Huma settlements 2013 (2013)
4. Pucher, J., Korattyswaropam, N., Mittal, N., Ittyerah, N.: Urban transport crisis in India. Transp. Policy **12**(3), 185–198 (2005)
5. Rangarajan, C., Mahendra Dev, S., Sundaram, K., Vyas, M., Datta, K.L.: Report of the expert group to review the methodology for measurement of poverty, New Delhi (2014)
6. Padam, S., Singh, S.K.: Urbanization and urban transport in India: the search for a policy. SSRN 573181 (2004)
7. Singh, S.K.: Urban transport in India: issues, challenges, and the way forward (2012)

8. Ministry of Road Transport and Highways Basic Road Statistics of India, Government of India, Various Issues from 2004–05 to 2016–17 (2017). https://morth.nic.in/basic-road-statistics-india
9. Sen, R., Sevani, V., Sharma, P., Koradia, Z., Raman, B.: Challenges in communication assisted road transportation systems for developing regions. In: NSDR 2009 (2009)
10. Roy, S., Sen, R., Kulkarni, S., Kulkarni, P., Raman, B., Singh, L.K.: Wireless across road: RF based road traffic congestion detection. In: 2011 Third International Conference on Communication Systems and Networks (COMSNETS 2011), pp. 1–6. IEEE (2011)
11. Rao, A.M., Rao, K.R.: Measuring urban traffic congestion-a review. Int. J. Traffic Transport Eng. **2**(4), 286–305 (2012)
12. Dubey, P.P., Borkar, P.: Review on techniques for traffic jam detection and congestion avoidance. In: 2015 2nd International Conference on Electronics and Communication Systems (ICECS), pp. 434–440. IEEE (2015)
13. He, F., Yan, X., Liu, Y., Ma, L.: A traffic congestion assessment method for urban road networks based on speed performance index. Procedia Eng. **137**, 425–433 (2016)
14. Salazar-Cabrera, R., de la Cruz, Á.P., Molina, J.M.M.: Sustainable transit vehicle tracking service, using intelligent transportation system services and emerging communication technologies: a review. J. Traffic Transp. Eng. (Eng. Ed.) **7**(6), 729–747 (2020)
15. Cafiso, S., Di Graziano, A., Pappalardo, G.: In-vehicle stereo vision system for identification of traffic conflicts between bus and pedestrian. J. Traffic Transp. Eng. (Eng. Ed.) **4**(1), 3–13 (2017)
16. Kotagi, P.B., Raj, P., Asaithambi, G.: Modeling lateral placement and movement of vehicles on urban undivided roads in mixed traffic: a case study of India. J. Traffic Transp. Eng. (Eng. Ed.) **7**(6), 860–873 (2020)
17. Theofilatos, A., Yannis, G., Vlahogianni, E.I., Golias, J.C.: Modeling the effect of traffic regimes on safety of urban arterials: the case study of Athens. J. Traffic Transp. Eng. (Eng. Ed.) **4**(3), 240–251 (2017)
18. Maheshwary, P., Bhattacharyya, K., Maitra, B., Boltze, M.: A methodology for calibration of traffic micro-simulator for urban heterogeneous traffic operations. J. Traffic Transp. Eng. (Eng. Ed.) **7**(4), 507–519 (2020)
19. Kumar, A., Sharma, R.K., Bansal, V.K.: GIS-based comparative study of information value and frequency ratio method for landslide hazard zonation in a part of mid-Himalaya in Himachal Pradesh. Innov. Infrastruct. Solutions **4**(1), 1–17 (2019). https://doi.org/10.1007/s41062-019-0215-2
20. Shafabakhsh, G.A., Famili, A., Bahadori, M.S.: GIS-based spatial analysis of urban traffic accidents: case study in Mashhad, Iran. J. Traffic Transp. Eng. (Eng. Ed.) **4**(3), 290–299 (2017)
21. Jeihani, M., James, P., Saka, A.A., Ardeshiri, A.: Traffic recovery time estimation under different flow regimes in traffic simulation. J. Traffic Transp. Eng. (Eng. Ed.) **2**(5), 291–300 (2015)
22. Kumari, S., Poddar, A., Kumar, N.: Delineation of groundwater recharge potential zones using the modeling based on remote sensing, GIS and MIF techniques: a study of Hamirpur District, Himachal Pradesh, India. Model. Earth Syst. Environ. (2021). https://doi.org/10.1007/s40808-021-01181-w
23. Kumar, A., Sharma, R.K., Bansal, V.K.: Landslide hazard zonation using analytical hierarchy process along National Highway-3 in mid Himalayas of Himachal Pradesh, India. Environ. Earth Sci. **77**(20), 1–19 (2018). https://doi.org/10.1007/s12665-018-7896-2
24. Nautiyal, A., Sharma, S.: A model to compute service life of rural roads using present pavement condition and pavement age. Compusoft **8**(7), 3261–3268 (2019). https://ijact.joae.org/index.php/ijact/article/view/981

25. Kumar, A., Sharma, R.K., Mehta, B.S.: Slope stability analysis and mitigation measures for selected landslide sites along NH-205 in Himachal Pradesh, India. J. Earth Syst. Sci. **129**(1), 1–14 (2020). https://doi.org/10.1007/s12040-020-01396-y
26. Nautiyal, A., Sharma, S.: Scientific approach using AHP to prioritize low volume rural roads for pavement maintenance. J. Qual. Maint. Eng. (2021). https://doi.org/10.1108/JQME-12-2019-0111
27. Kumar, A., Sharma, R.K., Bansal, V.K.: GIS-Based landslide hazard mapping along NH-3 in mountainous terrain of Himachal Pradesh, India using weighted overlay analysis. In: Singh, H., Garg, P., Kaur, I. (eds.) Proceedings of the 1st International Conference on Sustainable Waste Management through Design: IC_SWMD 2018, pp. 59–67. Springer International Publishing, Cham (2019). https://doi.org/10.1007/978-3-030-02707-0_9
28. Chavhan, S., Venkataram, P.: Prediction based traffic management in a metropolitan area. J. Traffic Transp. Eng. (Eng. Ed.) **7**(4), 447–466 (2020)
29. MoRTH: Road Accidents in India - 2019. Transport Research Wing, Ministry of Road Transport and Highways Government of India, New Delhi (2019)
30. Amadi, E.C., Ahaiwe, J., Etus, C., Akpabio, N.: Deployment of ICT for effective transportation system in Nigeria: a design model of Nigerian intelligent transportation system. Int. J. Sci. **3**(11), 24–32 (2012)
31. Agarwal, P., Singh, A.: Performance improvement of urban bus system: issues and solution. Int. J. Eng. Sci. Technol. **2**(9), 4759–4766 (2010)

Experimental Study on Overlying Strata Deformation and Failure Using Distributed Intelligent Sensing

Gang Cheng[1,2,3], Zhenxue Wang[1], Hong-Hu Zhu[2,3](✉), Dongyan Li[1], Wentao Xu[2], and Lei Zhang[2]

[1] School of Computer Science, North China Institute of Science and Technology (National Safety Training Center of Coal Mines), Beijing 101601, China

[2] School of Earth Sciences and Engineering, Nanjing University, Nanjing 210023, Jiangsu, China
zhh@nju.edu.cn

[3] Nanjing University High-Tech Institute at Suzhou, Suzhou 215123, Jiangsu, China

Abstract. With the continuous expansion and depth of coal mining in China, the disturbance of mining activities on the geological environment is unprecedented, which greatly aggravates the occurrence of various mine geological disasters and mining accidents. Therefore, real-time and intelligent sensing of overlying strata deformation and failure process has important practical significance. This paper carried out laboratory and on-site experimental research on mining-induced rock deformation based on distributed intelligent sensing and close-range photography technologies. The pre-implantation method was used in the laboratory test to lay the sensing cables inside the physical model. The continuous and distributed data of overlying strata deformation during coal mining were obtained. In the on-site test, the stress and deformation of overlying strata during coal mining were monitored and analyzed by laying four types of sensing cables in the borehole. The monitoring results showed that the two layers of igneous rock were the key layers to control the deformation and instability of rock, which should be monitored carefully. The research results had an important reference for revealing the deformation and failure process of deep rock mass in essence, improving early warning level of coal mining accidents, and guiding the green restoration of mining subsidence areas.

Keywords: Overlying strata deformation and failure · Separation evolution · Distributed intelligent sensing · Experimental research · Key layer

1 Introduction

With the increasing depth and intensity of coal mining, the difficulty is increasing gradually. Once the mining design method was improper, it would induce all kinds of serious disasters directly, such as rockburst, roadway deformation, ground subsidence, coal and gas outburst, mine water inrush, and et al. (Fig. 1). These disasters are usually the result of the deformation, movement, and failure of overlying strata during coal mining.

H.-H. Zhu et al. (Eds.): BRWSG 2021, LNCE 230, pp. 315–328, 2022.
https://doi.org/10.1007/978-981-16-9963-4_25

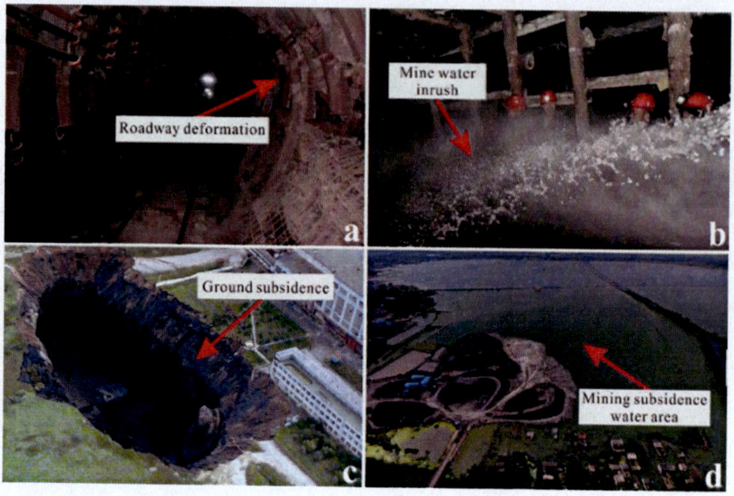

Fig. 1. Mine accidents and geological disasters caused by deep mining.

Underground coal mining often causes deformation, separation, fracture, and collapse of the overlying strata above the coal seam. With the movement of strata, mine pressure appears. To reveal the essence of the mine pressure phenomenon in the mining process, global scholars have put forward many stope hypotheses and theories [1–5]. Through decades of theoretical and experimental research, a variety of research methods for mining overburden deformation have been formed: engineering geological analogy method, numerical simulation analysis method, laboratory test method, on-site test method, et al. Due to mine engineering has the characteristics of large scale, complex influencing factors, strong concealment, high real-time monitoring requirements, and long monitoring period [6–8], the existing methods are difficult to monitor the deformation of overlying rock on the subsidence of roadway and ground in real-time during coal mining. Therefore, it is urgent to break through the technical bottleneck and realize a comprehensive understanding of the evolution process from the deformation and failure of overlying rock to the occurrence and development of ground subsidence. DFOS technology has become the research focus in the field of mine engineering monitoring [9, 10] owing to its full distribution, high precision, real-time, wide range, and anti-interference.

A mining area in Huaibei, China was invaded by a large area of magmatic rocks, which made the coal seam covered with igneous rocks, forming a unique mining condition and serious engineering geological problems. To study the deformation, failure, and separation evolution process of overlying strata during coal mining. This paper combined DFOS and close-range photography technologies and obtained the distributed monitoring results of rock mass stress and deformation through experimental research. It realized the monitoring of overlying strata deformation from macro to micro, from point to surface, from static to dynamic, and provided the theoretical basis for the stability of deep rock and the environmental carrying capacity of coal mining.

2 Technological Principles

Compared with traditional monitoring methods, distributed fiber optic sensing technology has unique advantages, including full distribution, long-distance, high sensitivity, and anti-electromagnetic interference. It is widely used in engineering applications such as bridges, tunnels, foundation pits, and slopes [11–14]. At present, the mainstream technologies are FBG (Fiber Bragg Grating Sensing), BOTDR (Brillouin Optical Time-Domain Reflectometer), BOTDA (Brillouin Optical Time-Domain Analysis), BOFDA (Brillouin Optical Frequency Domain Analysis), and DAS (Distributed Acoustic Sensing) [15]. In this paper, based on BOTDR, BOTDA, and CRP, laboratory model tests and on-site tests are carried out to obtain distributed strain data of overburden deformation in real-time. Table 1 is the technical performance comparison between BOTDR and BOTDA.

Table 1. Performance comparison between BOTDR and BOTDA

Technique classification	Demodulated signal	Technological principle	Equipment type	Performance parameter
BOTDR	Spontaneous Brillouin frequency shift	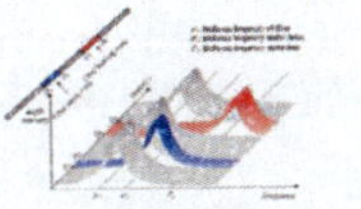	N8511 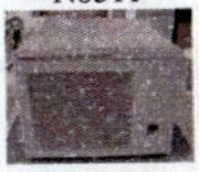	Maximum sampling interval: 0.05 m Frequency scan range: 10~12 GHz Strain test range: -1.5%~1.5% Strain test repeatability: <±100 με Measuring range: ≤80 km
BOTDA	Stimulated Brillouin frequency shift	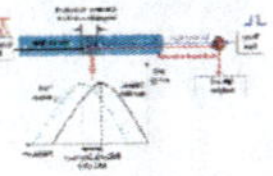	6050A 	Space resolution: 0.05 m Frequency scan range: 9~13 GHz Strain test accuracy: ±15 με Strain test range: -3%~4% Strain test repeatability: <±10 με Sampling resolution: 0.01 m

2.1 BOTDR/A

BOTDR generates spontaneous Brillouin scattering by the interaction between incident light and acoustic phonons in the fiber; BOTDA forms stimulated Brillouin scattering through the change of pump light pulse. When there is an axial strain (temperature) change along with the fiber, the Brillouin scattering light frequency will drift relative to the incident light frequency, and the frequency shift amount has a linear relationship with the axial strain (temperature) change of the fiber [16]. The axial strain (temperature) and Brillouin frequency shift of optical fiber can be expressed as:

$$\varepsilon = C_s(\nu_B - \nu_{B0}) + \varepsilon_0 \tag{1}$$

$$T = C_T(\nu_B - \nu_{B0}) + T_0 \tag{2}$$

where ε is the strain of the fiber; T is temperature; C_s is Brillouin frequency shift - strain coefficient; C_T is Brillouin frequency shift-temperature coefficient; ν_B is Brillouin frequency shift for optical fibers;ν_{B0}, ε_0 and T_0 are initial fiber Brillouin frequency shift, strain, and temperature, respectively [17]. When the temperature difference is less than five °C, the influence of temperature on the results is usually eliminated.

2.2 CRP

Close-Range Photography (CRP) uses a high-definition digital camera to record the location of speckles on the specimen's reference image and target image, select and match the target sub-region of the image, and obtain the displacement field of the measured region [18]. The measurement principle is shown in Fig. 2. By observing the change of coordinates before and after mining, the vertical displacement ΔH and horizontal displacement ΔU of measuring point M can be obtained:

$$\Delta H = hx_i - hx_0 \tag{3}$$

$$\Delta U = lx_i - lx_0 \tag{4}$$

where h is the vertical moving distance of point M; l is the horizontal moving distance for point M; x_i is the final position of point M; x_0 is the initial position of M point.

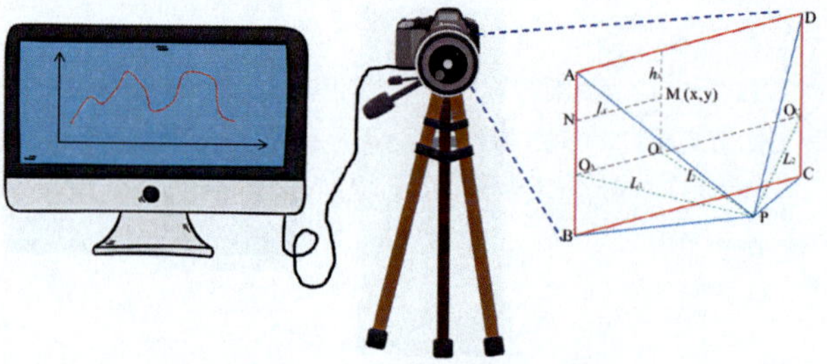

Fig. 2. CRP device and its technical principle.

3 Experimental Research on Deformation and Failure of Mining Rock Based on DFOS

3.1 Laboratory Test

The coupling between sensing fiber and rock mass significantly impacts the monitoring results, which directly determines the reliability of monitoring data. Zhu, Cheng, et al. carried out the pull-out test of fiber-sand coupling performance with BOTDA technology. The coupling process and the strain distribution of fiber were analyzed in detailed. The coupling process was divided into three stages: unity coupling, half coupling, and relative sliding. The displacement relationship model was established [19] (Fig. 3), which provided an important basis for the data validity evaluation of this experimental research.

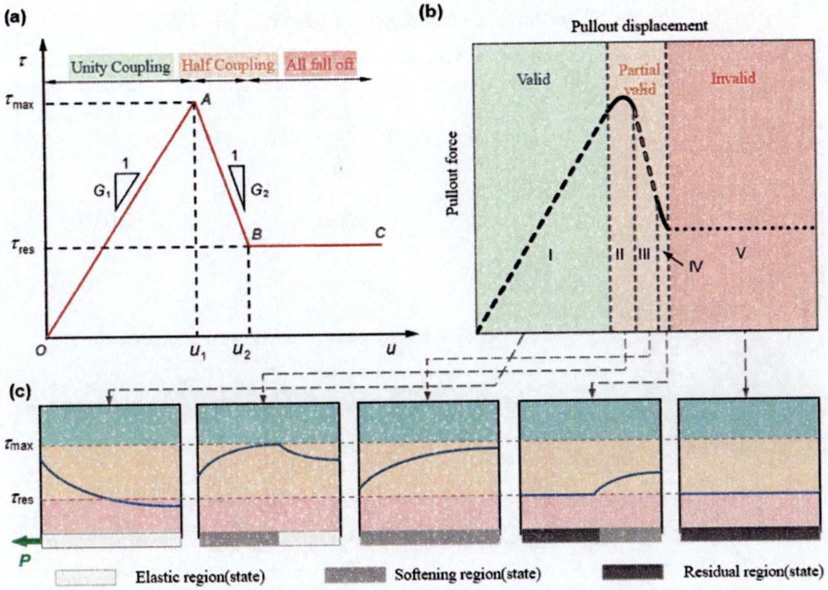

Fig. 3. Coupling process model of fiber-sand [19].

(1) Experimental process

The laboratory model test is carried out, taking the coal mining working face of a mine in Huaibei as the simulation object. The test steps are as follows:

1) Model design

Combined with the actual situation of the simulated coal face, the geometric ratio was 1:200, the time ratio was 17, and the stress ratio was 333. According to the load formula, the overburden load 18kN was calculated.

2) Model construction

The layout of the sensing cable is carried out by layered filling method, and the specific process was as follows: (a) Paste sensing cable; (b) Pre-pull cable; (c) Mixing similar materials; (d) Layered fill and compact; (e) Fix template.

3) Layout technique of sensing cable

According to the requirements of sensing cables layout and simulated formation deformation monitoring, 2 mm polyurethane sensing cable and fixed-point cable were selected. The main performance parameters of the cables are shown in Table 2, and the cable structures are shown in Fig. 4.

Since the deformation and failure of the overlying strata are mainly vertical compression deformation, tensile deformation and shear deformation, the vertical layout method is adopted. Three vertical plane U-shaped loops were arranged along the mining direction of the simulated coal seam, which are A12, B12, and C12, respectively. Two U-shaped

Table 2. Performance parameters of two types of cables

Cable type	Outer diameter (mm)	Maximum tension force (N)	Gauge factor (MHz/με)
Polyurethane sensing cable	2.0	200	0.04998
Fixed-point cable	2.0	300	0.04800

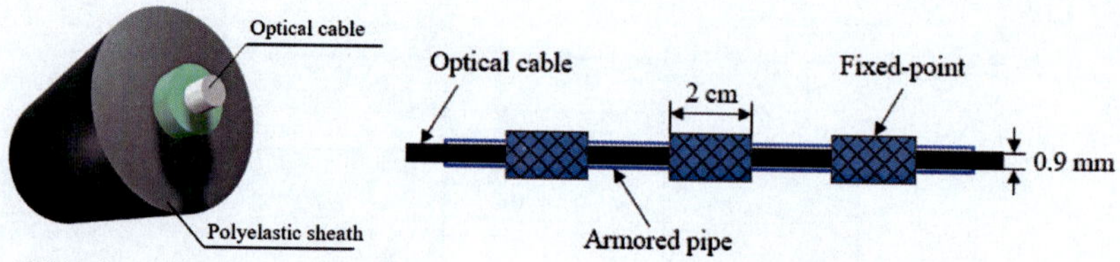

(a) Polyurethane sensing cable (b) Fixed-point cable

Fig. 4. Structural diagram of sensing cables.

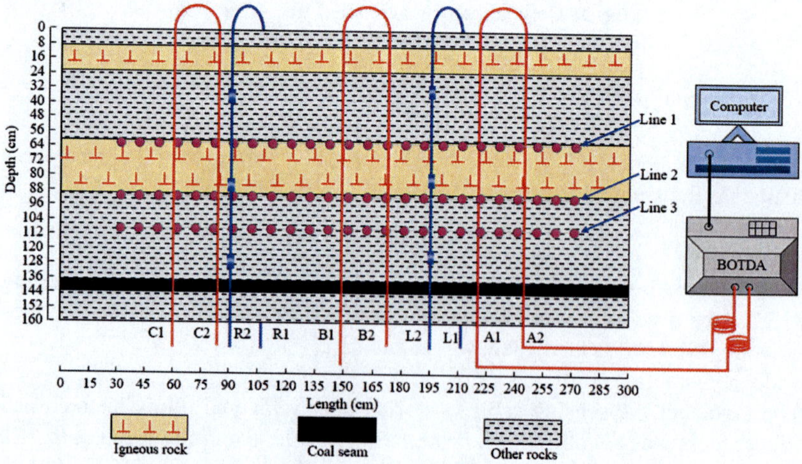

Fig. 5. Layout of the sensing cables in laboratory test.

loops were arranged in the vertical coal seam, which were L12 and R12, respectively. The layout position of the sensing cables is shown in Fig. 5.

4) Layout of close-range photography monitoring points

In order to compare the optical fiber monitoring data with the conventional monitoring data, close-range photography technology was used to observe the displacement of the overburden rock during coal mining. The layout steps were as follows: three horizontal measuring lines (depth: 61 cm, 91 cm, 111 cm) were laid on the model's surface. Each line had 25 measuring points, and the distance between each point was 10 cm. The starting and final measuring points are 270 cm and 30 cm, respectively.

(2) Experiment results and analyses

Limited to the length of this paper, A2, B2, C2 lines were selected for analysis. Figure 6 shows the strain test results of three cables. According to the A2 line, the rock mass below the lower igneous rock showed a continuous compression trend with the constant advancement of the coal seam, and the compression amount was increasing, the maximum value is about 7800 με. The upper rock mass of the lower igneous rock was compressed firstly. With the continuous mining, some cracks and separations were generated in the underlying rock mass, and the tensile state was presented at this time. With the constant advancement, the separation of the underlying rock mass was closed, and the upper rock mass of the lower igneous rock showed a trend of compression gradually. Since line B is located in the middle of the whole model, the stress mechanism in this area is the most complex, which can be explained by the test data of the cable. It could be found from the C2 line that the data showed a continuous compression state, indicating that the overall compression of the overlying strata along the vertical direction under coal mining.

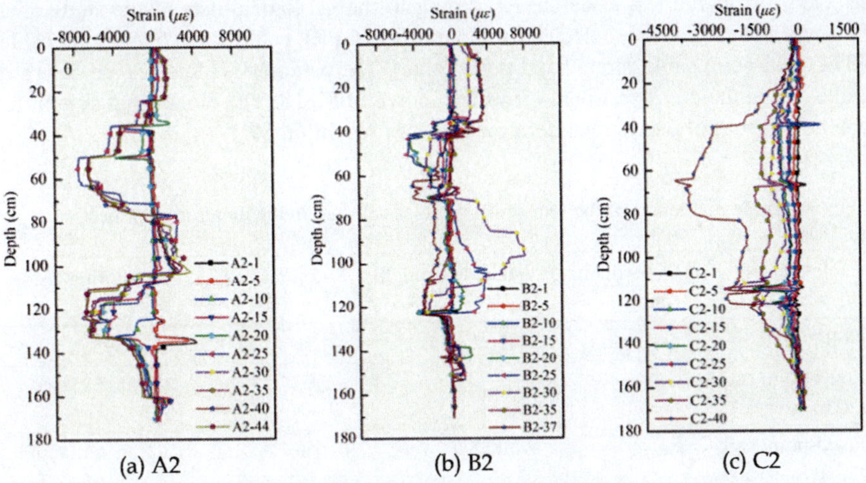

Fig. 6. Test results of sensing cables.

Take every 20 cm of coal seam excavation as a node to analyze the displacement of monitoring points. The displacement variations of each line are shown in Fig. 7. The displacement increased with the advancement of the working face and reached the largest in the middle of the model; the range of rock movement continued to expand with the depth increasing. In pace with the advancement of the coal seam, the influence of mining on the deformation of the front rock mass was increasing, showing an increasing trend. When coal mining was finished, the maximum displacement of each measuring line occurred in the middle of the model. The position of line 1 was far away from the coal seam, and the influence of mining was minimal. Line 3 was located in the caving zone, and the deformation law was complex, but the trend is the same as line 1 and line 2. The maximum displacement was at the height of 240 cm, about −4.36 cm, which is due to the maximum pressure at the right coal pillar after excavation.

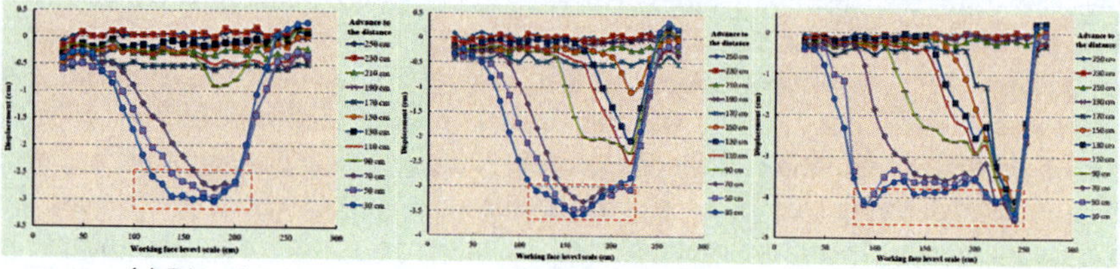

(a) Line 1 (b) Line 2 (c) Line 3

Fig. 7. Test results of CRP.

3.2 On-site Test

In order to verify the accuracy of the laboratory test, a borehole with 600 m depth was selected for on-site test research. According to the geological data of the study area, four types of cables were selected for monitoring (GFRP cable, 20 m fixed-point cable, 50 m fixed-point cable, strata fixed-point cable). The parameters of cables are shown in Table 3, and the cross-sectional views are shown in Fig. 8. The monitoring system for borehole layout of overburden deformation is shown in Fig. 9.

Table 3. Details of the four sensing cables selected in the monitoring project

Cable type	Outer diameter (mm)	Tensile strength (N)	Strain range (%)	Monitoring range
GFRP cable	5.8	4500	−1–1	0 m–595 m
Fixed-point cable (20 m interval)				175 m–595 m
Fixed-point cable (50 m interval)	5.0	3000	0–5%	308 m–595 m
Fixed-point cable (fixed at five strata)				274 m–595 m

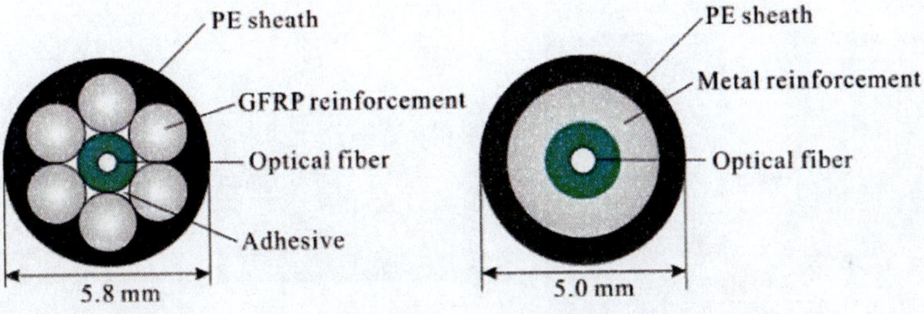

(a) GFRP cable (b) Fixed-point cable

Fig. 8. Cross-sectional view of the sensing cables [20].

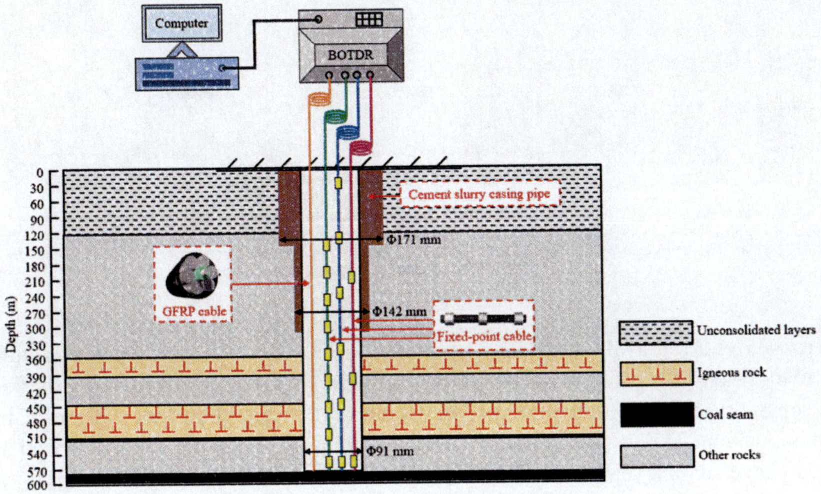

Fig. 9. The sensing cable layout of the monitoring system.

(1) Experimental process

Firstly, the ground borehole was carried out, and then the cable (device) was selected according to the monitoring target and geological conditions for borehole layout (Fig. 10). The process was as follows: (a) Making the guiding hammer; (b) Implantation of sensing cables; (c) Borehole backfill; (d) Sensing cable protection. The lowering speed of the drill pipe should be strictly controlled in the above process, and the layered filling method should be used for borehole backfill.

(2) Experiment results and analyses

Based on the measuring results of cables, the igneous rock was the key layer (Fig. 11). Before the working face was advanced through the borehole, the sensing cable strain in the rock layer above the bottom of the upper igneous rock was small. That is, this part of

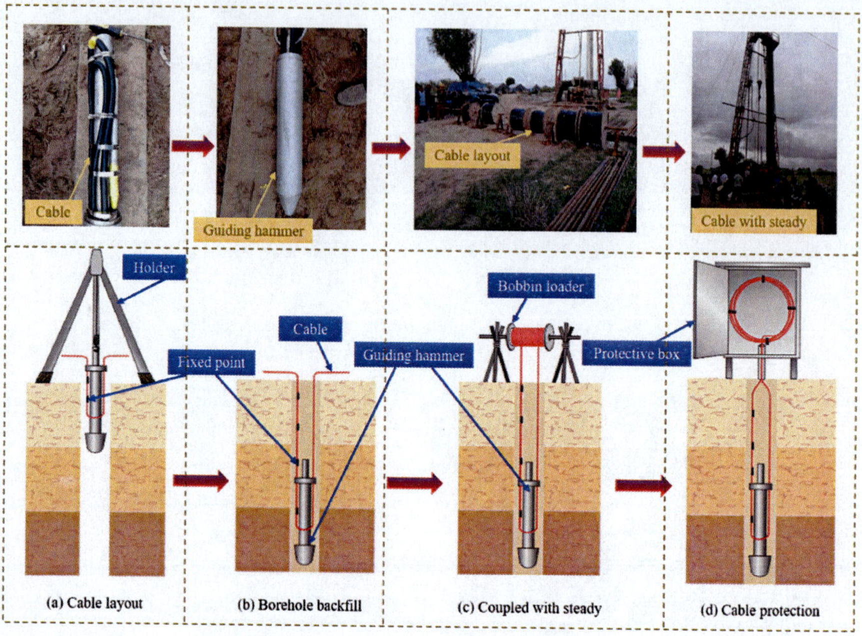

Fig. 10. Installation process of sensing cable.

the rock layer is relatively stable. There was a broken section at 390 m, which was judged as the position of separation development; with the advancement of the working face, the separation became more obvious, which led to the fracture of four types of cables at this position. The test results were consistent with the theoretical analysis results of the key layer, which verified the effectiveness of the application of DFOS technology in monitoring the deformation of overburden rocks. According to the test data of three fixed-point cables, when the working face was advanced through the borehole, the sensing cables between the depth of 450 and 580 m was at a continuous compression state, and the value was increasing, which showed that the advanced support pressure of the rock mass was also increasing. The cable deformation at 490 m was relatively larger, indicating that the stress deformation of rock strata caused by mining was mainly distributed between the fault zone and 490 m.

It could be seen from the monitoring data of 50 m fixed-point cable and strata fixed-point cable. There were two compression concentration areas at a depth of 574 m and 530 m. Therefore, the predicting heights of the caving zone and fracture zones were about 12.7 m and 56.7 m; according to the empirical formula, the height of the caving zone is 13.4 ± 2.5 m, and the height of the fracture zone is 53.6 ± 5.6 m (the mining thickness is 3 m). It showed that the monitoring results of cables were consistent with the empirical formula.

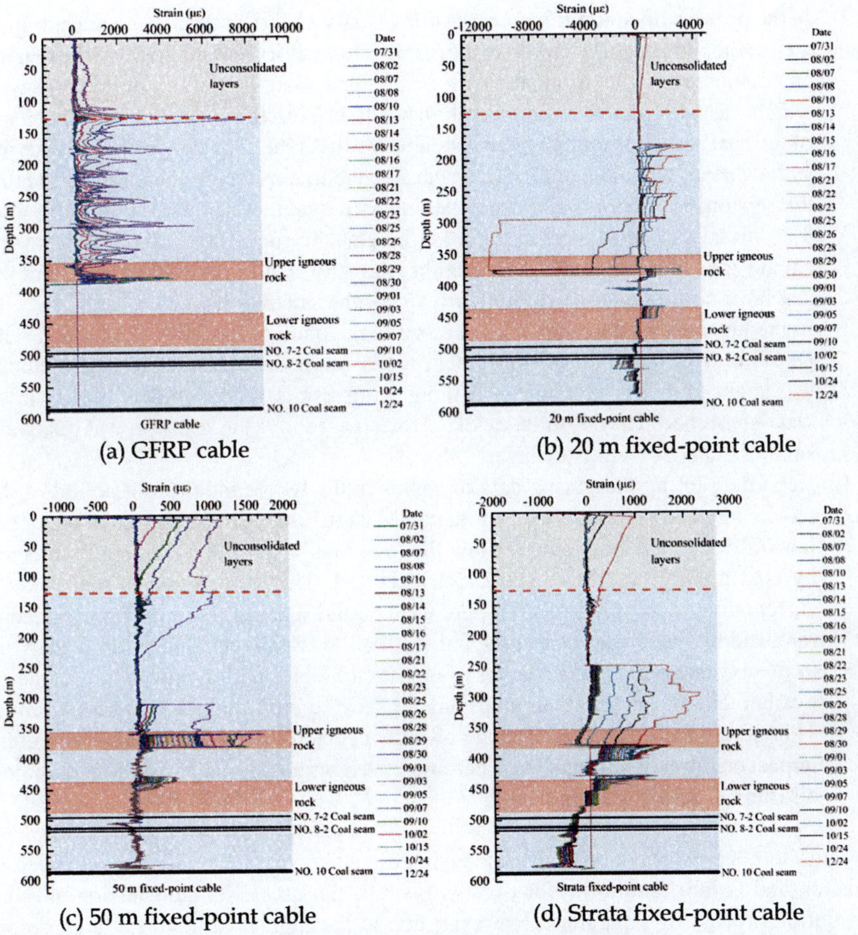

Fig. 11. Monitoring results of different cables.

4 Conclusion

The deformation, failure, and instability of the overburden rocks caused by coal mining is the direct cause of many serious mining accidents and also the direct root of ecological environment problems in mining areas. Therefore, it is of great significance to carry out effective and real-time monitoring. Based on the distributed intelligent sensing technology, the deformation and failure law of overlying strata and the evolution process of separation were studied through laboratory and on-site tests. The following conclusions were obtained:

(1) The intelligent sensing technology based on DFOS can effectively monitor the deformation and failure process of overlying rock mass during coal mining and obtain the continuous and distributed deformation data of rock mass in real-time.

(2) In the process of coal mining, separation usually occurs in the stress concentration area or the rock stratum interface; the result of laboratory test showed that the tension and compression mutation area was mostly located at the bottom of the key layer, which was determined as the separation position. With the continuous advancement of the coal seam, it gradually developed upward. As the subsidence of the overlying strata closed, when the deformation caused due to excavation increased gradually, the separation occurred and developed upward again. When the failure of mining rock mass was developed to a certain height, the rock fragmentation led to the caving zone and longitudinal separation gradually stabilizing. The existence of the key layer controlled the evolution process of the separation.
(3) In the laboratory test, the results of close-range photography were consistent with DFOS, and the maximum displacement of coal seam roof measured by both methods was about 4 cm. Through the cable monitoring results, the location of the key layer was determined. The deformation of the rock mass above the key layer was relatively smaller than below the key layer.
(4) According to the measuring data of cables in the whole monitoring period of the on-site test, it was found that the four cables were broken at the bottom of the upper igneous rock. The rock strata below the lower igneous rock were greatly affected by coal mining, and large separation occurred. The lower igneous rock and its surrounding rock were broken. There were partial separations in the upper igneous rock and its overlying rock mass, but the rock mass was relatively complete. Due to the existence of the key layer, it impacts the deformation and failure mode of overburden, and the rock stratum may be sheared at the bottom of the key layer. This failure form has a wide range of influence and fast speed, which has a great impact on the coal mining face. Therefore, the monitoring and prevention of mining disasters should be strengthened.

In short, the introduction of distributed intelligent sensing technology into the deformation and failure monitoring of overlying strata can reveal its deformation and the evolution process of separation. However, due to the high in-situ stress, large deformation, strong discontinuity, and vulnerability to multiple fields such as stress field, temperature field, and seepage field, the following aspects should be strengthened in the future:

(1) In the research of cable applicability for large deformation monitoring, the investigation of sensor monitoring technology with strong robustness, good durability, excellent stability, and high reliability should be carried out, as well as the development of special sensors should be strengthened to meet the adaptability of sensors under extreme conditions;
(2) In the study of the construction method of the multi-source and multi-field monitoring system, the guiding role of multi-field interaction relationship and additional effect in the deformation of rocks should be strengthened, and the multi-source and multi-field monitoring technology with diversified, multi-parameter, intelligent and full-period monitoring should be comprehensively developed.

Acknowledgments. This research was funded by the National Key Research and Development Program of China (Grant No. 2018YFC1505104), China Postdoctoral Science Foundation (Grant No. 2019M651786), State Key Laboratory of Geohazard Prevention and Geoenvironment Protection (Grant No. SKLGP2021K005), National Natural Science Foundation of China (Grant No. 41702347), and Hebei IoT Monitoring Engineering Technology Research Center (Grant No. 3142018055).

References

1. Helmut, K.: Mining Subsidence Engineering. Springer, New York (1983). https://doi.org/10.1007/978-3-642-81923-0
2. H. Kratzsch.: Mining damage and its protection. China Coal Industry Publishing House, Beijing, China (1984). (in Chinese)
3. Qian, M.G.: Equilibrium conditions of strata on mining site. J. China Univ. Min. Technol. **2**, 31–40 (1981). (in Chinese)
4. Song, Z.Q.: Practical Mine Pressure Control. China University of Mining and Technology Press, Xuzhou, China (1988).(in Chinese)
5. Qian, M.G., Miao, X.X., Xu, J.L.: Theoretical study on key strata in strata control. J. China Coal Soc. **21**(3), 225–230 (1996). (in Chinese)
6. Shi, B.: On field and their coupling in engineering geology. J. Eng. Geol. **21**(5), 673–680 (2013). (in Chinese)
7. Liu, Y., Li, W.P., He, J.H., Liu, S.W., Cai, L.Y., Cheng, G.: Application of Brillouin optical time domain reflectometry to dynamic monitoring of overburden deformation and failure caused by underground mining. Int. J. Rock Mech. Min. Sci. **106**, 133–143 (2018)
8. Wang, H., Qin, W.M., Jiao, Y.Y., He, Z.: Transitions and opportunities of geotechnical engineering monitoring in coming big data era. Rock and Soil Mechanics **35**(9), 2634–2641 (2014). (in Chinese)
9. Zhu, H.H., Shi, B., Yan, J.F., Zhang, J., Wang, J.: Investigation of the evolutionary process of a reinforced model slope using a fiber-optic monitoring network. Eng. Geol. **186**, 34–43 (2015)
10. Chai, J., et al.: Detecting deformations in uncompacted strata by fiber Bragg grating sensors incorporated into GFRP. Tunn. Undergr. Space Technol. **26**(1), 92–99 (2011)
11. Zhao, X.F., Lu, J., Han, R.C., Kong, X.L., Wang, Y.H.: Application of multiscale fiber optical sensing network based on Brillouin and fiber Bragg grating sensing techniques on concrete structures. Int. J. Distrib. Sens. Netw. **8**(10), 310797 (2012)
12. Gómez, J., Casas, J.R., Villalba, S.: Structural health monitoring with distributed optical fiber sensors of tunnel lining affected by nearby construction activity. Autom. Constr. **117**, 103261 (2020)
13. Wang, X., Shi, B., Wei, G.Q., Chen, S.E., Zhu, H.H., Wang, T.: Monitoring the behavior of segment joints in a shield tunnel using distributed fiber optic sensors. Struct. Control. Health Monit. **25**(2), e2056 (2017)
14. Zhu, H.H., Wang, Z.Y., Shi, B., Wong, J.K.W.: Feasibility study of strain based stability evaluation of locally loaded slopes: insights from physical and numerical modeling. Eng. Geol. **208**, 39–50 (2016)
15. Gao, L., Chen, H.D., Yu, X.J., Zhou, B.: Advance of the distributed optical fiber technology in geotechnical and geological engineering. Hydro-Sci. Eng. **32**(2), 93–99 (2013)
16. Sasaki, T., et al.: Distributed fibre optic strain sensing of an axially deformed well model in the laboratory. J. Nat. Gas Sci. Eng. **72**, 103028 (2019)

17. Zhang, L., Cheng, G., Wu, J.H., Minardoe, A., Song, Z.P.: Study on slope failure evolution under surcharge loading and toe cutting with BOTDA technology. Optical Fiber Technol. **66**, 102644 (2021)
18. Xu, W.T., et al.: Extraction and statistics of discontinuity orientation and trace length from typical fractured rock mass: A case study of the Xinchang underground research laboratory site. Chin. Eng. Geo. **269**(3), 105553 (2020)
19. Zhang, C.C., Zhu, H.H., Liu, S.P., Shi, B., Cheng, G.: Quantifying progressive failure of micro-anchored fiber optic cable–sand interface via high-resolution distributed strain sensing. Can. Geotech. J. **57**(6), 871–902 (2020)
20. Cheng, G., Shi, B., Zhu, H.-H., Zhang, C.-C., Wu, J.-H.: A field study on distributed fiber optic deformation monitoring of overlying strata during coal mining. J. Civ. Struct. Heal. Monit. **5**(5), 553–562 (2015). https://doi.org/10.1007/s13349-015-0135-6

Analysis of Influencing Factors on Bearing Characteristics of PCC Energy Pile Group

Lei Gao[1](✉), Yunhao Gong[1,2], Hui Zhou[1], Yumin Chen[1], and Zhongquan Xu[1]

[1] Key Laboratory of Ministry of Education for Geomechanics and Embankment Engineering, Hohai University, Nanjing, China
gaoleihhu@hhu.edu.cn

[2] Research Institute of Complex Engineering and Management School of Economics and Management, Tongji University, Shanghai, China

Abstract. PCC energy piles are a new type of energy pile technology in the field of geotechnical engineering in recent years, which has the technical advantages of convenient construction, good economy and energy savings. The three-dimensional numerical analysis model of PCC energy pile groups is established by finite element software. The influence of the pile number and pile spacing on the pile top displacement, pile tip resistance and pile body stress of the PCC energy pile group under two working conditions of temperature action and temperature-load combined action are studied, and the pile group and single pile are compared and analyzed under each working condition. The results show that the pile top displacement and pile end resistance of pile groups increase with increasing pile number and decrease with increasing pile spacing under the action of temperature. The pile body stress of the pile groups decreases with increasing pile number and increases with increasing pile spacing. When the load is added on the pile top, the displacement of the pile top caused by the temperature decreases, and the resistance of the pile end and the stress of the pile body increase.

Keywords: PCC energy pile · Group piles · Bearing characteristics · Number of piles · Pile spacing · Numerical simulation

1 Introduction

With the implementation of the "one belt, one road" strategy, more and more high-speed railways and highways have been built in the related countries. An energy pile is the pile foundation form combining ground source heat pump technology. Energy piles can efficiently utilize shallow geothermal energy along high-speed railways and highways to achieve the effect of energy conservation and emission reduction; they have attracted much attention during "one belt, one road" construction and have good application prospects [1–5]. The heat exchange between the superstructure and surface constant temperature layer is realized through the heat exchanger embedded in the energy pile [6]. According to the bearing characteristics of energy piles, Gui et al. [7] carried out field tests on energy piles. The research results show that the temperature change

H.-H. Zhu et al. (Eds.): BRWSG 2021, LNCE 230, pp. 329–344, 2022.
https://doi.org/10.1007/978-981-16-9963-4_26

of the energy pile will cause an additional temperature load. The size and distribution characteristics of the vertical stress of the pile body are related to the working load at the pile top and the constraints between the pile body and pile end. Its bearing characteristics are different from those of a pile foundation under a single working load. Based on the conventional model test, Wang et al. [8] found that the influence of pile temperature and soil moisture content around piles on the mechanical properties of pile-soil contact surfaces is related to the type of soil. Liu et al. [9–11] conducted field tests and numerical simulations on energy piles with different buried pipe forms, such as single U, double U and spiral pipes, and found that there are differences in the thermal-mechanical effects of energy piles with different buried pipe forms. Saggu et al. [12–14] carried out research on different sand compactnesses, different pile spacings, different pipeline embedding positions, different pile end constraints and different pile top constraints and compared them with the centrifuge test results of Rotta et al. [15]. Salciarini et al. [16] established a 3D numerical model by COMSOL software to study the influence of the layout of energy piles on the thermal-mechanical response of energy pile groups; the results show that when part of the energy piles in the energy pile group are heated, additional compressive stress will be generated in the energy pile, and additional tensile stress will be generated in the adjacent nonheated piles. With increasing heating time, the difference between the mechanical response of energy piles and nonheated piles will gradually decrease. The effects of temperature loads make energy piles different from other general pile foundations, and many scholars at home and abroad have used experimental means to monitor the temperature and moisture fields [17, 18].

Liu et al. [19] summarized the research status and engineering application of energy pile technology at home and abroad in recent years. He proposed the new PCC energy pile technology in view of the main problems existing in the engineering application of traditional energy piles and briefly analyzed the application prospects and research directions of the new energy pile technology. Huang et al. [20] carried out a model test on the bearing characteristics of PCC energy piles in air-dried sand to study the effect of cyclic temperature loading on the side friction characteristics and load transfer mechanism of PCC energy piles under heating conditions. The results show that the vertical bearing capacity of the PCC energy pile in air-dried sand increases slightly with increasing pile temperature, and the variation law of the side friction of the PCC energy pile is consistent with that of the conventional energy pile.

Currently, research on PCC energy piles is in its infancy, and most of the relevant research is focused on single piles. The research content mainly includes the thermal, mechanical and thermal-mechanical coupling characteristics of a single PCC energy pile, and there is relatively little research on the bearing characteristics of the PCC energy pile group. At the same time, the variation law of the bearing characteristics of the PCC energy pile group under temperature action and temperature-load combined action are not clear, and the influences of the number and spacing of piles for the pile group on the bearing characteristics of the PCC energy pile group are not clear. This paper intends to use the numerical simulation method to study the influences of pile number and pile spacing on the bearing characteristics of PCC energy pile groups under temperature action and temperature-load combined action. This research can provide a reference for the optimal design, construction and use of PCC energy pile groups in engineering.

2 Establishment of Numerical Model

This numerical simulation uses the finite element analysis software to simulate real physical phenomena. In the process of numerical simulation, it is assumed that the pile and soil are isotropic linear elastic materials, and the pile-soil contact surface is in complete contact. It is considered that the liquid phase is incompressible under isothermal conditions, without considering the creep of soil and the influence of thermal convection caused by solid heat transfer.

The model is 2.65 m in length, 2.00 m in width and 2.30 m in height, the upper layer of the model is 0.85 m thick clay, and the lower layer is 1.35 m thick sand. The outer diameter of the pile is 0.11 m, the inner diameter is 0.05 m, the length of the pile body is 1.65 m, and the upper part of the pile is 0.1 m exposed to the soil surface. The mesh generation is mainly divided into three parts, namely, soil mesh, pile mesh and air mesh. The mesh generation unit is a free tetrahedral (3D) element and a triangular (2D) element on the surface. A dense mesh is used in the temperature-sensitive area around the pile group, and a sparse mesh is used in the area far away from the pile group. Taking the pile group model of four PCC energy piles as an example, the mesh division of the model is shown in Fig. 1.

The physical, mechanical and thermal parameters of materials are comprehensively taken in combination with geotechnical tests, similar documents and relevant specifications [21, 22]. The parameters are listed in Table 1 and Table 2. The parameters of air are selected according to the built-in parameters in the software material library. The heat conduction parameters of air in the material library are variables with temperature and they are not listed in Table 2.

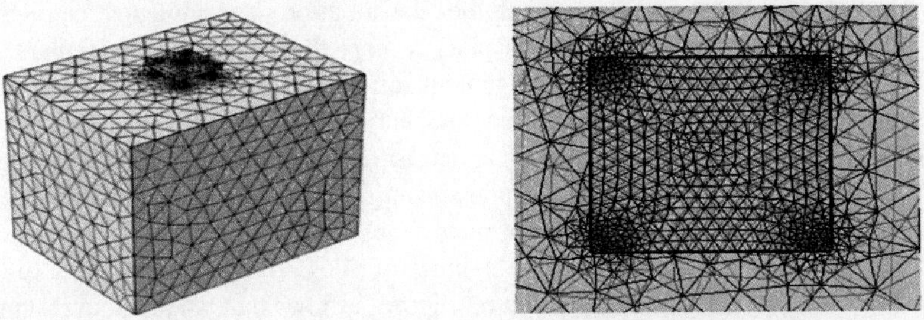

(a) (b)

Fig. 1. Meshes of numerical model: (a) Overall meshing, (b) Local meshing

In the heat transfer module of the model, the temperature of the environment and soil (including the pile and air in the soil layer) at the initial time is set to 9.5 °C, the water inlet temperature of the energy pile group is set to 30 °C, the cycle time is 120 h, the surrounding and bottom of the soil are thermally insulated, and the surface of the soil dissipates heat naturally.

In the solid mechanics module of the model, the bottom of the soil body is a fixed constraint, the periphery is roll support (fixed in the horizontal direction, only displacement in the vertical direction is allowed), and the surface of the soil body is a free boundary.

Table 1. Physical and mechanical parameters of materials

Material	P (kg/m^3)	E (MPa)	ν	C (kPa)	φ (°)
Clay	1930	24	0.3	28.6	23.1
Sand	1700	15	0.3	—	29.7
Concrete	2300	30300	0.2	—	—

Table 2. Thermal parameters of materials

Material	Cp (J/(kg·K))	α_T/K^{-1}	K (W/(m·K))
Clay	1050	1×10^{-5}	1.532
Sand	876	1×10^{-5}	1.826
Concrete	880	1×10^{-5}	1.4

To ensure that the stress of each pile in the pile group is the same as that of a single pile under different conditions, there is no cushion cap in the pile group model, the load is directly applied to the top of each pile in the pile group, and the load size is set to 5 kN.

3 Influence Analysis of the Number of Piles

When considering the influence of the number of piles on the bearing characteristics of the PCC energy pile group, this paper studies the situation of two piles, three piles and four piles because when the number of piles is more than five, the arrangement form of piles will be involved, and the arrangement form of piles is an important factor for the effect of the pile group and will affect the bearing characteristics of the pile group. To reduce the interference of the arrangement form, pile groups with a small number of piles are studied this time, and the pile spacing three times the pile diameter (3D) remains unchanged while the number of piles changes.

Figure 2 shows the influence of the number of piles on the change in pile top displacement under the action of temperature. Figure 2 shows that when the cycle time is the same, the pile top displacement of the pile group is greater than that of the single pile. For the pile group, the displacement of the pile top increases with an increasing number of piles because the pile spacing remains unchanged, the number of piles in the pile group is greater, and the temperature superposition phenomenon is obvious. In addition, the single pile in the pile group is also affected by the temperature field of the other piles.

Figure 3 is the diagram of pile groups with different pile numbers affected by the temperature of adjacent piles. The shaded part is the part affected by the temperature field of adjacent piles, and the double shaded part is the part affected by the superposition of the temperature of adjacent piles.

Figure 4 shows the influence of the number of piles on the change in pile top displacement under the temperature-load combined action. It can be seen from Fig. 4 that

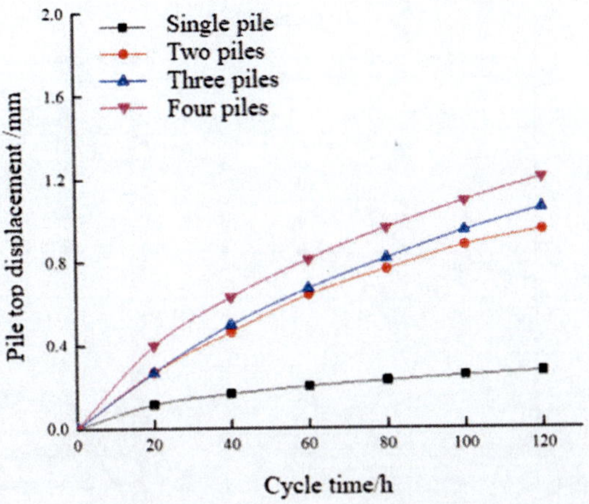

Fig. 2. Influence of different pile numbers on pile top displacement under temperature

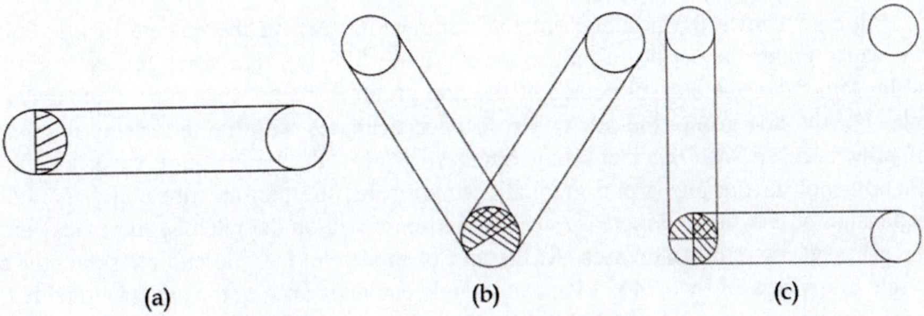

Fig. 3. Pile groups with different pile numbers affected by adjacent pile temperature: (a) Two piles, (b) Three piles, (c) Four piles

at the initial time, the pile top displacement of the single pile and pile group is negative because the effect of temperature has not been played at the initial time, and the pile body has a downward displacement under the action of vertical loading. With the cycle time, the pile top moves upward gradually due to the thermal expansion of the pile. Compared with Fig. 2, the variation law of the curves in the two figures is consistent, but at the end of the cycle, the displacement of the pile top of single pile decreases by 32.48% (0.0873 mm), the displacement of the pile top of pile group with two, three and four piles decreases by 15.64% (0.1484 mm), 8.32% (0.0877 mm) and 2.21% (0.0266 mm), respectively, it indicates that the load on the pile top restricts the thermal expansion of the pile, the restriction degree of single pile is greater than that of pile group, the restriction degree of pile group decreases gradually with the increasing of pile number.

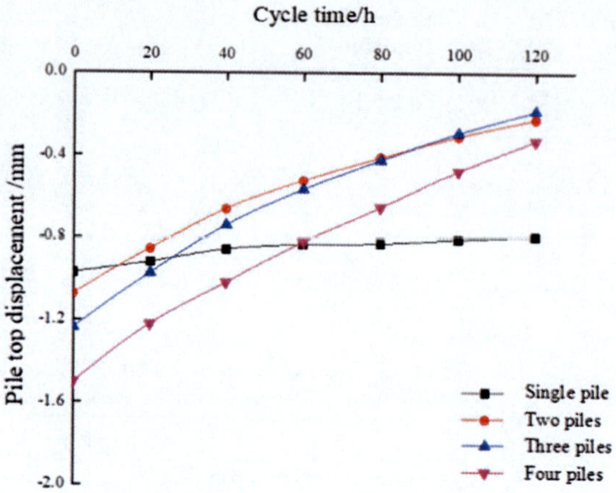

Fig. 4. Influence of different pile numbers on pile top displacement under temperature-load

Figure 5 shows the influence of the number of piles on the change in pile end resistance under the action of temperature. Figure 5 shows that when the cycle time is the same, the pile end resistance of the pile group is greater than that of the single pile. For the pile group, the pile end resistance increases with the increasing number of piles because with the increasing number of piles, the temperature superposition phenomenon in the pile group gradually strengthens, the thermal stress and thermal expansion of the pile end increase, the restriction of soil on the pile end increases, and the pile end resistance increases. At the end of the cycle, the pile end resistance of a single pile increases by 0.1442 kPa, and the pile end resistance of the pile group with 2, 3 and 4 piles increases by 4.6699 kPa, 6.2292 kPa and 12.2925 kPa, respectively, which are 32.38 times, 43.20 times and 85.25 times that of a single pile, respectively.

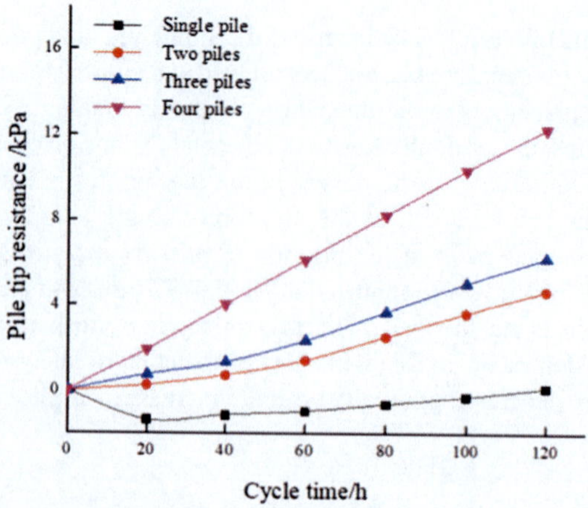

Fig. 5. Influence of different pile numbers on pile end resistance under temperature

Figure 6 shows the influence of the number of piles on the change in pile end resistance under the combined action of temperature and load. It can be seen from Fig. 6 that the change law of the pile end resistance under the combined action of temperature and load is consistent with that under the action of temperature, but the value increases. At the end of the cycle, the pile end resistance of a single pile increases by 15.95% (0.023 kPa), and the pile end resistance of the pile group with 2, 3 and 4 piles increases by 18.42% (0.86 kPa), 19.75% (1.23 kPa) and 28.72% (3.53 kPa), respectively, which indicates that the load plays an increasing role in the pile end resistance. Because the pile end moves downward and compresses the soil under the load, it makes the soil denser, and the constraint of the soil on the pile end increases. Under the same thermal expansion of the pile end, the soil is denser at the pile end, and the upward resistance to the pile end is greater.

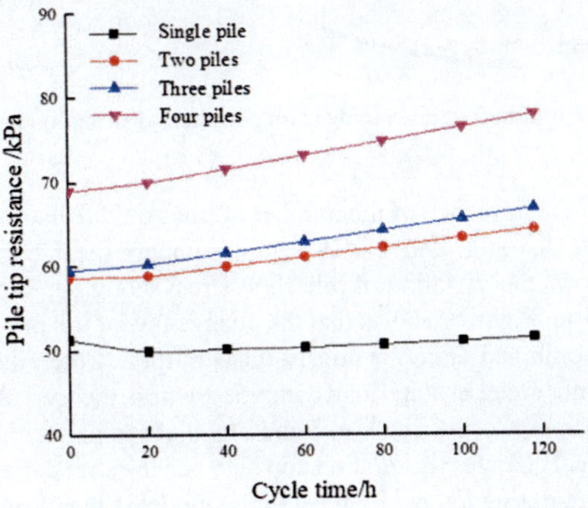

Fig. 6. Influence of different pile numbers on pile end resistance under temperature-load action

Figure 7 shows the influence of the number of piles on the distribution of pile stress along the depth under the action of temperature. Figure 7 shows that under the action of temperature, at the same depth, the pile stress of a single pile is greater than that of the group piles. For group piles, the pile stress decreases with an increasing number of piles because the existence of the pile group effect reduces the pile side friction of the pile group. As the number of piles increases, the reduction in pile side friction is more obvious. The decrease in pile side friction weakens the constraint around the pile. When the pile temperature increases to the same value, the constraint around the pile is smaller, and the thermal strain is greater. At the same time, the number of piles is increasing, the temperature superposition phenomenon is more obvious, the temperature rise of the pile body is faster, and the stress of the pile body is greater. It can be seen from the results in Fig. 7 that the decrease in pile body stress caused by pile side friction is greater than the increase in pile body stress caused by the increase in pile number (increasing pile body temperature). The maximum stress of the single pile and group piles is at a depth of 105 cm (69.69% of the pile length from the pile top), the maximum stress of the single pile is 124.78 kPa, and the maximum stress of the group piles is 66.87 kPa, 75.45 kPa and 89.63 kPa.

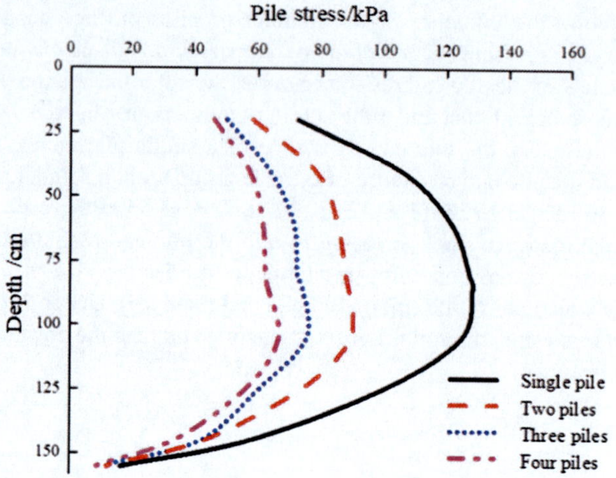

Fig. 7. Influence of different pile numbers on pile stress distribution under temperature

Figure 8 shows the influence of the number of piles on the distribution of pile stress along the depth under loading. Under different pile numbers, the pile stress decreases with depth. Figure 9 shows the variation of pile shaft stress along depth under temperature-load combined action. Figure 9 shows that the change law of the pile stress curve under temperature-load combined action is consistent with that of the pile body under load, which shows that under temperature-load combined action, the load plays a major role in the change in pile stress. Comparing Fig. 8 and Fig. 9, it can be seen that under the load, the temperature effect changes the relationship between the stress of a single pile and the group piles. The circulating temperature set in the model is low. When the temperature is high, the degree and status of the influence of temperature and load on pile stress may

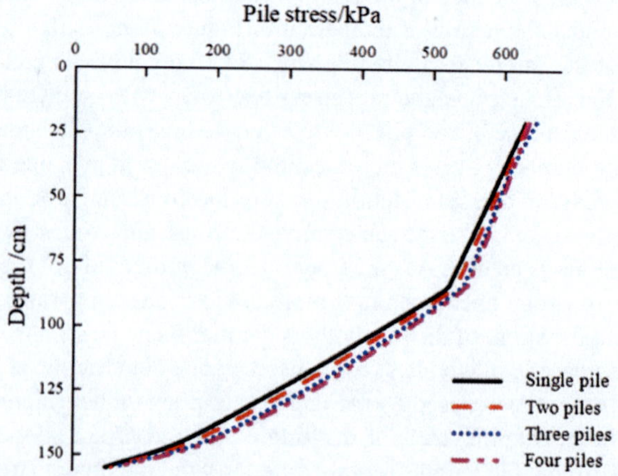

Fig. 8. Influence of different pile numbers on pile stress distribution under load

change, which should be given enough attention in the design and use of PCC energy piles.

Figure 10 shows the variation of pile stress under temperature-load combined action compared with that situation under load. Compared with Fig. 7, the variation law of pile stress along the depth in Fig. 10 is consistent, but the value increases. Taking the stress value at a depth of 105 cm as an example, the single pile increases by 5.07% (6.32 kPa), and the group piles with two, three and four piles increase by 4.86% (3.25 kPa), 4.02% (3.03 kPa) and 3.45% (2.31 kPa), respectively. The existence of the pile top load increases the pile top constraint, reduces the thermal strain of the pile body, and increases the thermal stress of the pile body when the pile body temperature rises.

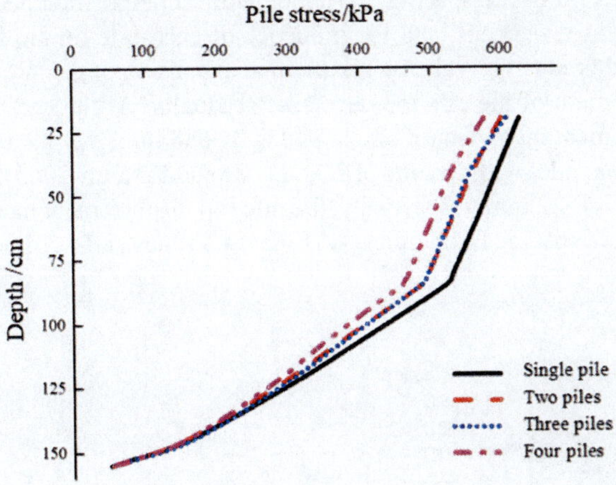

Fig. 9. Influence of different pile numbers on pile stress distribution under temperature-load

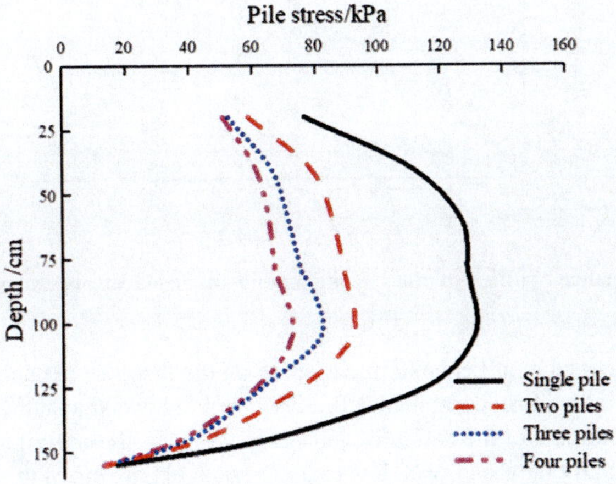

Fig. 10. Stress variation of pile body with different pile numbers under temperature-load

4 Influence Analysis of Pile Spacing

When considering the influence of pile spacing on the bearing characteristics of the PCC energy pile group, the four PCC energy pile models are established by finite element software. The pile spacing is set as two times the pile diameter (2D), three times the pile diameter (3D), four times the pile diameter (4D) and five times the pile diameter (5D), and the pile length remains unchanged.

Figure 11 shows the influence of pile spacing on the change in pile top displacement under the action of temperature. Figure 11 shows that when the cycle time is the same, the pile top displacement of the group piles is greater than that of the single pile. For the group piles, the displacement of the pile top decreases with increasing pile spacing because with increasing pile spacing, the temperature superposition phenomenon in the pile group gradually weakens, and the influence of each pile on the surrounding pile temperature field gradually weakens. The thermal expansion of the pile body decreases, and the displacement of the pile top decreases gradually. At the end of the cycle, the pile top displacement increment of single pile is 0.2688 mm, the pile top displacement increment of group piles with spacing of 2D, 3D, 4D and 5D are 1.2068 mm, 1.2016 mm, 1.1735 mm and 1.1532 mm, respectively, the pile top displacement increment of group piles with different spacing is 4.5 times, 4.5 times, 4.4 times and 4.3 times of single pile, respectively.

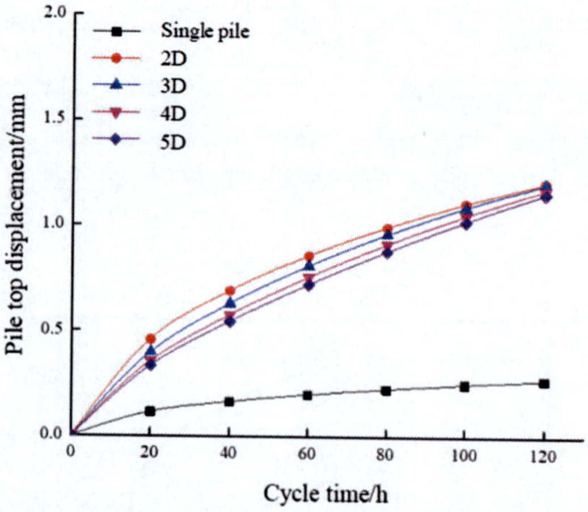

Fig. 11. Influence of different pile spacing on pile top displacement under temperature

Figure 12 shows the influence of pile spacing on the displacement change of the pile top under temperature-load combined action. Figure 12 shows that at the initial time, the pile body will settle under the action of vertical loading. With the progress of the cycle, the pile top gradually moves upward due to the thermal expansion of the pile. Compared with Fig. 11, the curves in the two figures have the same change law, but at the end of the cycle, the displacement of single pile top decreases by 32.48% (0.0873 mm), the displacement of group piles top with pile spacing of 2D, 3D, 4D and 5D decreases by

0.81% (0.0098 mm), 2.21% (0.0266 mm), 5.26% (0.0617 mm) and 5.71% (0.0659 mm), respectively, it indicates that the pile top load limits the thermal expansion of pile body. The restriction degree of a single pile is greater than that of group piles, and the restriction degree of group piles increases gradually with increasing pile spacing.

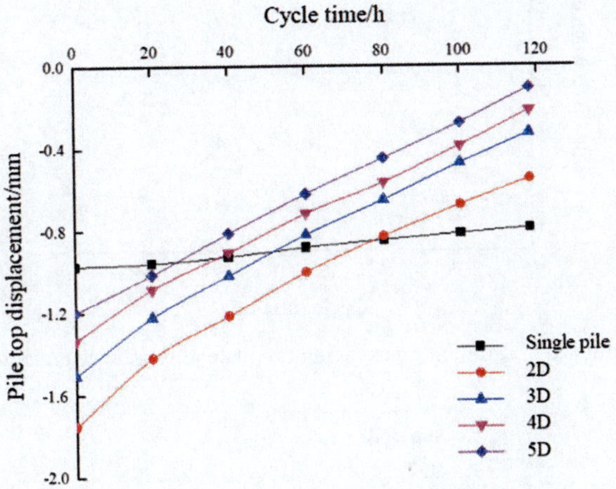

Fig. 12. Influence of different pile spacing on pile top displacement under temperature-load

Figure 13 shows the effect of pile spacing on the change in pile end resistance under the action of temperature. Figure 13 shows that when the cycle time is the same, the pile end resistance of the group piles is greater than that of the single pile. For group piles, the pile end resistance decreases with increasing pile spacing because with increasing pile spacing, the temperature superposition phenomenon for the pile group gradually weakens, the thermal stress and thermal expansion of the pile end decrease, the restriction of soil on the pile end decreases, and the pile end resistance decreases. At the end of the cycle, the pile end resistance of a single pile increases by 0.1442 kPa, the pile end resistance increments of the group piles with pile spacings of 2D, 3D, 4D and 5D are 13.4056 kPa, 12.2925 kPa, 8.9624 kPa and 4.9320 kPa, respectively, and the pile top displacement increments of the group piles with different pile spacings are 92.97 times, 85.25 times, 62.15 times and 34.2 times that of a single pile, respectively.

Figure 14 shows the influence of pile spacing on the change in pile end resistance under temperature-load combined action. Compared with Fig. 13, the change law of the pile end resistance curve under temperature-load combined action is consistent with that under the action of temperature, but the value increases. At the end of the cycle, the pile end resistance of a single pile increases by 15.95% (0.023 kPa), and the pile end resistance increments of the group piles with pile spacings of 2D, 3D, 4D and 5D are 30.66% (4.11 kPa), 28.72% (3.53 kPa), 27.00% (2.42 kPa) and 21.49% (1.06 kPa), respectively, which shows that the load plays an increasing role in the pile end resistance. Because the pile end soil is compressed and denser under the load, the constraint of the soil on the pile end is increasing. Under the same degree of thermal expansion of the pile end, the soil is denser at the pile end, and the upward resistance to the pile end is greater.

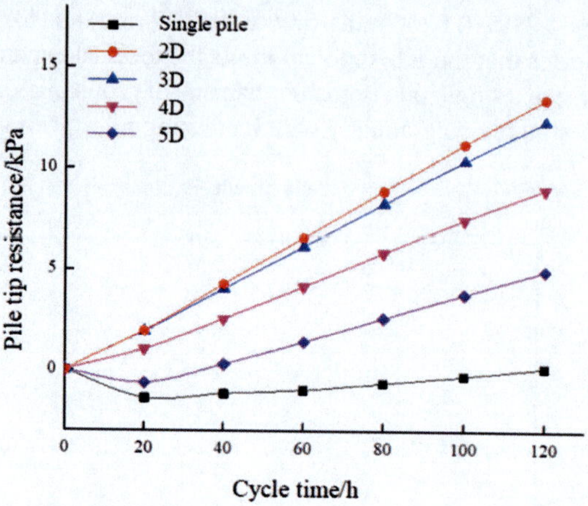

Fig. 13. Influence of different pile spacing on pile end resistance under temperature

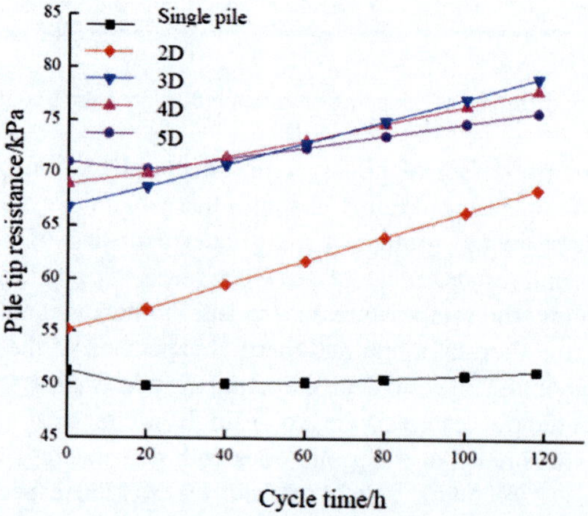

Fig. 14. Influence of different pile spacing on pile end resistance under temperature-load

Figure 15 shows the effect of pile spacing on the distribution of pile stress along the depth under the action of temperature. Figure 15 shows that under the action of temperature, at the same depth, the pile stress of a single pile is greater than that of the group piles. For group piles, the pile stress increases with increasing pile spacing because the existence of the pile group effect reduces the pile side friction of the pile group. The pile spacing is smaller, and the pile side friction is reduced more obviously. The decrease in pile side friction weakens the constraint around the pile. At the same temperature rise of the pile body, the constraint around the pile is smaller, and the thermal strain and the smaller the thermal stress are greater. At the same time, the pile spacing is smaller, the temperature superposition phenomenon is more obvious, the temperature rise of the pile body is faster, and the stress of the pile body is greater. The results show that the decrease in pile stress caused by pile side friction is greater than the increase in pile stress caused by the decrease in pile spacing. The maximum stress values of the single pile and group piles are at a depth of 105 cm, and the maximum values are 42.09 kPa, 66.87 kPa, 89.63 kPa, 109.98 kPa and 124.78 kPa.

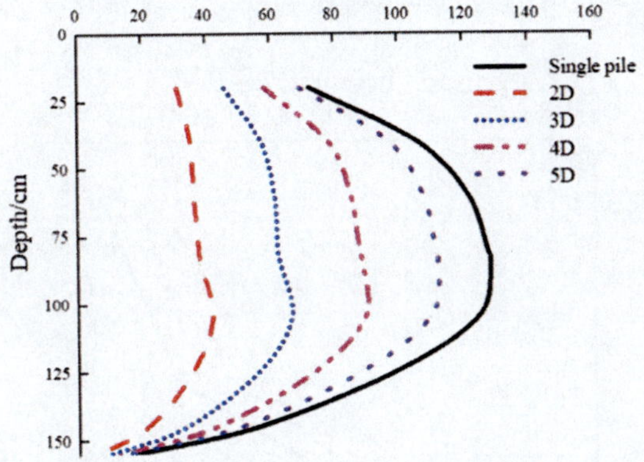

Fig. 15. Influence of different pile spacing on pile stress distribution under temperature

Figure 16 shows the effect of pile spacing on the distribution of pile stress along the depth under loading. Figure 17 shows the distribution of pile stress with depth under temperature-load combined action. The change law of the pile stress curve under temperature-load combined action is consistent with that along the pile body under load, which shows that under temperature-load combined action, the load plays a major role in the change in pile stress.

Figure 18 shows the variation of pile stress under temperature-load combined action compared with that under the action of load. Compared with Fig. 15, the variation law of pile stress along the depth in Fig. 18 is consistent, but the value increases. Taking

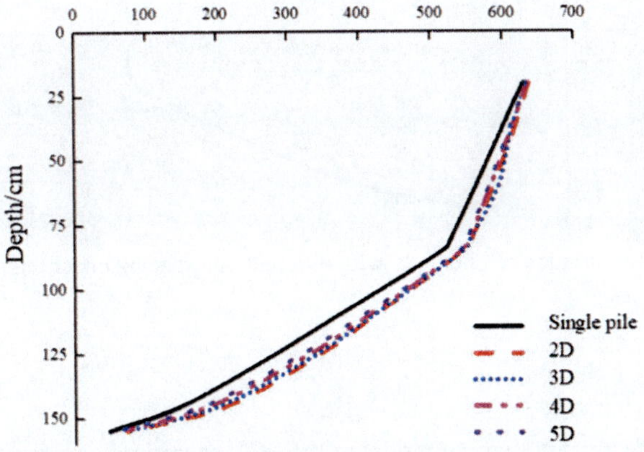

Fig. 16. Influence of different pile spacing on pile stress distribution under load

the stress value at a depth of 105 cm as an example, the group piles with pile spacings of 2D, 3D, 4D, 5D and a single pile increase by 2.92% (1.23 kPa), 3.45% (2.31 kPa), 4.06% (3.64 kPa), 4.23% (4.65 kPa) and 5.07% (6.32 kPa), respectively. This shows that the load increases the pile stress because the existence of the pile top load increases the pile top constraint and reduces the thermal strain of the pile. When the pile temperature rises, the thermal stress of the pile increases.

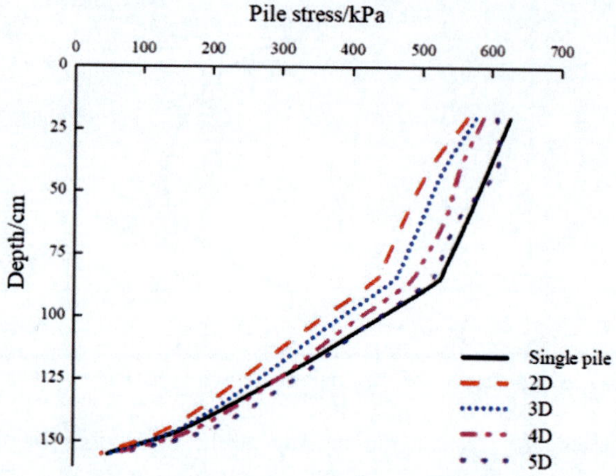

Fig. 17. Influence of different pile spacing on pile stress distribution under temperature-load

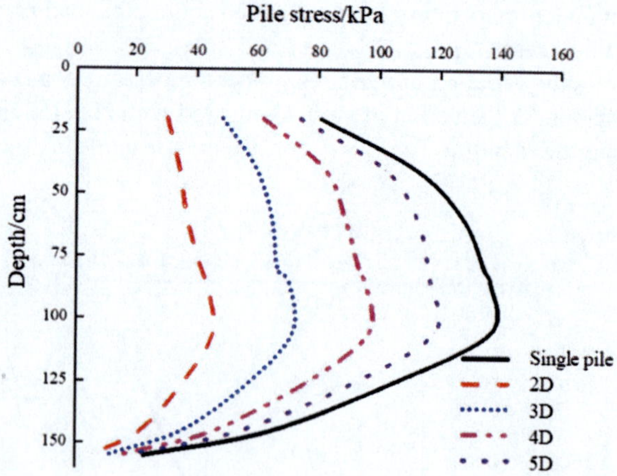

Fig. 18. Stress variation of pile body with different pile spacing under temperature-load

5 Conclusions

The three-dimensional numerical model of the PCC energy pile group is established by finite element analysis software, and the effects of the pile number and pile spacing on the bearing characteristics of the PCC energy pile group under temperature action and temperature-load combined action are studied. The results show that:

(1) Under the action of temperature, the pile top displacement and pile end resistance of group piles are greater than those of single pile, they increase with the increasing of pile number and decrease with the increasing of pile spacing, the strain of group piles is less than that of single pile, it decreases with the increasing of pile number and increases with the increasing of pile spacing.
(2) The vertical load on the pile top increases the pile restraint and limits the thermal expansion of the pile. Compared with no load on the pile top, when the pile top is loaded, the pile top displacement caused by temperature decreases, the pile end resistance increases and the pile stress increases.
(3) The pile group effect of temperature reduces the pile side friction of group piles and the constraint of soil on pile deformation. Compared with a single pile under the same working conditions, the decrease in pile stress caused by the decrease in pile side friction is greater than the increase in pile stress caused by the increase in temperature.

Acknowledgments. Financial support comes from the National Natural Science Foundation of China (No. 52027812), the Fundamental Research Funds for the Central Universities of Hohai University (No. B210202047, B210204020) are gratefully appreciated.

References

1. Lu, H.W., et al.: Combined load temperature field test and bearing behavior analysis of friction energy pile. Chin. J. Geotech. Eng. **39**(2), 334–342 (2017)
2. Wang, J., Zhu, H.H., Mei, G.X., Xiao, T., Liu, Z.Y.: Field study on bearing capacity efficiency of permeable pipe pile in clayey soil: a comparative study. Measurement **186**, 110151 (2021)
3. Brandl, H.: Energy foundations and other thermo-active ground structures. Geotechnique **56**(2), 81–122 (2006)
4. Gao, L., Ji, B.Q., Liu, H.L., Chen, Y.M., Cao, Y.: Numerical simulation study on deformation of PCC energy pile with spiral heat source. J. Disaster Prev. Mitig. Eng. **39**(04), 628–635 (2019)
5. Zheng, X., Shi, B., Zhu, H.H., Zhang, C.C., Wang, X., Sun, M.Y.: Performance monitoring of offshore PHC pipe pile using BOFDA-based distributed fiber optic sensing system. Geomech. Eng. **24**(4), 337–348 (2021)
6. Huang, X., Kong, G.Q., Liu, H.L., Peng, H.F., Hao, Y,H.: Study on negative friction characteristics of PCC energy pile under cooling cycle in summer. J. Disaster Prev. Mitig. Eng. **37**(04), 511–517 (2017)
7. Gui, S.Q., Chen, X.H.: In situ Test for structural responses of energy pile to heat exchanging process. Chin. J. Geotech. Eng. **36**(6), 1087–1094 (2014)
8. Wang, C.L., Liu, H.L., Kong, G.Q., Wu, D.: Analysis of influence of temperature cycle on deformation and stress of pile foundation under working load. Rock Soil Mech. **37**(s1), 317–322 (2016)

9. Liu, H.L., Wu, D., Kong, G.Q., Wang, C.L., Wu, H.W.: Study on heat transfer characteristics of energy piles with embedded and bound pipes. Rock Soil Mech. **38**(2), 333–340 (2017). [in Chinese]
10. Wang, C.L., Liu, H.L., Kong, G.Q., Ng, C.W., Wu, D.: Model tests of energy piles with and without a vertical load. Environ. Geotech. **3**(4), 203–213 (2016)
11. Gao, J., Zhang, X., Liu, J., Li, K.S., Yang, J.: Numerical and experimental assessment of thermal performance of vertical energy piles: an application. Appl. Energy **85**(10), 901–910 (2008)
12. Saggu, R., Clatkraborty, T.: Cyclic thermomechanical analysis of energy piles in sand. Geotech. Geol. Eng. **33**(2), 321–342 (2015)
13. Chen, D., Mcarmey, J.S.: Parameters for load transfer analysis of energy piles in uniform non plastic soils. Int. J. Geotech. ASCE **17**(7), 04016159 (2016)
14. Yang, T., Hua, Y.S., Liu, L.Z.: Numerical simulation of thermal-mechanical basic characteristics of suspended energy piles. J. Disaster Prev. Reduct. Eng. **37**(4), 518–524 (2017)
15. Rotta Loria, A.F., Donna, A.D., Laloui, L.: Numerical study on the suitability of centrifuge testing for capturing the thermal-induced mechanical behavior of energy piles. J. Geotech. Geoenviron. Eng. **141**(10), 04015042 (2015)
16. Salciarini, D., Ronchi, F., Cattoni, E., Tamagnini, C.: Thermomechanical effects induced by energy piles operation in a small piled raft. Int. J. Geomech. **15**(2), 04014042 (2015)
17. Cao, D., Shi, B., Zhu, H.-H., Wei, G., Bektursen, H., Sun, M.: A field study on the application of distributed temperature sensing technology in thermal response tests for borehole heat exchangers. Bull. Eng. Geol. Env. **78**(6), 3901–3915 (2018). https://doi.org/10.1007/s10064-018-1407-2
18. Cao, D.F., Zhu, H.H., Wu, B., Wang, J.C., Shukla, S.K.: Investigating temperature and moisture profiles of seasonally frozen soil under different land covers using actively heated fiber Bragg grating sensors. Eng. Geol. **290**, 106197 (2021)
19. Liu, H.L., Kong, G.Q., Wu, H.W.: Research progress in application of energy pile engineering and development of PCC energy pile technology. J. Geotech. Eng. **36**(1), 176–181 (2014). [in Chinese]
20. Huang, X., Kong, G.Q., Liu, H.L., Wu, H.W.: Experimental on thermal-mechanical characteristics of PCC energy pile under circular temperature field. Rock Soil Mech. **36**(3), 667–673 (2015)
21. Hao, Y.H., Kong, G.Q., Peng, H.F., Liu, H.L., Chen, F.: Simulation analysis of the influence of pile end constraints on the thermodynamic characteristics of pile. J. Disaster Prev. Reduct. Eng. **37**(04), 532–539 (2017)
22. Mehrizi, A.A., Porkhial, S., Bezyan, B., Lotfizadeh, H.: Energy pile foundation simulation for different configurations of ground source heat exchanger. Int. Commun. Heat Mass Transfer **70**, 105–114 (2016)

Author Index

H.-H. Zhu et al. (Eds.): BRWSG 2021, LNCE 230, pp. 345–346, 2022.
https://doi.org/10.1007/978-981-16-9963-4